ID0768806

CONTROL OF
ELECTRIC MACHINES

CONTROL OF
ELECTRIC MACHINES

IRVING L. KOSOW, PH.D.

Staten Island Community College
City University of New York

PRENTICE-HALL, INC., ENGLEWOOD CLIFFS, NEW JERSEY

Kosow, Irving L
  Control of electric machines.

  (Prentice-Hall series in electronic technology)
  Includes bibliographies.
  1. Electric controllers. 2. Electric driving.
3. Automatic control. I. Title.
TK2851.K66   621.31'7   72–3359
ISBN 0–13–171785–5

© 1973 by
PRENTICE-HALL, INC.
Englewood Cliffs, N.J. 07632

10

Printed in the United States of America

PRENTICE-HALL INTERNATIONAL, INC., *London*
PRENTICE-HALL OF AUSTRALIA, PTY LTD., *Sydney*
PRENTICE-HALL OF CANADA, LTD., *Toronto*
PRENTICE-HALL OF INDIA PRIVATE LIMITED, *New Delhi*
PRENTICE-HALL OF JAPAN, INC., *Tokyo*

To my wife
RUTH
and my children
SONIA, MARTIN, and **JULIA**

contents

preface

Control of Electric Machines is an outgrowth of the author's *Electric Machinery and Control*, published in 1964. In revising, supplementing, and updating that work, it became clear that two volumes were necessary to present the material properly and keep pace with the current state-of-the-art.

A variety of reasons dictated this choice. The original work was already fairly large (over 700 pp) and the contemplated new material would inevitably result in a most unwieldy, large, and expensive single volume. Further, a logical division between electric machinery theory and the control applications of electric machinery already exists in the literature. In the area of control of electric machinery, numerous works are found which do not deal with electric machine theory, so there is ample precedent for such a dichotomy.

The first volume, *Electric Machinery and Transformers*, is intended for the student who requires a background in the theory of electric machinery, transformers, and their characteristics. This volume, *Control of Electric Machines*, covers the industrial and commercial control applications of those machines introduced in the first volume. It is intended as a second text for students who have had a first course in the theory of electric machinery. It may also serve as a reference work for engineers and technicians currently engaged in the practice of selecting electric machinery for a variety of drive applications.

The contents of this volume reflects feedback from teachers and students who used the earlier *Electric Machinery and Control*. In response to numerous requests, the following modifications appear in this work:

1. The material which originally was covered in 5 chapters has been expanded considerably, as a result of new material, into 11 chapters.

2. Separate chapters have been added on automatic dc starters, ac starters, manual and automatic speed control of dc motors, manual and automatic speed control of ac polyphase motors, speed control of ac single phase motors, static (solid-state) control, ratings, selection and maintenance of electric machinery.

3. New sections have been added on electronic (solid-state) packaged drives for control of dc motors, performance of dc motors with electronic rectification, the brushless dc motor, electronic slip power control of polyphase induction motors, and electronic control of single phase motors.

4. The separate chapter on static control reflects the lastest devices used and is presented in such a way that the reader is familiarized with both noninverting and inverting logic systems.

5. In the chapter on automatic feedback control systems, an attempt has been made to introduce servomechanisms from both a qualitative and a quantitative point of view including an introduction to transfer functions, cascade compensation, gain-factor compensation and feedback compensation. The chapter is intended to serve as a broad introduction to and preparation for a separate course on servomechanisms.

The earlier rationale cited for the study of *Electric Machinery and Control* has been accentuated by two major worldwide problems: pollution (of our lands, waters and atmosphere) and overpopulation. The latter has resulted, in part, in tremendously increased demands for electric power and personalized transportion and this inevitably has produced the former. The electric car, cited by the author as a possibility in the earlier volume (1964) is rapidly becoming a reality. Inevitably, the techniques for control, braking, and regeneration of both dc and ac machines covered in this volume will be employed as use of electric cars becomes widespread, in response to the demand for a pollution-free vehicle.*

*J.T. Salihi, "The Electric Car - Fact and Fancy," *IEEE Spectrum,* June 1972, pp. 44-48.

The emphasis of the writing, based on the author's teaching and editorial experience, has been directed toward self-study. This has resulted in somewhat greater detail in text material, illustrative examples and many specific questions designed to motivate reading. It has the advantages of decreasing the teacher's workload and placing more responsibility on the student in the learning process, thus freeing the teacher to place greater stress on those aspects of the subject he feels require emphasis or in-depth study, and on those particular topics on which students require help.

Thanks and appreciation is expressed to the Prentice-Hall staff, generally, and to Mr. Steven Bobker, particularly, for his careful supervision of the production of the manuscript and the many helpful suggestions which resulted in the present format of the book. The author also acknowledges the support and help of Mr. Edward Francis, Editor, Career and Professional Education, Prentice-Hall.

As in the case of my other books and editorial work, my wife, Ruth, has contributed directly in the proofreading and indexing of the entire ms and indirectly by her encouragement, patience, and understanding throughout the many months of loneliness and isolation required to produce this book.

IRVING L. KOSOW

New York City, N.Y.

CONTROL OF
ELECTRIC MACHINES

overload, protective, and relay devices

1-0.
INTRODUCTION

Because of the wide availability of electric energy, electric motors are almost universally used in modern commercial and industrial occupancies to furnish the required mechanical motive power to drive mechanical machinery and control various industrial processes. Such machinery or other mechanical devices (valves, mechanical linkages, etc.) connected to motor shafts (either directly or coupled through gears, belts or pulleys) are called the *motor loads* or simply *loads*. In many cases, a load must be driven at a variety of speeds in either direction, in accordance with some desired preset sequence (e.g., an elevator in a high-rise apartment). Frequently, several motors are required in combinations in more complex sequences to control interrelated loads (e.g., processes in chemical plants and steel mills).

The energy supplied to the motor, depending on the nature of the load requirements, is usually programmed and controlled to obtain the desired load torque, speed and direction of rotation, at any given time, by a device called a *controller*. A controller may be *manually* operated (if the programming and sequence of events is controlled by a human operator) or *automatic*. The degree of automation is dictated by the requirements of the process or load to be controlled. If the program sequence or process to be controlled is simple, a motor stater (Chs. 3, 4 and 5) may be all that is required. Where the driven machine, process or load requirements dictate a more complex sequence and method of control, more complex controllers are needed. If the load or process is undisturbed by external variables or precise control of the load is not required, the controller and its process may be driven by an *open-loop* system (e.g., a preset timer controlling the various motor cycles of the motor of an automatic clothes washer). Where more precise control of the process or load speed and torque is indicated, *closed-loop* control (Ch. 10) is used. In the latter cases, the controller and its associated control devices are somewhat more complex.

In part, the selection of the type of motor used is dictated by the nature of load requirements, the type of energy available and the types of controllers commercially manufactured to adequately meet the load requirements. Some of these decision factors governing motor selection are discussed in Ch. 12.

It is assumed that the reader is generally familiar with the operating characteristics of various types of dc and ac motors.* This volume is concerned primarily with the techniques of stopping and starting various motors, controlling their acceleration and the various types of motor protection devices (overload, short-circuit, phase reversal, overspeed, undervoltage, etc.). Since the initial cost of a motor may be as high as 50% of the total overall control system initial cost, adequate protection is required to insure both long life and motor reliability. For this reason, a well-designed control system incorporates features to insure both motor and control system device protection, regardless of variations or disturbances of either supply or load conditions.

This chapter is devoted to those techniques used to "clear" (i.e., disconnect) both the motor and the control equipment from the supply in the event of such disturbances which may damage either.

1-1.
GENERAL

The National Electric Code (NEC) is deliberately specific about the purpose of branch circuit protection used for motor circuits. The intent of the code is to prevent fires of electric origin in motor branch circuits and feeders. Both feeder and motor branch circuit overload *and* short-circuit protection are specified clearly, as well as the minimum size of wires to be used for a single motor or for groups of motors. In the event of a *short circuit* within the motor, the branch circuit short-circuit protection will prevent damage also to the motor, its

* See I.L. Kosow, *Electric Machinery and Transformers,* (Englewood Cliffs, N.J.: Prentice-Hall, Inc., 1972).

starter, and its control equipment. The *branch circuit overload* protection, determined in part by the starting current and the type of motor, is designed to protect the *feeders* against continuous overloads. This feeder protection is somewhat higher, however, than that which is necessary to protect the *motor* against sustained operating overloads. It is necessary in addition, therefore, to protect the motor itself against operating overloads, using overload devices which are integral to the motor housing or contained in either a starter or a controller. The discussion of this chapter, therefore, will be limited specifically to those protective devices particularly related to electric machinery rather than to the feeders which supply it.* Other types of protective devices, which will be considered in addition to overload devices, include undervoltage and overvoltage protection, shunt-field failure, phase reversal and phase failure, and temperature and frequency-drift protection.

1-2.
FUSES

Perhaps the simplest motor overcurrent protective device is the fuse. Fuses are divided into two major categories: *low-voltage fuses* (600 V or less) and *high-voltage fuses* (over 600 V). Three types of fuses are shown in Fig. 1-1. The cartridge or ferrule-contact type, shown in Fig. 1-1a, is available in voltage ratings of 250 and 600 V in either a nonrenewable or a renewable type. The nonrenewable type shown in the figure contains an insulating powder (talc or suitable organic insulator) surrounding the fusible element. In the event of a short circuit, the powder is intended (1) to cool the vaporized metal, (2) to absorb the condensed metallic vapor, and (3) to quench any arc which may be sustained by the conductive metallic vapor. It is the presence of this powder that gives the fuse its high interrupting capacity in the event of sudden short circuits.

If the conductive metallic vapor created at the instant of short circuit were to sustain the short-circuit current, not only would the motor be permanently damaged but the fuse itself might explode because of the buildup of short-circuit current in the vapor. For this reason, the non-renewable type containing powder is preferred to the renewable type, which does not. The latter provides the advantage of reducing cost where frequent overloads occur and fuses must be replaced frequently.

Figure 1-1c shows the *plug fuse*, which is normally rated at 125 V, and available in low current ratings up to 30 A. These fuses have a medium-screw Edison base and are designed to be used in small starter or safety-switch boxes with 125-V, low-current motors.

As a general rule, fuses provide short-circuit rather than overload protection. With currents of approximately 25 to 50 times their rating,

* The subject of the selection and sizing of feeders and branch circuits (in accordance with the NEC or municipal codes) for light and power lines is normally considered the province of a course in electrical design; see R. C. Johnson, *Electrical Wiring: Design and Construction* (Englewood Cliffs, N.J.: Prentice-Hall, Inc., 1971).

3

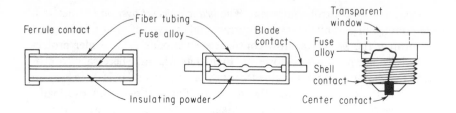

| (a) Cartridge type. | (b) Knife-blade type. | (c) Plug type. |

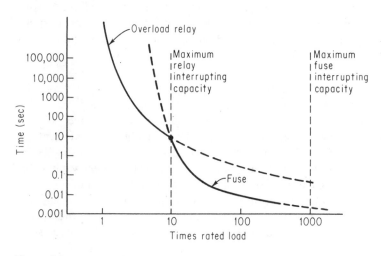

Figure 1-1

Low voltage fuse types and inverse time-current
curves for motor protection.

fuse elements will melt within one cycle (at 60 Hz) and interrupt the circuit within the next cycle, or in approximately $\frac{1}{30}$ second. Appendix Tables A-7 and A-8 specify fuse ratings from 150 per cent for *dc* motors up to 300 per cent of full-load current for single-phase and squirrel-cage induction motors. These ratings are of necessity greater than the normal overload occurring during either starting or braking operations and, as such, are unable to protect any motor against a continuous overload of from 100 to 140 per cent (or higher) of full-load current.

Attempts have been made to improve fuse characteristics for motor applications so that lower ratings may be used which will afford some overload as well as short-circuit protection. One type of fuse, called the *time-lag fuse*, which is available in knife-blade, cartridge, and plug types, provides a greater delay in the case of either momentary or sustained overloads before clearing (opening) the circuit. These fuses contain two

CHAP. ONE / *Overload, Protective and Relay Devices*

elements in series (or parallel): (1) a standard fuse link for short-circuit protection (25 to 50 times the rated current), and (2) an overload or thermal-cutout feature providing an inverse-time-delay characteristic for up to five times rated current. The inverse-time feature, for example, means that the circuit will be cleared by the latter element in anywhere from 3 minutes (at five times the rated current) to about 10 seconds (at approximately 20 times the rated current), since the heating effect varies as the square of the current. Thus, a fuse of relatively small rating may be employed to provide some overload protection and still not clear the circuit during starting or braking period surges of current. In the event of a short circuit, the instantaneous standard fuse link immediately clears the circuit to prevent damage.

Another type of fuse which has been developed attempts to improve the current-limiting ability* of these devices before the short-circuit current reaches its maximum or steady-state value. Ordinary cartridge fuses have some current-limiting ability in that they clear the circuit almost instantaneously before the short circuit has an opportunity to build up and fuse or weld the contacts of circuit breakers or overload relays. The *current-limiting power fuse*† contains silver alloy fusible elements surrounded by powdered quartz. When the fusible elements are vaporized by a short circuit, the silver-quartz vapor produces an extremely high resistance arc which serves to limit the current and its rise until magnetic protective equipment (such as a circuit breaker) has an opportunity to operate and clear the circuit.

Above 600 V, special *high-voltage fuses* are employed containing various techniques to extinguish the arc which may be sustained, particularly at high voltage, when the fusible element vaporizes because of excessive current. Some of these fuses are used particularly with low-interrupting-capacity switches or circuit breakers to obtain the advantage of the high interrupting capacity which these devices possess. The most common types of high-voltage fuses are (1) the liquid deion boric acid power fuse, (2) the expulsion fuse, and (3) the solid-material fuse.

The *liquid fuses* operate on the principle of the deionizing action of water vapor produced by the reaction between the arc and the boric acid liquid. Renewable "links" are available for these fuses in ratings of 7.5, 15, 23, and 34.5 kV, with interrupting capacities up to 0.45, 0.75, 1, and 1.2 million kVA, respectively.

---

* The use of series-connected reactance coils is still advocated as the best method of limiting rapidly rising short-circuit currents, not only in alternators but in motors as well, to prevent the short-circuit current from exceeding the interrupting capacity of protective equipment. A design by H. D. Short (*Elec. Eng.*, April 1950) using three reactors with not more than 1.57 W loss per hp per coil will limit any short circuit severely, regardless of the capacity of the supply system, until the normal overload equipment has an opportunity to operate.

† C. L. Schuck, *Transactions AIEE*, Vol. 69 (1950), pp. 770–776.

Smaller *high-voltage expulsion fuses* operate on the principle of the pressure, created by the formation of vapors within the fuse tube, actuating a release mechanism to expel the vapor and thus extinguish the arc.

Solid-material fuses use contact-spring mechanisms, held by the fuse element, which increase the air gap between contacts when the fuse melts, thereby extinguishing the arc.

1-3.
COORDINATION OF
FUSE AND
OVERLOAD RELAY

Although fuses lend themselves quite naturally to short-circuit or interrupting-capacity protection, they are somewhat limited in regard to overload protection for the reasons previously discussed. Various types of overload relays employing a number of basic principles will be discussed. These devices are designed to operate at anywhere from 110 to 250 per cent (depending on short-circuit protection) overload with (normal) interrupting capacities of up to 10 times the rated current. Figure 1-1d shows the coordinated use of a combination of fuse and overload relay to provide both overload and short-circuit protection. The inverse-time characteristics of both the overload relay and the fuse are shown in the figure. Normal motor operation occurs in the range of no load up to rated load, and the overload relay is not actuated. The operating time of the overload relay varies inversely as the overload current. As shown in the figure, an overload of approximately 10 times the rated current will produce an overload-relay-operating time of 10 s. A short circuit of 500 times the rated current will produce a fuse-operating time of less than 10 ms. The coordinated combination of fuse and overload relay permits the motor to operate within its normal range from no load to full load and, at the same time, permits small temporary overloads to occur without interrupting motor operation. Overloads of short duration are not sufficient to overheat or damage the motor if they do not occur too frequently. Various types of overload relays, employing various principles of operation, are used in dc and ac motor starters and controllers. Some of the more important types will be discussed in the following sections.

1-4.
INSTANTANEOUS
MAGNETIC
OVERLOAD RELAY

A typical overload relay employing a magnetic principle of operation is shown in Fig. 1-2a. This type of relay may be used on dc circuits and, with minor modification, on ac circuits (by inclusion of a stationary copper or brass sleeve surrounding the armature). As shown in Fig. 1-2a, the stationary contacts are *normally closed* when the *magnetic overload relay* is deenergized. At rated currents or those below rated, the spring pressure is sufficient to oppose motion of the armature. When the current reaches or exceeds a particular overload (say 125 per cent of the rated load), sufficient mmf is

6

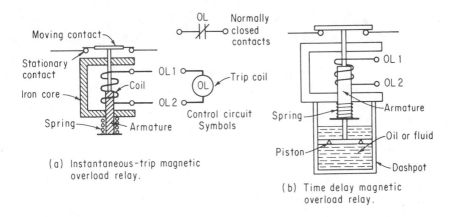

(a) Instantaneous-trip magnetic
overload relay.

(b) Time delay magnetic
overload relay.

Figure 1-2

Magnetic overload relays.

created to produce motion of the armature and the normally closed (n.c.) contacts are opened, serving to deenergize the motor. The instantaneous operation is usually sought in applications where the motor is used to drive a mechanical gear train, a conveyer, or drilling or woodworking machinery in which the overload represents a load condition requiring instant removal of the motor from across the line. The coils of magnetic overload relays are usually heavy copper conductors, having few turns and low inductance, connected in series with the armature of a dc motor or in series with the stator of an ac motor. The low inductance results in a short time constant (τ is proportional to L/R), and instantaneous relay operation when the current rises rapidly because of overload.

1-5.
TIME-DELAY
MAGNETIC
OVERLOAD RELAY

Figure 1-2b shows the same relay with the added provision of a dashpot or fluid damper to slow down the motion of the armature. The dashpot contains oil or a special fluid of the viscosity required to produce the necessary delay. Attached to the armature is a piston containing several holes through which the fluid passes. The piston rises when the coil mmf is sufficient to exert a pull on the armature. The size of the aperture holes may be varied, to vary the time delay of the relay, by removing the cup and turning a valve disc on the piston. The design produces an inverse-time characteristic (see Fig. 1-2). As the overload increases, the upward force on the piston is greater and the armature rises in a shorter time to trip the normally closed contacts.

The advantage of the magnetic *time-delay* relay, unlike the instantaneous type, is that sudden or momentary overloads are insufficient to

7

cause the motor to be disconnected from the line. If the overload continues for a given period, however, the piston slowly rises through the liquid to trip the closed contacts.

It should be noted that, as the armature rises, the magnetic reluctance is decreased (smaller air gap) and the armature pull is increased. If the air gap is great enough at the start, the retarding tension of the spring plus gravity should act in conjunction with the high reluctance to produce a slow rising motion of the armature at first. Magnetic time-delay relays incorporating this feature are called *vari-time* relays. The time delay is varied by changing the initial position of the armature air gap. When a short circuit or heavy load occurs, the relay acts almost instantly, providing the characteristic inverse-time feature required of all overload devices.

It should be noted that, in both the time-delay and the instantaneous types of relays, tripping the normally closed contacts deenergizes the motor and the relay overload coil simultaneously. This permits the relay to *reset* itself *automatically* and *instantly* because of the spring tension and gravity. These relays may be modified, however, by means of mechanical latches to catch and hold the armature. A mechanical manual release (or reset) button permits the armature to drop back to its released position before the motor can be restarted. This provision is usually made on motor starters in specific applications where the operator must be made aware that an overload has occurred or in small fractional horsepower motors (such as are used in connection with oil burners), in which an automatic reset might result in repeated cyclic overloads with possible damage to the motor, the overload device, the burner, and the occupancy because of an explosion.

Magnetic overload relays have the advantage of providing accurate adjustment of time-delay trip time to suit the individual motor. The operating coil (OL 1-2 in Fig. 1-2) either carries the motor line current directly on dc motors or, as in the case of larger ac motors, a current transformer is used. Adjusting the turns ratio provides a coarse trip adjustment. The fine adjustment requires adjustment of plunger position, spring, and/or piston orifices in the dashpot. Despite these advantages, however, magnetic overload relays are not as extensively used as thermal overload relays, for all the reasons described in Sections 1-6 through 1-10.

<table>
<tr><td>1-6.
FUSIBLE-ALLOY
THERMAL-MELTING
OVERLOAD RELAYS</td><td>A thermal relay, specifically designed for manual reset, shown in Fig. 1-3a, is the *fusible-alloy thermal-melting* relay. As shown in the figure, a high-wattage electric heater is connected in the load circuit (of a dc or ac motor). Under conditions of overload,</td></tr>
</table>

the heat is sufficient to melt the low-temperature (eutectic) fusible alloy and cause the spring to pull the latch and rotate the moving contact away from the stationary contacts. At first it might appear that, compared to the design of the magnetic overload relay, the fusible-alloy

8

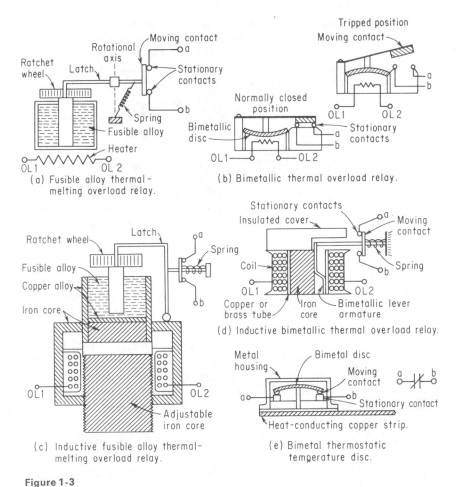

Figure 1-3

Thermal overload relays.

type is a "Rube Goldberg" affair. Actually, it is a practical and very popular overload relay for the following reasons.

1. Resetting of the relay requires both refreezing of the alloy and therefore a positive "waiting period" (which varies directly as the magnitude of the overload) before the motor may be reconnected across the line.

2. The relay must be *reset manually*, making the operator aware that an overload has occurred.

3. The relay will operate equally well on direct as on alternating current, since 1 A of alternating current produces the same heating effect as 1 A of direct current, and a given *heater rating* in amperes is *universal for all frequencies*. When operated on ac, the heater may carry the line current or some fraction of it, if a current transformer or inductive shunt is used. (see Section 1-10).

4. The same relay mechanism may be used interchangeably with *any* number of heaters (within a limited range), and overload "adjustments" may be obtained merely by changing the size of the heater (a wide and varied assortment of heater sizes in amperes is available) and/or by use of current transformers for larger ac motors.

5. The relay will work independently of gravity in either a vertical or a horizontal position, and all friction is overcome through the use of a fairly strong latch spring whose action is independent of gravity or current.

6. The accuracy and reliability of the relay is fairly high, since it is a function of the relatively constant melting point of the alloy.

7. The relay possesses the characteristic inverse-time heating required for the protection of all motors; i.e., since heat varies as the square of the current, a heavy overload will cause operation of the relay in a much shorter time.

1-7.
BIMETALLIC
THERMAL
OVERLOAD RELAYS

The advantages of using a separate heater (stated in the preceding paragraph) to actuate the normally closed overload contacts encouraged the development of other simpler and less expensive designs such as the *bimetallic thermal relay* shown in Fig. 1-3b.

An ordinary bimetallic rectangular strip will curve with the application of heat because of inequalities in the expansion of the two metals. This type of deflection is slow, and its slow action could lead to burning of the contacts in breaking a high-current highly inductive motor circuit. The design shown in Fig. 1-3b employs a circular bimetallic disc in which the upper side has a higher coefficient of expansion when heated. The forces of unequal expansion developed on a disc because of the heat are such that the disc must reverse its convexity suddenly rather than gradually. The quick-acting snap occurring at the instant of reversal has sufficient force to open (trip) the stationary contacts *a* and *b*, as shown in the tripped position of Fig. 1-3b. The tripping time of the bimetallic thermal overload relay is inversely proportional to the magnitude of the sustained overload current. Like the thermal-melting and the inductive time-delay relays, it permits instantaneous overloads to occur for short durations without disconnecting the motor from the line. Similar to the action of fusible-alloy types, the cooling time required to perform the transition from the tripped position to the normally closed position is a function of the magnitude of the overload or the heat (I^2R) developed. Thus, a very heavy overload will require a longer reset time than a slight overload. Although the design shown in the figure is automatically resetting, it is possible to modify the design to include a latching mechanism which catches and holds the moving contact when it is tripped, requiring manual reset or release.

Because bimetallic relays employ heaters, their mechanisms are interchangeable with fusible-alloy relays. The bimetallic relay has two

10

advantages not found in fusible-alloy types: It can reset itself automatically and, by means of a *compensating element*, adjustments may be made for ambient (room)-temperature variations.

1-8.
INDUCTIVE
FUSIBLE-ALLOY
THERMAL-MELTING
OVERLOAD RELAY

One of the disadvantages of fusible-alloy and bimetallic thermal overload relays, shown in Figs. 1-3a and b, is that the use of a separate heater (despite its various advantages) can permit overload adjustments only in discrete increments, based on the heater-current sizes available. While replacement heaters are (occasionally) available, since spares are normally supplied with the starter (if they can ever be found), it is hardly customary to have on hand an assortment of heaters of various values above and below the overload current setting at which the heater is rated. For a given motor application, it is sometimes desired to set the overload rating slightly higher during a given run or a particular operation. A design which makes this possible is the *inductive fusible-alloy overload relay* (the so-called *inductotherm* type), shown in Fig. 1-3c. This relay operates on the induction principle of eddy currents induced in a copper-alloy well and in the low-temperature fusible alloy contained within it. The relay, therefore, operates only on alternating current and is used for the overload protection of ac motors only. Since the heat produced in the fusible alloy is proportional to the flux density created by the current in the induction coil, an inverse-time characteristic (heat proportional to the square of the flux density) is produced by the eddy-current loss* for any given setting of the adjustable iron core.

When adjusted for an increased air gap at any given load, less heating effect is produced because of the reduction in flux density. There are two major advantages of the inductive fusible-alloy overload relay: (1) for a coil of a given current-carrying capacity, the overload trip setting is adjustable within limits; and (2) in conjunction with tapped current transformers (Fig. 1-4b), the same overload relay may be used with a wide variety of ac motors of higher or lower overload current rating with the same advantages of adjustment stated in (1).

1-9.
INDUCTIVE
BIMETALLIC
THERMAL
OVERLOAD RELAY

The extension of the overload range of a given overload coil is possible also with bimetallic relays as shown in Fig. 1-3d, which illustrates an *inductive bimetallic thermal overload relay*. This relay contains an iron core which is usually fixed (although a few designs have appeared which permit adjustment as well), inserted in a brass or copper sleeve to which a bimetallic lever armature has been welded. Eddy currents generated in the copper tubing

---

* The heat produced is a function solely of the flux density and not the current. Since the eddy-current loss varies as the square of the flux density, the exponential inverse-time characteristic is produced. A large air gap and a high current may not produce as much heating effect as a small air gap and a low current.

11

heat the bimetallic lever. An overload current produces sufficient heat to deflect the bimetallic lever and to trip the normally closed load-circuit contacts. Normally, this relay is automatically resetting; but a manual resetting feature may be added as described in Section 1-7. The advantages of this relay are the same as those stated in the preceding paragraph, with the added advantage that it is a simpler and more compact mechanism.

1-10.
AUXILIARY
THERMAL DEVICES

The bimetallic principle is also used in a device called a thermostatic or *temperature disc*, shown in Fig. 1-3e. The contacts of the disc are normally closed at the usual ambient temperature, and the disc may be riveted or welded to the *frame* or *bearing* of a motor. In the event of an increase in the ambient temperature because of ventilation failure, abnormally excessive line voltage, sustained or frequent overloads, defective bearings, lack of lubrication, or too frequent starting, the consequent heat developed is conducted to the disc and opens the normally closed contacts *a* and *b* connected in the line circuit. This device, therefore, operates as a result of the actual motor heat developed at the motor surface rather than because of the heat produced by the load current. Consequently, it is less sensitive to current overloads but more sensitive to ventilation, ambient temperature, and mechanical heating difficulties resulting from a lack of lubrication or defective bearings. On larger motors, the opening of contacts *a* and *b* may be used to actuate an alarm circuit. Larger dynamos employ *both* temperature discs and current-overload devices for maximum protection.

It should be noted that *all* overload relay contacts are normally closed since the relay of the coil is in series with the armature of a dc motor or the stator armature of an ac motor. The relay contacts and actuating force must be sufficiently large to produce the required interrupting capacity to break the circuit. The bimetallic temperature disc employs no relay coil whatever, but its contacts must be sufficiently large in smaller motors to interrupt the line or armature current. It should also be noted that the disc of Fig. 1-3e uses the higher-coefficient-expansion alloy on its lower side to increase its sensitivity and to reduce its operating time in the event of current overload. In larger ac motors, either normally closed or normally open contacts may be used to trigger an overload relay circuit or a trip coil in the circuit breakers supplying current to the motor. Temperature discs may be used as overload and/or temperature protection on either dc or ac motors.

The range of inductive-type relays used with ac motors may be extended by means of iron-core saturating devices. The *inductive saturating shunt*, shown in Fig. 1-4a, provides a limited extension of range by increasing the tripping time of any ac thermal overload relay during a starting or braking period or any other period of momentary current surge. It is usually during these periods that a thermal overload may trip its normally closed contacts even though its overload heater setting is the correct value for the motor under sustained overload conditions. Rather than purchase a larger heater (or use a current transformer) which increases the sustained overload rating, a saturating inductive shunt is employed. When the thermal overload relay coil or heater carries the rated current or less, the ac voltage drop across the heater or coil is below the voltage required to saturate the reactance coil. The reactance therefore has a high impedance, and very little load current is diverted from the heater or the highly resistive (low inductive) series-connected relay coil. During the momentary surge of braking or starting periods, however, the increased load current saturates the reactor, reducing its impedance and causing a larger portion of the load current to be diverted from the heater or (magnetic or inductive) relay coil. The tripping time of the relay is thus extended during these periods. The capacity rating of the inductive shunt is determined by the size and number of turns of its winding. The required rating is usually recommended by the manufacturer for use in conjunction with specific relays.

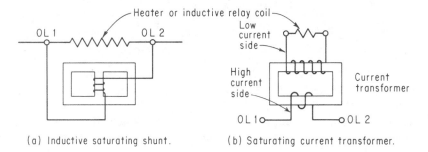

(a) Inductive saturating shunt.   (b) Saturating current transformer.

Figure 1-4

Thermal overload saturating devices used with thermal heaters.

A *saturating current transformer* may also be used in the same manner as an inductive shunt (although it is somewhat expensive for this purpose exclusively). As shown in Fig. 1-4b, when the primary current through OL_1–OL_2 is small, the transformer is unsaturated and the heater receives the full ratio of transformed current. During a current surge, on the other hand, the primary current is high, saturating the transformer (B remains constant) and decreasing the ratio of transformation. This means that an increase in the overload surge current does not produce a proportionate increase in the heater or relay current. Thus, the heater is also protected from momentary current surges by a saturating current transformer. At the same time, the current transformer may be used to extend the range (either up or down) of any ac thermal overload relay. A typical application is shown in Fig. 2-2. To summarize, the overload current transformer has *three* functions: (1) to protect the heater or relay from high overload

13

current surges (by self-saturation), (2) to increase the tripping time during momentary surges of current (by self-saturation), and (3) to extend the range of the relay heater or relay coil (by current transformation).

The thermal *heaters* used with all thermal relays are normally tested and rated at a 40°C *ambient temperature*, the standard allowable temperature rise of currently manufactured electric machinery.* If it is desired to determine the inverse-time-current characteristics of any thermal relay for any *other* ambient temperature, say 25°C, a new current curve may be drawn for the overload relay in Fig. 1-1 using Eq. (1-1):

$$I_{25°C} = I_{40°C}\left(\frac{T_{op} - 25°C}{T_{op} - 40°C}\right)^{1/2} \qquad (1\text{-}1)$$

where $I_{40°C}$ = current which will cause the element to trip at a given time, as rated by the manufacturer, at the operating temperature, T_{op}

$I_{25°C}$ = current that will trip the element in the same time at the lower ambient temperature of, say, 25°C

T_{op} = operating temperature of the thermal element in °C (it may be the melting point of a fusible-alloy element or the flexure temperature of a bimetallic disc)

Equation (1-1) shows that the *sensitivity* of any thermal relay mechanism used with heaters is a function of the operating temperature of the thermal element. When T_{op} in Eq. (1-1) is relatively high compared to the ambient temperature, a thermal relay is relatively insensitive to a change in ambient temperature, and the current rating is approximately the same for a wide range of ambient temperature values.

1-11.
OVERVOLTAGE
AND
UNDERVOLTAGE
PROTECTION

In general, overvoltage protection is not as common as undervoltage protection; and, since the former is simpler, it will be taken up first. There are two reasons why overvoltage protection is normally *not* provided in either manual or automatic starters and controllers: (1) voltages well above rated voltage rarely occur, and (2) small increases in voltage (applied across the armature) above the rated voltage for either a dc or ac machine result in an increase in the torque and a consequent reduction in the armature current.† Where, however, an increase in voltage will produce a change

---

* I. L. Kosow, *Electric Machinery and Transformers* (Englewood Cliffs, N.J.: Prentice-Hall, Inc., 1972), Sec. 12–17.

† The reason for this stems quite naturally from the theory of the doubly excited dynamo. If the stator excitation of an induction motor is increased, the rotor and stator currents must decrease to produce the same torque for the same load. Similarly, increasing the field excitation of a dc motor automatically decreases the armature current (despite the small increase in armature voltage) because of the increase in counter emf ($E_c = K\phi S$).

in the torque and the speed of a system in such a way as to affect adversely its output, or to damage electric and mechanical apparatus, it is necessary to use some form of overvoltage protection.

Figure 1-5a shows a typical instantaneous inductive relay having normally closed contacts *a* and *b* when normal (or less than normal) line voltage is applied to terminals V_1 and V_2. The normally closed contacts are connected in series with the line, or in a control circuit which energizes the motor. In the event of an overvoltage across V_1–V_2, the current in the relay coil is sufficient to pull up the iron armature (against gravity and sometimes against a spring) to trip the normally closed contacts *a* and *b*. The relay may be adjusted by means of an adjustable stop to trip at either higher or lower overvoltages, as desired.

This type of relay may be used on either direct or alternating current; the only modifications required in the case of the latter is a copper or brass sleeve (to reduce chattering) and a laminated core (to reduce eddy currents). In the configuration shown in Fig. 1-5a the relay contacts will remain open for the duration of the overvoltage, and will reset *automatically* when the voltage returns to normal.

In some types of operation this may not be advantageous, however, and a latching mechanism may be added to catch the moving armature and latch it until it is reset manually. Although an instantaneous-type overvoltage relay may be desirable in some applications, there are situations in which the *temporary* fluctuation of line voltage (above and below the rated value) may produce annoying machine disconnects and work stoppage. In these situations, an *inverse-time-delay overvoltage relay* of the type shown in Fig. 1-5b may be used. This type of relay is designed for ac operation only; it is similar in appearance to an ac induction watt-hour meter. Two sets of quadrature field poles and windings (not shown in the figure), similar to those of a two-phase induction motor, are housed below and above the solid aluminum disc. The main pole windings below the disc induce an emf in and excite the auxiliary (quadrature) windings above the disc. The rotation of the disc depends on the magnitude of field excitation and the drag created by the position of the permanent magnets with respect to the disc. For a given permanent magnetic field strength, the outer edge position will prevent rotation more than an inner position. Since the fields are displaced in space (but not in time), a true two-phase rotating magnetic field does not exist, and the torque is small. Under conditions of overvoltage, however, sufficient torque may be produced to cause the disc to rotate slowly despite the eddy current damping of the permanent magnet. A sustained overvoltage, therefore, will cause the disc to rotate slowly (counterclockwise in Fig. 1-5b) and will also cause the latching pin to actuate the moving contact about the pivot, breaking the normally closed contact *a–b*. As long as the overvoltage is sustained, torque is exerted against the spring to maintain the contact open. When the voltage drops to normal, the spring countertorque against the latching

15

pin returns and resets the overvoltage relay to its normal position. The induction principle employed in the time-delay overvoltage relay is used as well in field protective relays, ratio differential relays, reverse current relays, undervoltage relays, and power-factor relays.*

Undervoltage protection is almost always employed in both dc and ac starters and controllers. The National Electric Manufacturers Association (NEMA) distinguishes between two forms of undervoltage protection, specifically:

1. *Low-voltage release*, in which the device operating on the reduction or failure of voltage will cause interruption of power to the main circuit but will not prevent the reestablishment of the main circuit when the line voltage is restored to its normal or rated value.
2. *Low-voltage protection*, in which the effect of a reduction or failure of voltage will cause and maintain interruption of power to the main circuit.

The distinction between low-voltage release and low-voltage protection is largely a matter of safety and depends on the nature of the motor load. In the case of moving machinery such as a circular saw or bandsaw, in the vicinity of an operator, the restoration of voltage to normal might cause serious injury to the operator should the motor start up automatically. On the other hand, remotely located equipment such as pumps, air compressors, air conditioners, food freezers, or fans should be restarted immediately and automatically after the voltage returns to normal.

Undervoltage is usually much more serious than overvoltage in most applications. An appreciable reduction in line voltage across the armature of a loaded and running dc or ac motor results in a corresponding appreciable increase in armature current to maintain the same excitation flux (as in a doubly excited dynamo) and the same torque, to drive the applied load. In ac motors, the reduction in torque varies as the square of the voltage. The consequent reduction in voltage, torque, and speed may even result in the operation of centrifugal devices and/or the breakdown of the torque to standstill in single-phase induction motors.

The relay shown in Fig. 1-5a may be used as an *instantaneous undervoltage inductive relay*, operating on either alternating or direct current. In this configuration, the normal or rated line voltage applied to contacts V_1–V_2 is sufficient to maintain contacts a'–b' in their normally closed (energized) position. When the voltage drops below a certain value, or in the absence of line voltage, contacts a'–b' are opened as the relay armature drops because of the combined force of gravity and added

---

* These relays look so much like watt-hour meters that they are often mistaken for them.

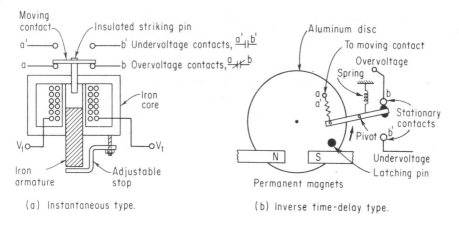

(a) Instantaneous type.　　　(b) Inverse time-delay type.

Figure 1-5

Overvoltage and undervoltage relays
(See text).

spring tension (not shown). Instantaneous undervoltage relays may be adjusted by means of series resistors in the coil circuit, or by a change of spring tension, to drop out at a predetermined voltage. The relay shown in Fig. 1-5a is automatically resetting. It may also be latched, if desired, for manual reset to release the armature and restore it to its normally closed position when the voltage returns to normal.

As in the case of the instantaneous overvoltage relay, temporary reductions or fluctuations in line voltage may produce annoying interruption of motor operation. A time-delay undervoltage relay is shown in Fig. 1-5b. In this figure, the undervoltage contacts $a'-b'$ are normally closed under conditions of rated normal voltage. The excitation of the main and quadrature fields produces sufficient torque to maintain the latching pin against the movable contact, and the spring is extended. A variable resistor is used in series with the lower main field winding (not shown) to adjust the torque of the latching pin against the pivot arm with the rated voltage. When the voltage drops below normal, the spring tension overcomes the disc torque and opens contacts $a'-b'$. Momentary reductions in voltage, however, are insufficient to cause sufficient rotation of the disc because of the retarding viscous friction of the (eddy current damper) permanent magnet, which also provides the required inverse-time characteristic. Normally closed contacts $a'-b'$ may be arranged to operate the trip coils of circuit breakers or to interrupt control circuits whenever the voltage drops below normal. Circuit arrangements are also used which simultaneously make it impossible to start the motor unless the line voltage is up to its rated value.

17

1-12.
REVERSE POLARITY
PROTECTION

In dc generator battery charging, a reversal of current or polarity will cause the generator to tend to operate as a motor. If it is unable to turn (its prime mover may be held stationary), the stalled generator attempting to act as a motor may be damaged. In many electrochemical processes, moreover, such as electroplating, a polarity reversal (even of short duration) may result in a defective product. In plugging braking (for example), a reversal of the armature current will cause reversed rotation, and some protection is necessary.

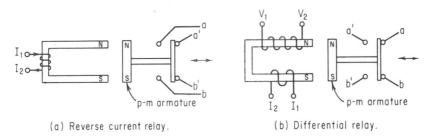

(a) Reverse current relay.   (b) Differential relay.

Figure 1-6

Polarity-control relays.

Figure 1-6a shows a simple *reverse-current relay* using a *permanent-magnet* armature. The circuit may be used as either a normally closed or a normally open relay. When current is flowing from I_1 to I_2 as shown, the relay opens contacts a–b and closes contacts a'–b'. When current is flowing from I_2 to I_1, the electromagnet reverses the polarity shown in the figure, and contacts a–b are closed while contacts a'–b' are opened. The reverse-current relay (sometimes called a *directional relay*) is tripped when the voltage across the coil produces sufficient mmf to move the permanent-magnet (p–m) armature. Thus, it may also be used as a protective device (1) to prevent a circuit from being energized if the polarity is incorrect, and (2) to shut down a generator or to disconnect a battery source if the polarity is incorrect or if the current tends to reverse.

The same principle may be employed in the *differential* or voltage-current *relay* used in the automotive voltage regulators, shown in Fig. 1-6b. The voltage and current coils are wound in the same direction around the iron core. The voltage developed by the generator sends current to the automobile battery, causing the p–m armature of the relay to take a given position. When the generator voltage drops below the battery voltage and the current reverses, the relay operates, opening one set of contacts (disconnecting the generator from the battery) and closing another (lighting a discharge lamp).

CHAP. ONE / *Overload, Protective and Relay Devices*

1-13.
PHASE-REVERSAL
AND
PHASE-FAILURE
PROTECTIVE
RELAYS

The "two-phase" induction disc principle may also be employed for either phase reversal or failure of one phase of a polyphase system. In these applications, the moving contact is welded or bolted to the rotating disc shown in Fig. 1-7a. The upper and lower quadrature windings, each connected to a different phase, are balanced (by means of potentiometers) to produce a clockwise torque against the retarding spring and to maintain moving contact *a* against stationary contact *b*. In the event of phase reversal, the direction of rotation is instantly reversed, and contacts *a–b* are opened,

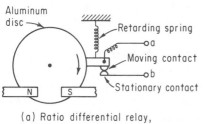

(a) Ratio differential relay,
winding temperature relay,
phase failure, phase reversal
or power factor relay.

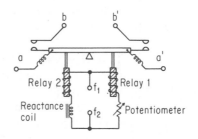

(b) Frequency control relay.

Figure 1-7

Phase, power-factor, and frequency-control relays.

disconnecting the motor from the line. In the event of the more serious loss of one phase because of a blown fuse in one of the lines of a three-phase motor,* the single-phase torque is insufficient compared to the retarding spring tension, and contacts *a–b* are also opened. Momentary voltage fluctuations are damped out by the permanent-magnet damper, which also serves to produce an inverse-time characteristic.

1-14.
DIFFERENTIAL
RELAY

Because the induction principle operates by means of a balance of mmf and current in the main and quadrature coils of the relay, the principle may be employed to detect slight unbalances in ac circuits. The two lower main coils (not shown in Fig. 1-7a) are a sum and difference coil, respectively, while the upper quadrature coils also have a sum and difference coil. The current in the sum coils is balanced against the current in the difference coils. If the currents are balanced and equal, no resultant field is produced and the disc will not rotate.

* A running, three-phase SCIM (squirrel-cage induction motor) in which one line is accidentally opened will continue to operate as a single-phase motor. Depending on the load, the line current, even though it is increased by $\sqrt{3}$, may still be insufficient to trip an overload coil, particularly when lightly loaded.

The current through the sum coils tends to maintain contacts *a–b* in Fig. 1-7a closed. The current through the difference coils is proportional to the unbalance of windings resulting from (1) an increase in the temperature of one of two windings, (2) leakage from the slip rings of synchronous motors or wound-rotor induction motors to ground, (3) leakage from one stator phase to another, or (4) leakage from one stator phase to ground. The current in the difference coils will produce counterclockwise torque, tending to open the contacts *a-b* in response to any of the aforementioned conditions. The differential relay operates with a relay for each phase and is connected to detect only an unbalance within the machine itself, rather than to detect line current or system unbalance (although it could easily do so and is sometimes used in that manner).

1-15.
FREQUENCY-
CONTROL RELAY

Two relays, each in the corresponding arm of a simple bridge network, may be used to detect either an increase or a decrease in frequency as shown in Fig. 1-7b. At the desired frequency, usually 60 Hz, the current in one relay is adjusted by means of a potentiometer to maintain both pairs of contacts open. As long as the desired frequency is maintained across terminals f_1–f_2, the armature will remain stationary. An increase in frequency will cause an increased impedance in the reactance coil and a decrease in current in relay 2, causing contacts *a–b* to close. Similarly, a decrease in frequency will result in an increase in current in relay 2, causing contacts *a'–b'* to close. The bridge circuit is thus a direction-sensitive error detector (frequency discriminator). Relay 1 acts as a retarding or balancing force against relay 2, so that frequency changes affect both relay coils equally. The potentiometer is unaffected by frequency changes, and the reactance coil is the frequency transducer. Since the speed of an induction or synchronous motor is proportional to frequency, this device may be used in applications where frequency-speed changes are critical (Section 5-17), and the motor will not start or will cease operating if the frequency is not precisely at its rated value. At the same time, the relay closure will actuate corrective networks to restore the frequency to its normal value.

1-16.
OVERLOAD CIRCUIT
BREAKERS

No discussion of motor protection would be complete without a description of the *overload circuit breaker* (OCB). A circuit breaker is an overload device intended to interrupt or clear a circuit (in the same manner as a fuse) but without injury to itself. Like fuses, circuit breakers act primarily as short-circuit protection in conjunction with any of the various overload devices in motor circuits discussed above. Also like fuses, they are rated as *low-voltage* circuit breakers (below 600 V) and *power* circuit breakers (above 600 V). There are numerous types of circuit breakers employing various methods for tripping a circuit and quenching the arc which forms between the stationary and moving contacts. Small low-voltage breakers employ bi-metallic heating elements having inverse-time characteristics and operate on the principle of the bimetallic overload relay (Sec. 1-9). Larger low-voltage breakers employ combinations of magnetic and thermal trip elements in which the former provide instantaneous tripping characteristics for short-circuit protection, and the latter inverse-time tripping for sustained overload currents.

Small *air circuit breakers* are generally enclosed in a Bakelite case and are available as single-, double-, or triple-pole molded-case air breakers in frame sizes of 50, 100, 225, 400, 600, and 800 A. Each frame size has a variety of standard continuous-current ratings determined by the rating of its trip elements. The interrupting capacities of low-voltage molded-case air circuit breakers may be as high as 35 kA. Molded-case breakers may be reset by means of a handle whose latching mechanism is independent of the handle speed. Thus, in closing or opening highly inductive circuits, the contacts will provide a quick make or break so as to avoid the pitting of contacts and to provide rapid extinction of the arc. The handle is generally *trip-free*, meaning that the contacts cannot be held closed manually during sustained overloads or a short circuit.

Larger air circuit breakers are available below 600 V in frame sizes up to 4000 A with interrupting capacities up to 100 kA. Figure 1-8a shows a *manual air circuit breaker* of the plunger or solenoid type. The inductive element shown is an instantaneous type, whereas those employing dashpots on the moving armature provide a variable time delay, as previously discussed. Figures 1-8a and b show, respectively, the series-connected operating coil OL_1-OL_2 and manner in which the moving contacts are electrically connected for circuit interruption. Some of the larger low-voltage air circuit breakers are electrically rather than manually reset by means of a closing solenoid, also providing the advantage of remote resetting in the event of overload.

High-voltage or *power circuit breakers* have voltage ratings from 2.5 to 350 kV and three-phase interrupting capacities from 25,000 to 25,000,000 kVA. Power breakers are invariably rated in kilovoltamperes (kVA) rather than amperes.

Power magnetic air breakers are available in voltages up to 18 kV and in standard three-phase capacities of 750,000 kVA. They are similar in construction to low-voltage air breakers, but they are electrically operated. In larger capacities, *air-blast* or *compressed air breakers* are used to extinguish the arc. Compressed air breakers are rated up to 34.5 kV at 2.5 million kVA.

Even larger capacities are available in *oil circuit breakers*, where the contacts are immersed in oil to quench the arc and to cool the contacts. Oil circuit breakers are invariably used for power distribution (alternator and transformer protection) rather than motor protection.

21

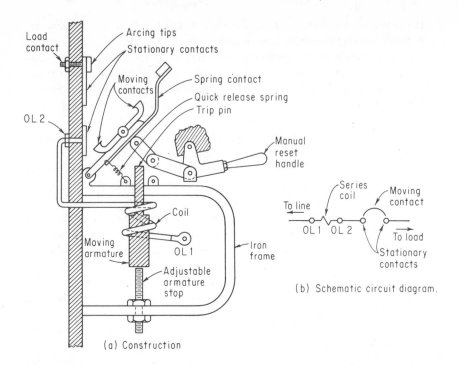

Load
contact

Arcing tips

Stationary contacts

Moving
contacts

Spring contact

Quick release spring

Trip pin

OL 2

Manual
reset
handle

Coil

Moving
armature

OL 1

Iron
frame

Adjustable
armature
stop

(a) Construction

Series
coil

Moving
contact

To line

OL 1   OL 2

To load

Stationary
contacts

(b) Schematic circuit diagram.

Figure 1-8

Instantaneous magnetic-trip overload
circuit breaker.

1-17.
ARC PROTECTION

Circuit breakers and overload relays invariably employ some method of protecting their contacts from the harmful effects created by the *electric arc* ($LI^2/2$) which forms when any highly inductive circuit is interrupted. In the air circuit breaker of Fig. 1-8a, for example, the circuit is interrupted at the main moving contacts *before* the arcing tips are opened. The combination of rising heat and rapid action of the spring contact holding the moving arcing tip serves to break contact quickly without damage to the main contacts.

In larger circuit breakers, magnetic fields are employed to react against ionized current-carrying air particles that sustain the arc; such a field concentrates and forces the arc upward into an insulated arc chute, thus physically removing the arc from the contacts. This principle is called the *deionization* principle, and the magnetic field is created by either a permanent magnet or an electromagnet called a *blowout coil*. By forcing the arc into a restricted and insulated chute, deionization of the air and extinction of the arc are accomplished. Multiple breaks are sometimes employed in arc chutes, dividing the arc into a number of series discharges and which tends to deionize the air more rapidly.

Figure 1-9a shows one type of protective relay employed to disconnect a dynamo in the event of flashover between insulated circuits (commutator, slip-ring, or stator) and the frame of the machine (ground). The flashover protection relay shown in Fig. 1-9a is an instantaneous inductive relay whose coil is well insulated and is designed to carry current from the frame of the machine to ground. In the event of flashover to ground, the relay disconnects instantly, requiring manual reset.

Figure 1-9b shows a typical application of a *blowout coil*. A shunt motor is started with its field rheostat short-circuited (full field current) in order to develop maximum starting torque $(T = k\phi_f I_a)$ and rapid acceleration. When the contacts shown in the figure are opened to reduce the field current and to permit speed control by means of the field rheostat, the sudden change in field current may produce sufficient energy to damage the rheostat. A high-resistance blowout coil (connected in parallel with the field circuit across the armature), or a permanent (alnico) magnet located directly below the contacts, creates a field which serves to extinguish the arc rapidly through arc chutes.

Figure 1-9c shows a method called the *line-arc principle*, in which a copper bus is connected in the arc circuit in such a direction as to produce an mmf which repels the arc (ionized air particles carrying current). The arc is forced rapidly upward, reducing the pitting of the contacts that interrupt the arc.

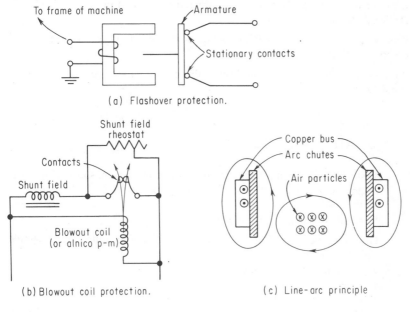

(a) Flashover protection.

(b) Blowout coil protection.

(c) Line-arc principle

Figure 1-9
Arc protection devices.

23

BIBLIOGRAPHY

Bellinger, T. F., and R. A. Gerg. "Closer Overload Protection for Polyphase Motors," *Allis-Chalmers Electrical Review* (Second Quarter 1960).

"Electronic Standards for Industrial Equipment" (General Motors Technical Center), *Electrical Manufacturing* (August 1958).

Graphical Symbols for Electrical Diagrams (ASA Y32.2) (New York: American Standards Association, 1954).

Gutzwiller, F. W. "Using the Silicon Controlled Rectifier for Protection," *Electro-Technology* (October 1961).

Harwood, P. B. *Control of Electric Motors* (New York: John Wiley & Sons, Inc., 1970).

Heumann, G. W. "Motor Protection," in *Magnetic Control of Industrial Motors*, part 2, Chap. 6 (New York: John Wiley & Sons, Inc., 1961).

Howell, J. K., and J. J. Courtin. "Temperature Protection for Induction Motors ... Today and Tomorrow," *Westinghouse Engineer* (November 1959).

Industrial Control Equipment (Group 25) (ASA C42.25) (New York: American Standards Association).

Industrial Control Equipment (UL508) (Chicago: Underwriters Laboratories, Inc.).

James, H. D., and L. E. Markle. *Controllers for Electric Motors* (New York: McGraw-Hill Book Company, 1952).

Karr, F. R. "Squirrel-Cage Motor Characteristics Useful in Setting Protective Devices," AIEE Paper 59–13.

Kosow, I. L. *Electric Machinery and Transformers* (Englewood Cliffs, N.J.: Prentice-Hall, Inc., 1972).

Wilt, H. J. "Circuit Factors in Motor Protection," *Electrical Manufacturing* (June 1959).

QUESTIONS AND PROBLEMS

1-1.   Distinguish between branch circuit overload protection, short-circuit protection, and motor overload protection as to
a. Purpose
b. Relative magnitude for a given motor

1-2.   Give one function and one application for each of the following protective devices:
a. Overload circuit breaker
b. Line circuit fuse
c. Undervoltage relay
d. Overvoltage relay
e. Shunt-field failure relay

24

      f. Phase-reversal relay

      g. Excessive-temperature-rise relay

      h. Frequency-drift relay

1-3. Distinguish between low-voltage and high-voltage fuses as to differences in construction and reasons for the differences. Give three types of high-voltage-fuse designs and consider the relative merits of each design.

1-4. Give one specific advantage of ferrule contact and of knife-blade and plug-fuse construction.

1-5. Describe the construction and purpose of a time-lag fuse and a current-limiting fuse.

1-6. Distinguish between fuse rating and interrupting capacity, including those factors which determine these ratings.

1-7. Give two advantages of the coordinated combination of overload and short-circuit protective devices.

1-8. An instantaneous magnetic overload relay has a voltage rating of 28-V dc, a resistance of 1.4 Ω, and an inductance of 0.28 H. A current of 15 A is required to operate the relay, used to protect a motor whose rated current is 12 A. Calculate

      a. The time required for the relay to clear the circuit in the event of short circuit

      b. How much longer the relay will sustain short-circuit current compared to a 10-ms fuse

Note: Use the equation: $i = \left(\dfrac{E}{R}\right)\left(1 - \epsilon^{\frac{-tR}{L}}\right)$ to solve Problems 1-8 and 1-9.

1-9. Repeat Problem 1-8 for a time delay relay having an inductance of 28 H used as an overload relay, all other values being the same.

1-10. List seven advantages of fusible-alloy, thermal-melting overload relays and two disadvantages, as compared to other types of overload relays.

1-11. List two major advantages for *inductive* fusible-alloy and inductive bimetallic overload relays compared to normal fusible-alloy and bimetallic types.

1-12. Give five applications for bimetallic temperature discs in which the disc is more responsive to motor heat than overload devices.

1-13 Give two iron-core devices used to extend the range of inductive-type relays.

1-14. Give three functions of an overload current transformer.

1-15. Give two reasons why overvoltage protection is normally not provided on magnetic starters or controllers.

1-16. Give two types of relays used for overvoltage protection and discuss the principles used for each.

1-17. Give two forms of undervoltage protection and define each.

1-18. Why is undervoltage protection required for most dc or ac motors?

1-19. Describe the device and give one application in terms of a circuit diagram for the following: a time-delay undervoltage relay, a reverse-polarity relay, a reverse-current relay, a differential relay, a phase-reversal relay, a phase-failure relay, and a frequency-control relay.

1-20. Give two ways in which circuit breakers differ from overload relays.

1-21. The bimetallic element in a thermal relay operates at a temperature of 100°C. The relay is rated to trip at 50 A at an ambient temperature of 40°C. Calculate the trip current at ambient temperatures of
a. 20°C
b. 60°C

ANSWERS

1-8(a) 0.277s (b) 28 times longer for relay to clear circuit. 1-9(a) 28s (b) 2.8 × 10³ times as long. 1-21(a) 57.7 A (b) 40.8 A.

standard control-circuit symbols and diagrams

2-1.
GENERAL

The various protective devices described in Ch. 1 and other control devices described throughout the work are represented, *graphically*, on a variety of electrical diagrams, in *symbolic* form. The use of symbols is a form of shorthand which saves space and writing in representing particular control-circuit elements on a variety of *electrical wiring diagrams*.

To the engineer, technician, and electrician, electrical diagrams serve a variety of purposes. They may serve as a record of the various devices used. They may serve to show how they are connected electrically. They also record the "logic" of the control circuitry. Depending on purpose, therefore, a variety of diagrams are used. An *interconnection diagram* only shows the terminal connections among control equipment, remote control stations, and the various electrical machinery and associated equipment in a particular installation. Such a diagram would be used by an electrician in making the necessary electrical interconnections between the various pieces of equipment, which are physically located apart from each other. An equipment *wiring* or *connection diagram* would show the various wiring connections between the circuit elements of a single piece of control (or other electrical) equipment. Such a wiring diagram would reflect the physical proximity of the various circuit elements within the equipment and show the connections between them. A connection diagram is useful to the technician in troubleshooting a specific item of control equipment and also to the assemblers who initially constructed it.

A third, and undoubtedly most useful, type of electrical diagram is the *schematic* or *elementary* wiring diagram. This diagram serves as a record of the logic of the control circuitry. It shows the wiring between the various control elements in the control circuitry, apart from the power circuits, arranged in such a sequence that the *logic* of the control circuitry is emphasized. Typical schematics are shown in Figs. 2-2 and 2-3. Schematics are used by engineers and technicians to comprehend the *logic* of the control circuitry and ensure that it functions properly. Schematic diagrams do not show the *physical* location of the various control elements within a control panel or the location of various remote equipments and electric machines. But they do show the wiring between all the control elements in a simple, graphic form so that anyone familiar with standard control-circuit symbols can comprehend the *scheme* or sequence of control-circuit operation.

Some schematic diagrams are supplemented by *contact sequence charts*, which are *tables* that summarize and facilitate understanding of control-circuit logic and also enable troubleshooting in the event of failure of particular relays to operate in the proper sequence. A typical contact sequence chart is shown in Fig. 2-3b.

In order to comprehend schematic or elementary wiring diagrams, it is first necessary to understand the symbols used on such diagrams and the designations used on these symbols. Section 2-2 describes standard control-circuit symbols and Section 2-3 covers the abbreviations for devices and functions used with these symbols.

2-2.
STANDARD
CONTROL-CIRCUIT
SYMBOLS

While an attempt has been made by the International Electrotechnical Commission (IEC) to develop an international set of symbol standards for electric circuits, generally, and control circuits, specifically; at this writing there is no single set of graphical standards used internationally. In the United States the American Standards Association (ASA) and the National Electrical Manufacturers Association (NEMA) have developed standards for electrical drawings as well as for industrial electrical control equipment. The IEEE has summarized ASA and NEMA standards for graphical electrical symbols* universally used throughout the United States and in other countries as well. The draftsman or engineer seeking guidance as to symbols should refer to these publications, particularly in regard to new or non-standard devices or equipment not specifically designated.

A typical set of standard control-circuit symbols used on schematic wiring diagrams is shown in Fig. 2-1. The symbol for a thermal device (Sections 1-6 to 1-10) is shown in Fig. 2-1a. Such a symbol might represent a bimetallic element, heated either directly or indirectly, and its normally closed (n.c.) contacts. Note that the contacts are represented in their *deenergized* or *nonactivated* state, as shown in Fig. 2-1a. In the event of overload or short circuit, the contacts of a thermal device are opened and the particular *control circuit* (in which these contacts are wired) is *cleared* or opened. Figures 2-1b and c, respectively, show fuse and air circuit breaker symbols. Such overload devices are used in series with power circuits supplying electrical power to electric motors. In the event of overload or short circuit, these devices clear the main *power* circuit *directly*

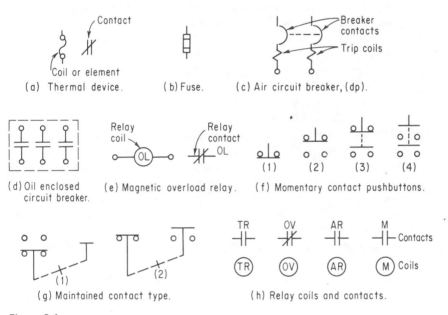

(a) Thermal device. (b) Fuse. (c) Air circuit breaker, (dp).

(d) Oil enclosed circuit breaker. (e) Magnetic overload relay. (f) Momentary contact pushbuttons.

(g) Maintained contact type. (h) Relay coils and contacts.

Figure 2-1

Standard control circuit symbols.

* ASA Y32.2, *Graphical Symbols for Electrical Diagrams* (New York: American Standards Association, 1954); *Industrial Control System Standards* (NEMA ICS 1–101) (National Electrical Manufacturers Association). See also IEEE standard 315-1971, Institute of Electrical and Electronic Engineers, N.Y.C., N.Y.

rather than a control circuit. (See Section 2-5 for distinctions between power and control circuits.)

The magnetic overload relay (Section 1-4 and 1-5) performs the same logic in Fig. 2-1e as that shown for the thermal element in Fig. 2-1a. The overload sensed by the relay coil in a power circuit will open the n.c. relay contact in a control circuit. Note that the symbol for a coil is a closed circle and its particular identification is designated by an abbreviation in capital letters, in accordance with the designation given in Table 2-1. When more than one relay performs the same function on a given schematic diagram, it is customary to follow (or prefix) the letter designation with a number. Note in Figs. 2-1e and h that both the coil and contact are designated with the same abbreviation.

Figure 2-1f shows *momentary* (monostable or not maintained) contact pushbuttons. The implication of such a symbol is that the pushbutton is only temporarily removed from its shown deactivated position. Thus, Fig. 2-1f(1) shows a n.c. pushbutton which *temporarily* breaks a line but returns to close the contacts when deactivated. Similarly, Fig. 2-1f(2) shows a double-contact pushbutton (PB) which temporarily breaks its upper contacts and makes (closes) its lower contacts, returning to its deactivated state shown in the figure, when the mechanical activating force is removed. Figure 2-1f(3) performs the same function as 2-1f(2) using a two-pole rather than a single-pole PB. Figure 2-1f(4) shows a double-pole, double-contact (DPDC) arrangement, which temporarily closes two separate lines or circuits.

Pushbuttons (PBs) of the *maintained* (not momentary) type are shown in Fig. 2-1g. Such PBs have two (or more) stable states, similar to a rotary switch. The single-pole PB in Fig. 2-1g(1), when activated, closes its upper contacts and clears its lower contacts. When reactivated mechanically, it is once more restored to the position shown in the figure. Figure 2-1g(2) shows a two-pole arrangement which accomplishes the same logic in opening one line and closing a second for a sustained period. Figure 2-1h shows a number of other types of relay coils and the deactivated state of their contacts, labeled in accordance with the abbreviations of Table 2-1. Note that the overvoltage relay (OV) clears a circuit when the voltage across its coil rises above a certain value, activating the relay, in agreement with Fig. 1-5a.

The control-circuit symbols shown in Fig. 2-1 are merely representative symbols used on schematic diagrams. A more complete list of symbols is given in Appendix A-2, showing those most commonly used. But the most complete list of symbols is found in ASA Y32.2 or NEMA ICS 1-101, and IEEE Std. 315-1971.

2-3.

ABBREVIATIONS
USED FOR DEVICE
DESIGNATIONS AND
NUMBERING

In connection with designations of control-circuit schematic diagram symbols, it was noted that certain standard *abbreviations* were used to designate relays, relay contacts, switches, or other devices, such as instruments, transformers, etc. The complete list of abbreviations used as standards to designate either electrical devices or functions is found in NEMA ICS 1-101.05. Table 2-1 is a partial list of such abbreviations as used in subsequent chapters.

TABLE 2-1 ABBREVIATIONS USED TO DESIGNATE DEVICES
          OR FUNCTIONS

FUNCTION OR DEVICE	ABBREVIATION	FUNCTION OR DEVICE	ABBREVIATION
Acceleration (accelerating)	A	Plugging	P
Ammeter	AM	Power factor meter	PFM
Braking	B	Pressure switch	PS
Capacitor(s)	C1, C2	Pushbutton	PB
Centrifugal switch	CS	Reactor(s)	X1, X2
Circuit breaker	CB	Resistor(s)	R1, R2
Control relay	CR	Reverse	REV
Current transformer	CT	Rheostat	RH
Diode(s)	D1, D2	Selector switch	SS
Disconnect switch	DS	Silicon-controlled rectifier	SCR
Field accelerating	FA	Solenoid valve	SV
Field loss	FL	Squirrel cage	SC
Float switch	FS	Squirrel-cage induction	
Forward	F	motor	SCIM
Fuse	FU	Starting contactor	S
Hoist	H	Tachometer	TACH
Jog	J	Transformer(s)	T1, T2
Limit switch	LS	Transistor(s)	Q1, Q2
Lower	L	Undervoltage	UV
Main contactor	M	Voltmeter	VM
Master switch	MS	Watt-hour meter	WHM
Overcurrent	OC	Wattmeter	WM
Overload	OL	Wound-rotor	WRIM
Overspeed	OS	induction motor	
Overvoltage	OV		

It is customary when designating devices that perform the same *control function* to *prefix* the number *before* the letter abbreviation. Thus, if three accelerating relays are used in a control circuit of a schematic diagram, both the relays and their contacts are designated as 1A, 2A, and 3A. Correspondingly, in designating electrical *components* or *devices*, used in conjunction with these functions, it is customary to *suffix* the letter abbreviation with the number. Thus, if three transistors are used, their designations are Q1, Q2, and Q3. (This widely used distinction, however, is

31

arbitrary since NEMA ICS 1-101 only specifies that function or devices designations may be prefixed *or* suffixed with appropriate numbers.) Since the convention described above is rather extensively used, it will be used throughout this work.

2-4.
SPECIAL DEVICE
ABBREVIATION
DESIGNATIONS

Occasionally, certain devices, particularly relays, have special functions. This is true of a multiple-contact relay, which may have certain auxiliary contacts and coils having specialized functions incorporated within the device. Thus, a given relay may contain a closing coil, a latch coil, and a trip coil. While the main coil and contactor of such a relay is designated as M, the holding coil may be designated HC and the trip coil TC, respectively.

The complete listing of standard abbreviations used to identify special functions of certain relay coils or contacts may be found in NEMA ICS 1-101.06, but Table 2-2 lists a few which are used in the subsequent chapters.

TABLE 2-2 ABBREVIATIONS USED TO DESIGNATE SPECIALIZED RELAY COILS AND CONTACTS

RELAY COIL FUNCTION	ABBREVIATION
Closing coil	CC
Holding coil	HC
Latching coil	LC
Time-delay contacts (closing)	TC or TDC
Time-delay contacts (opening)	TO or TDO
Trip coil	TC
Unlatch coil	ULC

2-5.
DISTINCTIONS
BETWEEN POWER
AND CONTROL
CIRCUITS

Power circuits are those which supply energy *directly* to the terminals of dc and ac motors. Designations for power-circuit markings are found in ASA Standard C6.1, 1956. In accordance with this standard, the armature terminals of a dc motor are designated A1 and A2, the (shunt) field terminals are designated F1 and F2, and the series field terminals are designated S1 and S2. Similarly, the stator terminals of a wound-rotor induction motor (WRIM) are designated T1, T2, T3, etc., and the rotor terminals of a WRIM are designated M1, M2, M3, etc. (as per IEEE Std. 58-1956).

The (source) lines of a power circuit are designated L1, L2, L3, etc., and are shown in a heavier weight than the (lighter) lines of control circuits. Again, while this convention is not standardized, it will be used throughout this text to distinguish readily between power and control circuits.

Control circuits are those circuits which control the flow of power from the source to the load (i.e., a motor). As indicated above, these are drawn as lighter-weight lines in the schematics used conventionally in industry and throughout this work. Various systems or schemes are used to designate control circuit wire numbers and terminals. The only ASA requirement is that such terminal markings follow a logical order and are consistently shown on schematics, equipment-wiring connection diagrams, and equipment interconnection diagrams. This consistency permits reference between the various diagrams so that the wiring terminals of individual pieces of equipment may be tagged in advance of delivery to facilitate both wiring installation and troubleshooting in the event of wiring errors between pieces of equipment.

In preparing control-circuit drawings, it is usually customary for the engineering designer or draftsman to draw the schematic first, followed by a timing sequence chart (if necessary), a connection wiring diagram showing the internal wiring of control equipment panels, and an interconnection diagram between remote-control stations, the control equipment, the power-source main disconnect, and the motor. The schematic diagram contains the logic of the control system employed, all the necessary control devices, and associated connections. Thus, a clear schematic will simplify layout and wiring of the control equipment diagram as well as later preparation of an interconnection diagram between pieces of equipment. The schematic also provides a clue to the method of designating terminal markings of power and control circuits.

For all these reasons, therefore, *only* schematic diagrams are shown throughout this text, with occasional control sequence charts where required. Such schematics will show both control and power circuits designated as described in this section, and as shown in the remainder of this work.

2-6.
TYPICAL
CONTROL-CIRCUIT
SCHEMATICS

The control and power circuits for a simple ac across-the-line starter* for a squirrel-cage induction motor are shown in Fig. 2-2. As described in Section 2-5, there are two basic circuits: (1) the single-phase (208-V) control circuit, in *light* line weight, supplied through fuses *FU*1 and *FU*2, and (2) the three-phase motor line or power circuit, in *heavy* line weight, supplied through the three-phase circuit breaker, *CB*, and disconnect switch, *DS*.

Note that the control circuit in Fig. 2-2 is not *energized* until the triple-pole disconnect switch, *DS*, and the manual circuit breaker, *CB*, are *each* closed. Even then the control circuit is inoperative until a remotely located (momentary) start button is depressed to energize the main relay, *M*, and close its main and auxiliary contacts in the power and control circuit lines, respectively. Thus, relay coil *M* serves as (1) undervoltage protec-

* See Section 5-2 for a more complete discussion of this starter.

33

tion (in the event of line voltages well below 208 V), (2) a triple-pole line-starting contactor, and (3) a means of stopping the motor, either manually, or, in the event of overload, automatically.

The control-circuit operation is initiated by depressing the momentary contact START button; this closes the large, normally open (n.o.), three-pole, main contactors M and the smaller, n.o. auxiliary contact M that shunts the START button. The function of the auxiliary contact M is to maintain the control circuit in an energized state when the START button is released. The motor starts across the line with all the M contacts closed.

The motor may be stopped by any one of the following methods:

1. Manually, by depressing the momentary n.c. STOP button.
2. By a sustained overload operating the inductive relay coils (through current transformers) OL_1 and OL_2, respectively, causing their n.c. contacts to open.
3. By a short circuit in the motor, causing the trip coils of the circuit breaker to disconnect the motor from the line, deenergizing coil M and resetting all M contacts to their n.o. position.
4. By a short circuit or overload in the control circuit, causing its low-rated fuses to open.
5. By a sustained undervoltage sufficient to cause coil M to become de-energized.
6. By opening the main switch disconnecting the motor from the three-phase supply.
7. By accidental single phasing of the motor, e.g., the opening of one of the M contacts in the line, or a defective breaker pole, causing an overload ($\sqrt{3}$ times the normal load current) and operating either OL_1 or OL_2, or both.

In accordance with the provisions of the National Electric Code (NEC), both short-circuit and overload protection are provided for the motor (as well as a separate switch disconnect, although in some circumstances the circuit breaker may serve this purpose). In addition, separate fuse protection is provided in both lines of the control circuit, as shown in Fig. 2-2.

A similar application is shown in Fig. 2-3a for a dc shunt motor. Since a shunt motor cannot be started directly across the line, the armature is accelerated by means of three steps of variable resistance in series with the armature. Short-circuit protection is provided in this application by fuses, and overload protection by a thermal overload relay. Both lines of the control circuit are fused, and the starting relay coil M also maintains the armature disconnected from both sides of the line when the main line two-pole switch is closed, in accordance with NEC standards.

The motor is started by depressing the momentary contact START button. The normally open main contacts M and the auxiliary contact M close, energizing lines 1 and 2 of the control circuit and the armature

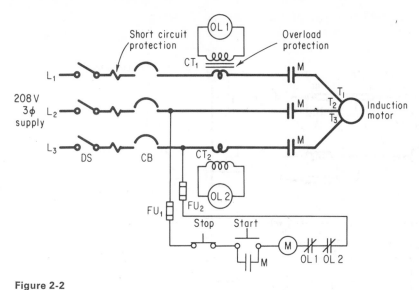

Figure 2-2

Control circuit protective devices used in an
induction motor starter.

circuit with full series resistance. The field circuit is energized when the
two-pole switch is closed to start the motor with full field (an inductive
field circuit takes longer to reach its maximum flux than the highly resistive armature circuit). The armature starts and accelerates for a period
determined by the dashpot-type time-delay acceleration relay $1A$. After
a brief period, relay $1A$ closes its n.o. contacts, shorting out one-third of
the series armature resistance and, at the same time, energizing line 3 of
the control circuit. After a similar brief period, relay $2A$ closes its n.o.
contacts, shorting out an additional third of the total series resistance,
and, at the same time, energizing line 4 of the control circuit. After an
equal interval of time, relay $3A$ closes, shorting the remaining third of
the armature series resistance, and the armature accelerates to its rated
speed across the line. Each time that additional resistance is cut out in
series with the armature, the armature develops more armature current
and more torque, and accelerates to a higher speed, producing more
counter emf. A convenient representation of the relay sequence with
respect to time is shown in Fig. 2-3b. The sequence shows that in the OFF
position, no relays are energized. When the START button is pressed, only
relay M is energized, followed by relays $1A$, $2A$, and $3A$, respectively.
In the RUN position, all four relays are energized. When the STOP button
is pressed, all relays are deactivated simultaneously.

Several interesting circuit protective devices are employed in this
starter: (1) the resistor TR, shunting the shunt field, is a *thyrite* resistor

35

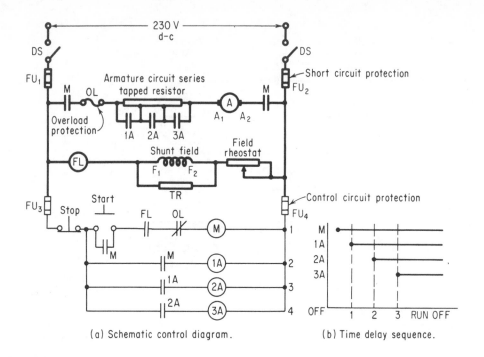

(a) Schematic control diagram.   (b) Time delay sequence.

Figure 2-3

Control circuit protective devices used in a
dc shunt-motor starter.

having a *high* resistance at *low* voltages but a low resistance at high voltages. Thus, when the main switch or the *M* contacts are opened, the high voltage across the shunt field is expended in the thyrite resistor rather than across the switch contacts; (2) a field-loss relay, *FL*, is included in the field circuit to protect the motor from runaway (in the event of an open field winding or a defective field rheostat) by deenergizing its n.o. contacts in line 1 of the control circuit, closed when main switch, *DS*, is closed.

The motor may be stopped:

1. By manually depressing the STOP button.
2. By opening the two-pole main line disconnect switch, *DS*.
3. By a sustained overload sufficient to operate the thermal overload relay *OL*, opening its n.c. contacts in line 1 of the control circuit.
4. By an open field circuit, deenergizing field-loss relay *FL*, opening its n.o. contacts in line 1 of the control circuit. (These contacts are closed when *DS* is closed.)
5. By a short circuit in the armature or field circuit, causing the main (*FU*1, *FU* 2) power fuses to open.

36

6. By a short circuit in any of the parallel-connected *control* lines, causing either of their low-rated fuses to open.

7. By an undervoltage sufficient to cause coil *M* to become deenergized.

8. By defective or inoperative contacts or relays, particularly coil *M*.

Subsequent chapters will show schematic diagrams employing the general protective devices described in Ch. 1, as well as other protective devices of a more specialized nature particularly germane to the type of motor or control employed.

BIBLIOGRAPHY

"Electronic Standards for Industrial Equipment" (General Motors Technical Center), *Electrical Manufacturing* (August 1958).

Ellenberger, J. "Start-Run Protection of Split-Phase Motors," *Electrical Manufacturing* (December 1959).

General Principles upon Which Temperature Limits Are Based in the Rating of Electrical Equipment (Publication No. 1) (New York: American Institute of Electrical Engineers).

Graphical Symbols for Electrical and Electronic Diagrams (ANSI Y32.2-1970) IEEE Standard 315-1971.

Harwood, P. B. *Control of Electric Motors*, 4th ed. (New York: John Wiley & Sons, Inc., 1970).

Heumann, G. W. "Motor Protection," in *Magnetic Control of Industrial Motors*, part 2, Chap. 6 (New York: John Wiley & Sons, Inc., 1961).

Howell, J. K., and J. J. Courtin. "Temperature Protection for Induction Motors . . . Today and Tomorrow," *Westinghouse Engineer* (November 1959).

Industrial Control Equipment (Group 25) (ASA C42.25) (New York: American Standards Association).

Industrial Control Equipment (UL508) (Chicago: Underwriters Laboratories, Inc.).

James, H. D., and L. E. Markle. *Controllers for Electric Motors* (New York: McGraw-Hill Book Company, 1951).

Johnson, R. C. *Electrical Wiring* (Englewood Cliffs, N.J.: Prentice-Hall, Inc. 1971).

Jones, R. W. *Electric Control Systems* (New York: John Wiley & Sons, Inc., 1953).

Karr, F. R. "Squirrel-Cage Motor Characteristics Useful in Setting Protective Devices," AIEE Paper 59–13.

Kosow, I. L. *Electric Machines and Transformers* (Englewood Cliffs, N.J.: Prentice-Hall, Inc., 1972).

Lebens, J. C. "Positive Over-Temperature Protection—With Heat Limiters," *Electrical Manufacturing* (January 1958).

Libby, C. C. *Motor Selection and Application* (New York: McGraw-Hill Book Company, 1960.)

Motor and General Standards (MG1) (New York: National Electrical Manufacturers Association; see also NEMA, ICS 1–101).

Raskhodoff, N. M. *Electronic Drafting and Design*, 2nd Ed. (Englewood Cliffs, N.J.: Prentice-Hall, Inc. 1972).

QUESTIONS AND PROBLEMS

2-1.   a. Name three kinds of electrical diagrams and give the function of each.
      b. Which of the above is most useful for comprehending control logic?

2-2.   Explain the meaning of the following terms:
      a. Normally closed contacts
      b. Normally open contacts
      c. Momentary contact
      d. Clearing a power circuit
      e. Clearing a control line
      f. Automatic disconnect
      g. Manual disconnect
      h. Activated pushbutton
      i. Energized relay

2-3.   For the fuse shown in Fig. 1-1,
      a. Draw the graphical symbol
      b. Show the device abbreviation above the symbol

2-4.   For the time-delay magnetic relay shown in Fig. 1-2b,
      a. Draw the graphical symbol for the relay coil and contacts in the deenergized state
      b. Label both coil and contacts with appropriate device abbreviation

2-5.   Repeat parts a and b of Question 2-4 for each of the thermal relays shown in Figs. 1-3a through d.

2-6.   Repeat parts a and b of Question 2-4 for the thermal overload heaters shown in Fig. 1-4 used with the bimetallic relay of Fig. 1-3b.

2-7.   Repeat parts a and b of Question 2-4 for the *undervoltage relay* shown in Fig. 1-5a.

2-8.   Define
      a. Power circuits
      b. Control circuits

2-9.   Draw a series circuit consisting of momentary START and STOP PBs in series with the undervoltage relay coil of Question 2-7. Explain
      a. Conditions under which contacts a'-b' are open
      b. Conditions under which contacts a'-b' are momentarily closed

c. Modify your drawing to show that relay coil *UV* contacts V_1-V_2 are *continuously* energized when the START button is momentarily depressed

d. On the basis of modification in part c, explain why START PBs are always shunted by auxiliary contacts of a main relay

e. Why STOP PBs are always n.c.

2-10. Using standard control-circuit symbols given in Appendix A-2, draw circuit diagrams a showing power and control-circuit schematic to energize a single 120-V, 500-W lamp from a 120-V single-phase ac supply by means of PBs and relays which will turn the lamp on and off. Include overload protection using

a. Overload relay

b. Circuit breaker with overload relay

c. Discuss the relative merits of parts a and b in terms of versatility, manual disconnect, etc.

2-11. Draw a properly labeled schematic diagram for automatically starting a single-phase capacitor start motor (terminals T_1-T_2) across the line showing

a. START and STOP momentary contact PBs

b. DPST main disconnect switch

c. Two main line contactors and main relay, *M*

d. Fuses for short-circuit and control-circuit protection

e. Two single-coil *OL* relays

2-12. Draw a schematic diagram for automatically starting and accelerating a loaded series dc motor in three steps using series armature protective resistance showing

a. Parts a through d of Question 2-11

b. One single-coil overload relay

2-13. In the event of reduced mechanical load on the series motor (and reduced line current), there is the danger that the motor might run away. What protection is afforded in the circuit of Question 2-12 to disconnect the motor from the line

a. Manually. Explain

b. Automatically. Explain

c. Modify the schematic drawn in Fig. 2-3 to include a n.c. overspeed switch to automatically disconnect the motor when the speed rises 200 per cent above rated speed.

manual
dc and ac starters

3-1.
GENERAL

This chapter and subsequent chapters will deal specifically with devices generally known as *controllers*. An electric controller may be defined as "a device (or group of devices) which serves to govern, in some predetermined manner, the electric power delivered to the apparatus to which it is connected."* The term "govern" is generally taken to imply variation, modification, or modulation of the power that is made available to the apparatus (motor). The term "predetermined" may be construed as the desired, intended, or conditioned sequence which the controller is capable of exhibiting or performing.

An electric *starter* is defined as a controller whose primary function is to start and accelerate a motor.

In this chapter, we will deal with *manual starters* using both direct current and alternating current, and in Ch. 4 we will deal with both

* "American Standard Definitions of Electrical Terms," ASA C42; and also "Standards for Industrial Control," NEMA Publication ICS 1–101.

dc and ac automatic starters. Subsequent chapters will cover manual and automatic controllers of a more sophisticated nature which will incorporate, in addition to starting functions, various other types of control features, such as memory, braking, speed control, and reversing.

The term "manual" implies the association of the hand of a human being (or its equivalent) in combination with an electric starter. This combination is, perhaps, the most sophisticated of all servomechanisms. It carries with it two implications: (1) that the human being possesses an intelligence, and (2) this intelligence is being directed in a purposeful way to operate a manual starter so that a motor may be properly started and accelerated under any conditions of load to reach its rated speed. It is almost impossible to build into any automatic device a versatility equivalent to that possessed by a human operator.* The senses of smell, touch, sight, hearing, and sound, properly used, may go a long way toward prolonging the life of the motor under extremely adverse starting and running conditions. Although there are certain advantages to automatic starting (see Section 4-1, the infinite superiority of *proper* manual control cannot be denied.

3-2.
STARTING AND
ACCELERATING DC
MOTORS

Figure 3-1 shows, in schematic form, the resistance in series with the armature circuit required for accelerating series, shunt, and compound motors, respectively, with a manual starting device for removing the series armature resistance in six steps. It should be noted that the shunt and compound motors are usually started with full field current, i.e., their field rheostat resistance is zero, and the series motor is always started under load.

At the instant of applying a voltage, V_a, across the armature terminals in order to cause a motor to rotate, the motor armature is not producing any counter emf since the speed is zero. The only current-limiting factors are the armature brush voltage drop and the resistance of the armature circuit, R_a. Since neither of these, under normal conditions, amounts to more than 10 or 15 per cent of the applied voltage, V_a, across the armature, the overload is many times the rated armature current, as indicated by Ex. 3-1.

Example 3-1 serves to illustrate that damage may be done to a motor unless the starting current is limited by means of a starter. (Commercial motor starters, both manual and automatic, will be covered in detail in Sections 3-3 through 3-6 and Ch. 4, but the following brief description is an introduction to the subject.) The current in Ex. 3-1 is excessive because of a lack of counter emf at the instant of starting.

* Having observed the manner in which numerous students in the laboratory and industrial workers in factories operate manual starters, the author is tempted to grim humor about these last statements; but perhaps they are better left to the imagination of the reader.

41

A 120-V dc shunt motor has an armature resistance of 0.2 Ω and a brush voltage drop of 2 V. The rated full-load armature current is 75 A. Calculate the current at the instant of starting and the per cent of full load.

Solution

$$I_{st} = \frac{V_a - BD}{R_a} = \frac{120 - 2}{0.2} = 590 \text{ A (counter emf is zero)}$$

$$\text{Per cent full load} = \frac{590 \text{ A}}{75 \text{ A}} \times 100 = \textbf{786 per cent}$$

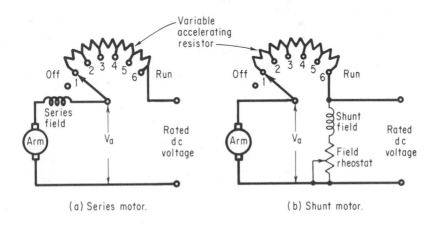

(a) Series motor.

(b) Shunt motor.

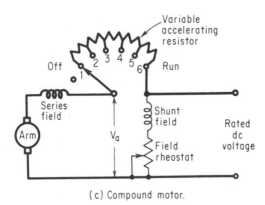

(c) Compound motor.

Figure 3-1

Schematics showing how accelerating armature resistance is connected and varied in armature circuit to accelerate series, shunt, and compound motors.

Once rotation has begun, counter emf is built up in proportion to speed. What is required, then, is a device, usually a tapped or variable resistor, whose purpose is to limit the current during the starting period and whose resistance may be progressively reduced as the motor gains speed. Given an external resistor, R_s, in series with the armature, the above equation must be modified for computing the armature current.

$$I_a = \frac{V_a - E_c - BD}{R_a + R_s} \qquad (3\text{-}1)$$

where V_a = voltage applied across the armature
E_c = counter emf generated in the armature
BD = voltage drop across the brushes
R_a = armature resistance

The value of the starting resistor at zero speed or any step along the way may be computed from Eq. 3-1, as illustrated by the following example:

EXAMPLE 3-2

Calculate the various values (taps) of starting resistance to limit the current in the motor of Ex. 3-1 to
a. 150 per cent rated load at the instant of starting
b. A counter emf which is 25 per cent of the armature voltage, V_a, at 150 per cent rated load
c. A counter emf which is 50 per cent of the armature voltage at 150 per cent rated load
d. Find the counter emf at full load, without starting resistance

Solution

$$R_s = \frac{V_a - E_c - BD}{I_a} - R_a \quad \text{[from Eq. (3-1)]}$$

a. At starting, E_c is zero; $R_s = \dfrac{V_a - BD}{I_a} - R_a = \dfrac{120 - 2}{1.5 \times 75} - 0.2 = 1.05 - 0.2$
$= \mathbf{0.85\ \Omega}$

b. $R_s = \dfrac{V_a - E_c - BD}{I_a} - R_a = \dfrac{120 - 30 - 2}{1.5 \times 75} - 0.2 = 0.782 - 0.2$
$= \mathbf{0.582\ \Omega}$

c. $R_s = \dfrac{120 - 60 - 2}{1.5 \times 75} - 0.2 = 0.516 - 0.2 = \mathbf{0.316\ \Omega}$

d. $E_c = V_a - I_a R_a - BD = 120 - (75 \times 0.2) - 2 = \mathbf{103\ V}$

Note that, in Ex. 3-2, a progressively decreasing value of starting resistance is required as the motor develops an increased counter emf as a result of acceleration. This is the principle of the armature resistance starter.

The manner in which a starter is used in conjunction with the three basic types of dc dynamos, used as motors, is shown in Fig. 3-1. The

43

techniques shown here for motor starting are schematic diagrams only; commercial forms of manual and automatic starters and controllers are covered in Sections 3-3 through 3-6.

The shunt and compound motors are started with full field excitation (i.e., the full line voltage is impressed across the field circuit) in order to develop maximum starting torque $(T = k\phi I_a)$. In all three types of dynamos, the armature starting current is limited by a high-power series-connected variable starting resistor. In commercial practice, the initial inrush of armature current is generally limited to a higher value than the full-load current (see Ex. 3-2), again to develop greater starting torque, particularly in the case of large motors which have great inertia and which come up to speed slowly.

Figure 3-2 is a graphical description of the variation in armature current and speed as the shunt motor is accelerated in steps to its rated speed. It should be noted that, because the motor starts from rest and because its armature has inertia, the first acceleration step (point 1) requires the greatest length of time before the acceleration becomes almost zero and the speed approaches that determined primarily by (1) the reduced voltage across the armature, (2) the load on the motor, (3) the field flux,

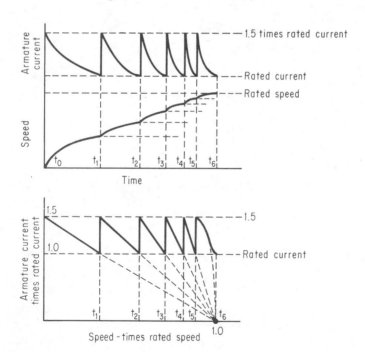

Figure 3-2

Acceleration of a shunt motor.

and (4) the counter emf ($S = k[V_a - I_aR_a]/\phi$). Advancing the arm to the second position (point 2), reduces the resistance and supplies a current surge of 150 per cent of the rated load (developing increased torque to accelerate the motor to the higher asymptotic speed determined by its new armature voltage).

Each time that the armature resistance is reduced at each step, the motor accelerates but requires less time in which to reach its asymptotic speed (zero acceleration) and reduce its current to approximately the rated load. The last step, however, may require somewhat more time than the intermediate steps because of the increased loading at increased speed, as shown in Fig. 3-2.

The two upper curves of Fig. 3-2 show armature current and speed, respectively, plotted against time. The lower curve shows armature current as a function of rated speed. If, at step 1, the motor could accelerate to higher speeds, the increased counter emf would reduce the armature current to almost zero at the rated speed. This is true of any other step as well; but, at each step, the shunt motor speed is limited ultimately by the equation ($S = k[V_a - I_aR_a]/\phi$), where V_a is the difference between the line voltage and the drop across the specific resistance in series with the armature.

Although the curves shown in Fig. 3-2 are given for a shunt motor, they hold true for series and compound motors as well, with slight differences in the variation of curvature. The number of acceleration steps is almost a function of the horsepower capacity of the motor. Larger motors having more inertia require more steps and a longer time interval to approach a given asymptotic speed.

Extremely small motors of fractional horsepower have so little inertia that no starter is required; they may be started directly across the line, since they accelerate and develop a (self-protective) counter emf almost immediately.

3-3.
COMMERCIAL
THREE-POINT
STARTER

Figure 3-3 shows the connection wiring diagram for a commercial *three-point manual starter* used to start a shunt motor. The figure incorporates four separate and distinct equipment units: (1) the three-point starter; (2) a safety switch box containing a switch, with short-circuit and overload protection; (3) the shunt motor, consisting of its armature and field circuits; and (4) an external field rheostat. The operation is the same as that described in the preceding section. Note that, at point 1 of the starter shown in Fig. 3-3, the full field voltage is applied to the field through a *holding coil*. When the starting arm reaches its last step (step 6), the current in the field circuit (through the low starting resistance, the holding coil, the field rheostat, and the field winding) produces sufficient mmf in the holding

45

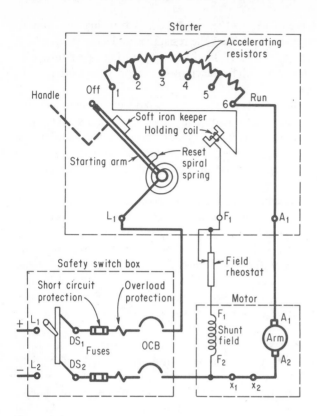

Figure 3-3

Manual three-point starter for shunt motor.

coil to maintain (the soft-iron keeper of) the starting arm magnetically attracted in this position against the countertension of the spiral reset spring.

The holding coil and the spring serve to provide (1) open or weakened field protection, and (2) undervoltage protection. In the event of either condition, the countertension of the spring (which is normally adjustable)* is sufficient to disengage the starting arm and return it to its OFF position. The three-point starter may also be used for a compound motor; the series field connection may be made at points X_1–X_2 in the motor, as shown in Fig. 3-3.

The motor may be stopped by any of the following conditions: (1) the opening of the DPST main line switch; (2) a short circuit (the main

---

* The counterforce may be "increased" also by employing nonmagnetic shims at the surface of the holding coil, by reducing the size of the keeper, or by placing a resistance in series with the holding coil. The last-named method is the most preferable.

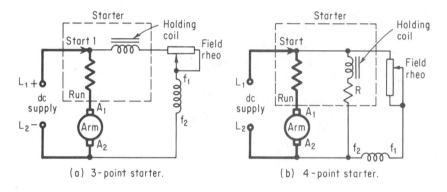

Figure 3-4

Schematic diagrams of three- and four-point starters.

fuse clearing the circuit) in the armature or the field circuit, including the starter; (3) an overload in the armature circuit (operating the overload circuit breakers, OCBs); (4) a weak or open-field circuit; or (5) a reduction in the line voltage.

As shown in Fig. 3-3, the primary purpose of the holding coil is to maintain the motor across the line, at the same time affording undervoltage and weak-field protection. The schematic diagram shown in Fig. 3-4a is a schematic representation of the commercial three-point starter. The starter consists of two parallel circuits, as shown. Its previously stated open-field protection advantage becomes a disadvantage when the motor is loaded and higher speeds are desired. In order to obtain higher speeds, it is necessary to weaken the field current and the flux (by increased field rheostat resistance). Whenever the field current is reduced, however, the starting arm is released and the motor stops.

Frequent and unexpected work stoppages of this nature led to the *four-point starter* shown in schematic form in Fig. 3-4b. The four-point starter differs from the three-point starter in that the four-point design provides *three* parallel paths across the supply rather than two. The additional path across the line has been created by the separate holding coil circuit in series with a protective resistor R. The resistor may be a tapped or semivariable type to counterbalance the tension of the spiral reset spring.

3-4.
COMMERCIAL
FOUR-POINT
STARTER

The *commercial four-point starter*, in which a fourth terminal is required for the holding coil circuit to L_2, is shown in Fig. 3-5. Because of its independent holding coil circuit, shown in Figs. 3-5 and 3-4b, it is possible to vary the current in the field circuit *independently* of the holding coil circuit. This is, of course, the greatest advantage of the four-point starter; but this advantage makes

47

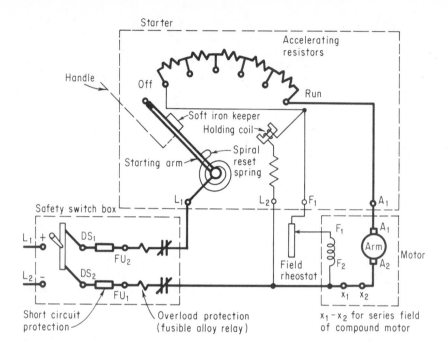

Figure 3-5

Manual four-point starter for shunt or
compound motor.

for three disadvantages:

1. If the field is weakened considerably or opened accidentally, the motor will race away to dangerous speeds; auxiliary protective devices, such as centrifugal overspeed circuits, are frequently used in industry.

2. In order to raise the speed when the motor is driving a heavy load, the field current is weakened; restarting the motor then causes the load to be accelerated much too rapidly and dangerously (the armature current increases to compensate for the weakened field flux in accordance with doubly excited dynamo theory).

3. The increase in the armature current may continue to trip the overload and even the short-circuit protective devices until the field rheostat resistance is decreased.

The last two disadvantages are overcome by incorporating an *automatic memory* feature into the four-point starter, as shown in Fig. 3-6. The field rheostat is built directly into the starter; and when the holding coil releases the starting arm (as a result either of undervoltage or of opening the line disconnect switch), the starting arm automatically *resets* the field rheostat arm to the OFF position. The operation of the starter

48

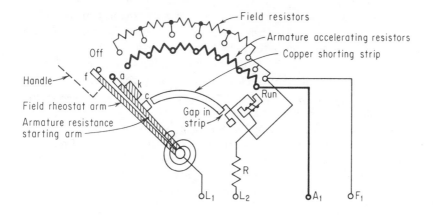

Figure 3-6

Manual four-point starter for shunt or compound motor with built-in speed adjusting field rheostat.

is as follows:

1. The handle moves both the field rheostat arm (contact f) and the armature resistance starting arm (contacts a and c) from the OFF position to the first armature resistance position; at this position, all the field rheostat resistance is shunted by the armature contact c, connected through a copper shorting strip to terminal F_1, and the motor is started with full field current.

2. Contact a on the armature starting arm provides maximum armature resistance on starting.

3. The armature starting resistance is reduced in the normal manner until the last or RUN position is reached, passing the gap in the copper shorting strip. In the RUN position, only the armature resistance arm is held by its keeper, k, and the holding coil; the field resistance arm is free to be moved by the handle.

4. The handle thus serves to start the motor and to control the setting of the field rheostat resistance by adjusting contact f of the arm to the desired increased speed above the basic speed (full field current).

5. In the event of undervoltage or of the opening of the line disconnect switch, the armature resistance starting arm is released by the holding coil, returning both arms to the OFF position.

The four-point starter (Fig. 3-6) with a built-in speed adjusting field rheostat (sometimes called a *compound four-point starter* or a *manual four-point speed controller*) possesses the advantages of (1) always starting the motor with full field current and (2) eliminating the requirement of a separate external field rheostat, since it may be used both for starting and for speed adjustment by control of the field rheostat.

49

It still has the disadvantage, however, of not providing open-field or weak-field protection. Another disadvantage is that, because the arm may not be at its customary fully clockwise RUN position, there is a temptation to stop the motor by returning the handle to the OFF position in a counterclockwise direction. This is highly dangerous because, in the process, the field is continuously weakened and the motor speeds up.

No starter, including the starter described above, should ever be stopped by manually "forcing" the arm away from its RUN position. In the starters shown in Figs. 3-3 through 3-7, such action will result in pitting of the starting contacts (step 1) when the heavy armature current and the highly inductive field circuits are interrupted. In addition, the starter of Fig. 3-6 will develop dangerously high speeds.

The proper way to stop a motor is to disconnect its line terminals, L_1-L_2, from the supply, using a line-disconnect switch. The arm will automatically be returned to its OFF position.

3-5.
SERIES-MOTOR
MANUAL STARTERS

A series motor is, fundamentally, a two-terminal or two-point circuit. The commercial three-and four-point starters previously discussed are used with shunt and compound motors, interchangeably, for the same starting current or horsepower rating. Special modifications are required for series-motor starters. Figure 3-7a shows a two-point manual series-motor starter in which the holding coil (a few turns of heavy wire) is connected in series with the series field and the armature. During the starting and the running period, a loaded series motor will have a sufficiently heavy armature current to excite the holding coil sufficiently to hold the starter arm in its RUN position. In the event of loss of load, or reduction of load to a point where dangerously high speeds develop, the reduction in armature current is sufficient to weaken the holding coil and to release the starting arm.

The two advantages of the series-motor two-point no-load release manual starter are (1) overspeed protection in the event of the removal or reduction of load, and (2) undervoltage protection.

In applications involving speed control, where the load is coupled to the motor in such a way that it cannot be removed (as in railway traction service, for example), the load on the motor may be reduced (at high speeds) sufficiently to disconnect the motor from the line, using the no-load release type of starter. The starter shown in Fig. 3-7b, the three-point undervoltage release manual series starter, is employed where there is no possibility of accidental runaway due to loss of load.

This starter is, to the series motor, what the four-point starter is to shunt and compound motors. Three terminals are available, L_1, L_2, and S_1. It provides undervoltage protection only, and does not protect the motor in the event of a weakened field. Generally, centrifugal overspeed protective devices are used with this starter. A variable protective resistor,

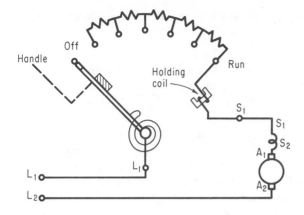

(a) 2-point starter, no load release.

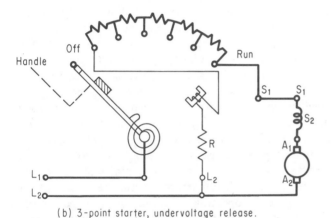

(b) 3-point starter, undervoltage release.

Figure 3-7

Manual series motor starters.

R, is connected in series with the holding coil (many turns of fine wire) across the line to adjust the release of the starting arm.

It should be noted that the three-point series undervoltage starter (with terminals L_1, L_2, S_1) is not normally interchangeable with the three-point starter (with terminals L_1, F_1, A_1), because the former has a fairly high protective resistance which would provide a weak field current for the shunt or compound motor. It is sometimes possible to short out this resistance, however, and, by renaming the terminals, modify the starter for shunt or compound operation. In the same way, by bringing out a field terminal, F_1, the starter shown in Fig. 3-7b may also be converted to a four-point starter (Fig. 3-5) with relatively little modification. This

51

simplifies the parts required for the manufacture of starters of a given horsepower.

3-6. MANUAL DRUM CONTROLLER

The manual, *flat* (faceplate) starters described above all possess certain electrical and mechanical disadvantages when used with shunt, compound, or series motors of greater horsepower. It is difficult to construct arcing and blowout protection for these starters, and the nature of the flat sliding contacts with gaps between them results in the pitting of contacts when starting larger motors.

These disadvantages are overcome in the rotary-type *drum* "controller," which has a set of rotatable heavy circular copper strip conductors, insulated from each other and positioned vertically as shown in Fig. 3-8a. The figure represents a 360° development of the contacts which are rotated by a handle at the top (not shown in the figure). A set of stationary spring-loaded contact fingers, separated by arc chutes and provided with magnetic blowout protection (Fig. 1-9b), are arranged vertically and concentric with the moving contacts a' through g' shown in Fig. 3-8b. The length of the movable contact segments may be varied to suit the desired predetermined sequence of operations, and the tension of the stationary finger contacts may also be adjusted. This construction, therefore, provides a greater flexibility in "governing the electric power in a predetermined manner" to start and control the speed of a motor.

The simple *nonreversing drum controller* shown in Fig. 3-8 is used in this application to start a series motor; it may be used for shunt and compound motors, as well, as indicated in Fig. 3-8b. A tapped accelerating resistor provides three steps of acceleration. When the rotary drum contacts are rotated to position 1 in Fig. 3-8a, contacts a' and b' are shorted, and the motor starts with full series armature resistance. After a suitable time delay, when the operator rotates the drum to position 2, contacts a', b', and c' are shorted, reducing the series armature resistance to two-thirds of its value and accelerating the motor to a higher speed. After a proper delay, the operator rotates the drum to position 3, where contacts d', b', and a' are shorted, maintaining direct contact from L_1 to d' and reducing the series armature resistance to one-third of its total resistance value. Finally, after another suitable delay, the drum is rotated to position 4 (RUN), where contacts a', b', and e' are shorted, thus shorting the entire series resistance and placing the series motor across the line.

It should be noted that there are two sets of common or shorted segments on the rotary drum: segments a through e and segments f, g. The first set serves to decrease the series armature resistance progressively and connect the motor to the line. The second set serves to energize main contactor relay M and its main and auxiliary contacts shown in Fig. 3-8b in the OFF position. The relay M thus maintains itself energized in all positions through its auxiliary M contact. When the drum is in any

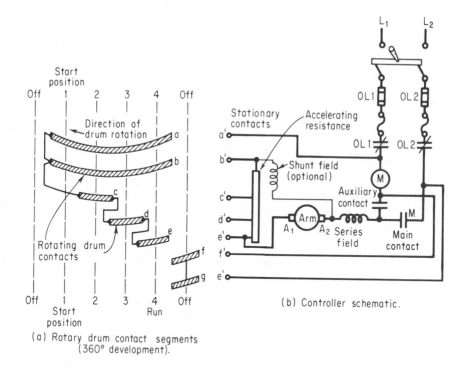

Figure 3-8

Manual drum controller (with overload, short
circuit, and undervoltage protection) for
starting series motor.

of the starting positions or in the RUN position (positions 1 through 4,
respectively), relay M will (1) serve as undervoltage protection, since it is
in parallel with the armature circuit across the line, and (2) necessitate
the return of the drum to the OFF position in order to restart the motor
in the event of a motor stoppage for any of the following causes.

The motor will stop under any of the following conditions: (1)
opening the main line switch; (2) a reduction in live voltage sufficient to
deenergize relay M; (3) an overload sufficient to cause thermal overload
relays OL_1 and OL_2 to trip their normally closed contacts; (4) a short
circuit in either the control relay circuit, the drum controller, or the motor
circuit, causing the fuses to clear the circuit; or (5) an open in the shunt
field circuit (shown as a dashed line) of either a compound or shunt motor
used with this controller.

If the resistor used in series with the armature of the motor has a
continuous rather than an *intermittent* duty rating, the drum controller
may be used as a method of armature resistance *speed control*, as well as

a manual dc motor starter, and the drum handle may be left continuously in any of positions 1 through 4.

Some of the advantages of drum controllers are as follows:

1. They are relatively low in price, in small and medium sizes, compared to automatic starters.
2. They are fairly compact in size, but the starting or speed control resistors must be externally and separately mounted.
3. The drum is completely enclosed and, if necessary, may be manufactured as gastight, waterproof, or explosion-proof.
4. Their operation is extremely simple.
5. They are mechanically very simple and rugged.
6. Too rapid acceleration may be prevented by built-in mechanical delay mechanisms.
7. They are extremely versatile electrically, because relatively complex internal connections may be made to permit reversing, braking, and other speed-control operations.

An improvement on the drum controller is the *cam* controller. The only essential difference between drum and cam controllers is the manner in which contact is made. In appearance and operation they are identical. In the cam controller, the rotating cylinder carrying movable contacts is replaced by a set (or decks) of layers of insulated cams which open and close sets of stationary contacts. The cam controller provides advantages over the drum controller in that (1) its contacts are relatively simple to maintain and replace, (2) new switching sequences and combinations may be readily created by replacing individual cams or decks, and (3) stationary contacts may be designed for extremely rapid opening and closing, not possible with drum controllers.

Smaller cam-operated drum controllers are sometimes called *master switches*. Indeed, both the drum and the cam controller are large-contact, multicontact, multipolar switches. Thus, master switches may be of the faceplate type, the drum type, or the cam-operated type, and they usually have fewer poles than the larger contact drum and cam controllers. One would be hard put, however, to distinguish between a *small* drum or cam controller and a *large* drum- or cam-operated master switch.

3-7.
MANUAL STARTING OF SQUIRREL-CAGE INDUCTION MOTORS (SCIMs)

Depending on the horsepower rating, small or large SCIMs may be started manually by using master switches or drum controllers. Where the capacity of the three-phase supply is sufficient and also in the case of small induction motors, across-the-line starting is employed as shown in Fig. 3-9a. Drum- or cam-operated two-position switches (OFF and RUN) would be used to close contacts a–a', b–b', and c–c', simultaneously, in the RUN position.

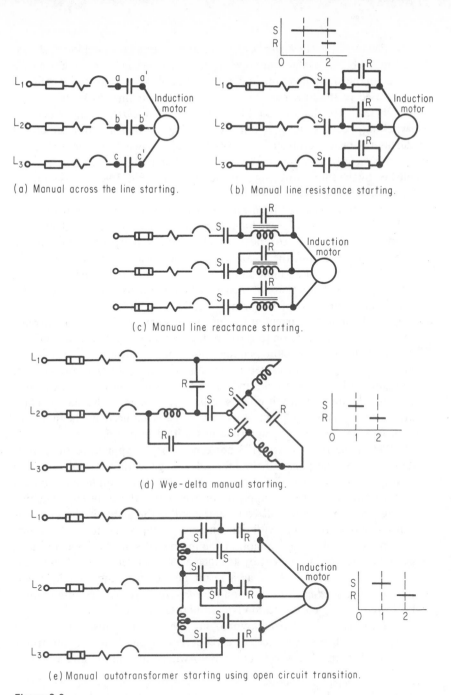

(a) Manual across the line starting.

(b) Manual line resistance starting.

(c) Manual line reactance starting.

(d) Wye-delta manual starting.

(e) Manual autotransformer starting using open circuit transition.

Figure 3-9

Manual starting of ac induction motors using
master switches, drum- or cam-operated controllers.

*Primary resistance starting** by means of manual switching is shown in Fig. 3-9b. All contacts are open in the OFF position. The S contacts are closed in the START position, and, after a suitable time delay, the R contacts are closed in the RUN position. This type of switching, as shown by the switch-sequence diagram in Fig. 3-9b, is called *closed-circuit transition* switching, because the S contacts energizing the motor stator are closed during the transition from the START to the RUN switch positions.

*Primary reactor starting** by means of manual switching is accomplished by the same switching sequence, shown in Fig. 3-9c, as was employed in primary resistance starting.

*Wye-delta starting** by means of manual switching is shown in Fig. 3-9d. This method of switching necessitates an open-circuit transition from the START to the RUN position. When the switch moves from its OFF to its START position, the S contacts are closed and the motor starts in wye. When, after a suitable interval, the switch is advanced to its RUN position, the S contacts are opened and the R contacts have not yet closed. This momentary loss of power may produce current transients of larger magnitude than the current at the instant of starting.† The stator currents in their transient decrease to zero produce a collapsing flux which induces a large voltage in the closed-circuit rotor rotating at a given speed. The induced voltage produces rotor currents which, by Lenz's law, attempt to maintain the air-gap flux; and, since the rotor is moving and carrying current, the air-gap flux it creates induces a voltage in the stator. Thus, at the very instant of power loss, a voltage is induced in the stator of approximately the same magnitude but shifted in phase.

When the full voltage (173 per cent more) is applied to the stator, depending on the phase of the induced voltage at that instant, the total stator applied voltage is the phasor sum of the instantaneous induced voltage and the increased applied voltage. Thus, the transient switching voltage may exceed the full voltage applied to the stator if the motor is started across the line with its windings delta-connected. The stator current is, of course, a function of the impressed stator voltage and may exceed the rated line current by as much as 600 per cent. Since these transients are produced directly as a result of the switching technique employed, the *closed-circuit* transition is *generally preferred*, because there is no instant at which the stator is disconnected from the line.

Autotransformer starting‡ employing open-circuit transition, is shown in Fig. 3-9e. The stator receives a reduced voltage (from the autotransformer when the S contacts are closed) whose magnitude and resulting

---

* For a complete discussion of the theory of starting methods of SCIMs see I. L. Kosow, *Electric Machinery and Transformers* (Englewood Cliffs, N. J.: Prentice-Hall, Inc., 1972), Chap. 9.

† S. B. Toniolo, "Behavior of Induction Motors after Short Interruptions of Supply," *Electrotecnica*, Vol. 30 (1943), pp. 181–194.

‡ See Kosow, *op. cit.*, Secs. 9–15 and 9–18.

56

starting current varies with the tap selected. After a suitable period, the switch may be brought to its RUN position opening the S contacts and closing the R contacts by open transition. Open transition is necessary here to avoid short circuiting a portion of the transformer winding. The autotransformer method produces the highest starting torque per line-ampere of starting current, compared to the resistance and reactance methods of reduced-voltage starting. Closed-transition autotransformer starting is also possible and is shown in Fig. 3-12.

*Part-winding starting,** in which the motor is started using half of its total three-phase stator winding with a consequent reduction in starting current (because of the higher impedance of the stator), is shown in Fig. 3-10a. The closed-circuit transition switching is shown in Fig. 3-10b and the sequence diagram in Fig. 3-10c. The S contacts are closed when the switch is in its START and RUN positions, while the R contacts are closed only in the RUN position.

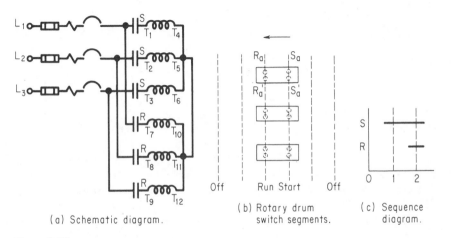

(a) Schematic diagram.

(b) Rotary drum switch segments.

(c) Sequence diagram.

Figure 3-10

Part-winding manual starting of an induction motor using drum-operated switch and showing sequence diagram.

**3-8.
WOUND-ROTOR
INDUCTION MOTOR
(WRIM) MANUAL
STARTING**

Like dc motors, smaller WRIMs may be started by using manual faceplate starters, as shown in Fig. 3-11a, and larger wound-rotor motors may be started by using drum controllers, as shown in Fig. 3-11b. In Fig. 3-11a, the stator is energized and protected by means of an overload circuit breaker (OCB). The motor does not start until the manual secondary resistance starter is rotated free from its open OFF position to its maximum rotor

* See Kosow, *op. cit.*, Secs. 9-15 and 9-18.

57

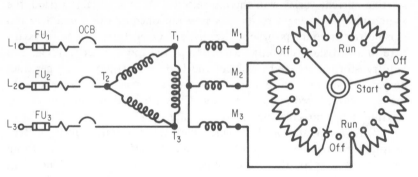

(a) Manual faceplate secondary resistance starter.

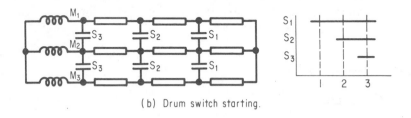

(b) Drum switch starting.

Figure 3-11

Two methods for manually starting a
wound-rotor induction motor.

resistance position. The motor accelerates to that value of speed and slip determined by its rotor resistance and will attain maximum speed and minimum slip in the RUN position, where the rotor winding is short circuited. The external starting resistances are usually intermittently rated for starting purposes. If it is desired to use the starter shown in Fig. 3-11a for speed-control purposes, then the resistors should be rated for *continuous* duty based on the continuous rotor current at any value of slip.*

Larger wound-rotor motors may be manually started by using drum- or cam-operated switches. These have the same advantages over faceplate starters listed in Section 3-6. Figure 3-11b shows the rotor circuit only, since the stator circuit is the same as in Fig. 3-11a. A closed-circuit transition sequence is employed, and the first set of starting resistors are shorted out by contacts S_1, the second set by S_2, and the last set by S_3. Depending on the size and severity of starting under load, additional sets of starting resistors and contacts may be used.

---

* A complete discussion of the theory of WRIMs is given in Kosow, *Electric Machinery and Transformers* (Englewood Cliffs, N.J.: Prentice-Hall, 1972). Sec. 9-11.

3-9.
POLYPHASE
SYNCHRONOUS
MOTOR MANUAL
STARTING

Since the stator of a synchronous motor is the same as that of an induction motor, the polyphase synchronous motor may be started as an induction motor on its damper windings* by any of the SCIM methods covered previously in Section 3-7. In all these methods, however, the field circuit is short-circuited (not energized during the starting period) until the rotor is close to synchronous speed. At this point, the dc field circuit is opened and dc voltage is applied to it, pulling the rotor into synchronism. As shown in Fig. 3-12b, the field is short-circuited through normally closed contact *R* to a discharge resistor. Thus, during the starting period, the field circuit acts in the same manner to aid the damper winding (in starting the motor as a squirrel-cage induction motor) until the rotor approaches synchronous speed.

The method of starting shown in Fig. 3-12a is a reduced-voltage auto-transformer manual starter using a closed-circuit transition; this is shown

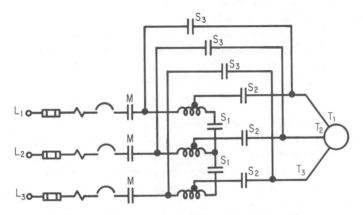

(a) Autotransformer start using closed circuit transition.

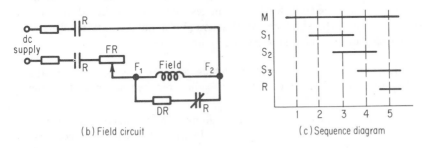

(b) Field circuit   (c) Sequence diagram

Figure 3-12

Manual starting of a synchronous motor.

* A complete discussion of starting methods for synchronous motors is given in Kosow, *op. cit.*, Sec. 8-5.

for purposes of comparison with Fig. 3-9e. In the first starting position, shown in the sequence table of Fig. 3-12c, contacts M are closed, but the rotor does not rotate in this position. In the second starting position, contacts M and S_1 are closed, energizing the autotransformers only but not the stator of the synchronous motor. In the third starting position shown in the sequence diagram of Fig. 3-12c, contacts M, S_1, and S_2 are closed, starting the synchronous motor under conditions of reduced voltage. After a suitable time interval, the switch is advanced to position 4, where contacts S_3 are closed and contacts S_1 are opened. The former contacts are closed before the latter are opened, to provide a closed-circuit transition to the higher stator voltage.

The rotor accelerates toward synchronous speed, and the switch is advanced to the fifth position, closing n.o. contacts R to the dc supply and opening n.c. contact R, thus short-circuiting the field. The field is thus excited, and the motor pulls into synchronism. At position 5, contacts S_2 are also opened. As shown by the sequence diagram, in the RUN position, n.o. contacts M, S_3, and R are closed, and n.c. contact R in the field discharge circuit is open. When advanced to the OFF position, the former group of contacts open and the latter contact closes. The motor is disconnected from the dc and ac lines, and the field discharge is absorbed by the discharge resistor DR, which may be a thyrite resistor in order to reduce the field power loss and to withstand the high voltage applied.

3-10.
MANUAL STARTING OF SINGLE-PHASE MOTORS

Manual switching techniques for single-phase series or universal motors may be the same as shown for the dc series motor. Induction-type single-phase motors requiring reduced-voltage starting are also connected in a similar manner as the polyphase induction motor previously considered.

3-11.
ADVANTAGES AND DISADVANTAGES OF MANUAL STARTERS

Indicated earlier in this chapter but summarized here for emphasis, the following advantages may be attributed to *manual* lever-handle faceplate starters and manual rotary-handle drum- or cam-type starters. These advantages are:

1. The time required for the switching sequence may be varied by the operator to suit the particular loading requirements placed on the motor.

2. Manual starters are simpler than automatic starters in both construction and maintenance. Neither electrical interlocks nor relays are necessary, and there is less possibility of breakdown.

3. Manual starters are more compact and weigh less than equivalent automatic starters of the same horsepower, since they contain less equipment. Where space and weight are serious factors, this may be of major importance.

4. Drum- and cam-type starters may be completely enclosed and rendered watertight or explosion-proof. Thus, they are less susceptible to damage caused by moisture and dust than automatic starters.

5. The initial and maintenance costs of manual starters are less than for automatic starters.

The following limitations or disadvantages of manual starters have led to the development of automatic starters (to be discussed in Ch. 4).

1. There is a physical limit to the size of manually operated starters. The larger starters have heavy contacts, requiring a good deal more effort to rotate the handle. Extremely large "manual" types have been "motorized" by means of stepping switches, offsetting this disadvantage somewhat. But there is no doubt that a large manual starter may tire an operator if frequently and continuously used.

2. Although the starter may be remote from the motor, the operator is in the immediate vicinity of the switching contacts where a great deal of power may be going through transition. In smaller starters, this is not serious; but in larger units, despite their complete enclosure, the possibility of rupture and internal explosion (however remote) still exists.

3. Where remote operation from a confined location occurs, the switching controller requires more space than a pushbutton station used for a magnetic automatic starter.

4. Because of its restricted enclosure, the possibility of a buildup of ionized gases through frequent switching within the sealed enclosure may lead to breakdown and extensive internal damage. Continuous-duty master or drum switches may require some means of ventilation to expel these ionized gases. Although this condition may also develop in an enclosed magnetic starter, there is less possibility of a buildup because the enclosure is larger (of necessity). Furthermore, the making and breaking of automatic contacts is of shorter duration, producing less ionized gas.

BIBLIOGRAPHY

Bellinger, T. F., and R. A. Gerg. "Closer Overload Protection for Polyphase Motors," *Allis-Chalmers Electrical Review* (Second Quarter 1960).

"Electronic Standards for Industrial Equipment" (General Motors Technical Center), *Electrical Manufacturing* (August 1958).

Ellenberger, J. "Start-Run Protection of Split-Phase Motors," *Electrical Manufacturing* (December 1959).

Fitzpatrick, D. "Reduced-Voltage Starting for Squirrel-Cage Motors," *Electrical Manufacturing* (March 1960).

Graphical Symbols for Electrical Diagrams (ASA Y32.2) (New York: American Standards Association, 1954). Updated in IEEE Std No. 315-1971.

Harwood, P. B. *Control of Electric Motors*, 4th Ed. (New York: John Wiley & Sons, Inc., 1970).

Heumann, G. W. *Magnetic Control of Industrial Motors*, 3 vols. (New York: John Wiley & Sons, Inc., 1961).

Industrial Control Equipment (Group 25) (ASA C42.25) (New York: American Standards Association).

Industrial Control Equipment (UL508) (Chicago: Underwriters Laboratories, Inc.).

James, H. D., and L. E. Markle. *Controllers for Electric Motors* (New York: McGraw-Hill Book Company, 1952).

Jones, R. W. *Electric Control Systems* (New York: John Wiley & Sons, Inc., 1953).

Kosow, I. L. *Electric Machines and Transformers* (Englewood Cliffs, N.J.: Prentice-Hall, Inc., 1972).

Motor and General Standards (MG1) (New York: National Electrical Manufacturers Association).

QUESTIONS AND PROBLEMS

3-1. Why must the first and last acceleration steps of a dc motor accelerated by means of a resistance starter require more time than any of the intermediate steps?

3-2. Why may fractional horsepower universal or dc motors be started across the line without requiring protective series resistance?

3-3. List five reasons why a commercial three-point manual starter will disconnect the motor from across the line.

3-4. Give one advantage and three disadvantages of a manual four-point starter compared to a three-point starter.

3-5. Show how the disadvantages of the four-point starter of Question 3-4 are overcome using a compound four-point manual starter.

3-6. a. Draw a diagram for a three-point manual *series* motor starter
   b. Show by diagram how this starter may be converted to a three-point manual starter which may be used with either a shunt or compound motor
   c. Repeat part b for conversion to a four-point manual starter
   d. Is it possible to convert a commercial three-point or four-point starter to a three-point *series* motor starter? Explain, using diagrams.

3-7. Give five advantages of cam or drum controllers over faceplate starters for manual starting of a dc motor.

3-8. List five advantages of manual starters over automatic starters.

3-9. List four disadvantages of manual starters compared to automatic starters.

62

3-10.   A 15-hp, 115-V, dc, 1750-rpm shunt motor has a rated armature current
        of 105 A, an armature resistance of 0.06 Ω, and a brush voltage drop of
        3 V. It is desired to start and accelerate the motor to rated speed in five
        equal steps using a drum switch and a separate tapped starting resistor.
        Calculate
        a. The starting resistance and each of its taps if the initial inrush motor
           current is to be limited to 175 per cent of rated armature current
        b. The speed to which the motor is accelerated at each step of the drum
           switch
        c. The internal torque developed at end of each step
        d. The internal torque developed at the beginning of each step
        e. Draw the complete diagram of the starter showing all calculated values
        Hint: See Ex. 3-2.

3-11.   Discuss in full detail the effect on the size and power rating of the resistors
        used in Problem 3-10 if
        a. The motor is to be started and stopped occasionally
        b. The motor is started and stopped frequently
        c. The "starter" is to be used as a speed controller.

3-12.   A 15-hp, 220-V, three-phase squirrel cage induction motor has a full-load
        line current of 40 A. The inrush current when started at full voltage is
        five times rated current at a locked rotor power factor of 0.35. Assuming
        that it is desired to limit the starting current to 2.5 times rated current
        using a line resistance starting, calculate
        a. The resistance of each line resistor
        b. The power factor at the instant of starting.

3-13.   Repeat Problem 3-12 using line reactance starters and calculate
        a. The resistance and reactance required
        b. The power factor at the instant of starting (assume that the reactors
           contain negligible resistance).

3-14.   A 125-hp, 220-V, three-phase squirrel-cage motor has an inrush starting
        line current of 1000 A when starting across the line. Calculate the inrush
        current when started by
        a. The wye-delta method
        b. The line resistance method where the impressed voltage is 60 per cent
           rated voltage
        c. An autotransformer using a 60 per cent tap (neglect transformer
           exciting current).

3-15.   Calculate the starting torque as a fraction of normal starting torque at
        full-line voltage for each of the three methods of starting in Problem
        3-14. Tabulate for ready reference and comparison.

ANSWERS

3-10(a) 0.56 Ω, 0.433 Ω, 0.319 Ω, 0.2045 Ω, 0.0895 Ω (b) 350 rpm, 700 rpm,
1050 rpm, 1400 rpm, 1750 rpm (c) 44.6 lb-ft (d) 66.9 lb-ft. 3-12(a) 0.9 Ω (b)
0.885 PF. 3-13(a) 0.655 Ω (b) 0.1745 PF. 3-14(a) 333 A (b) 600 A (c) 360 A.
3-15(a) $\frac{1}{3}T_s$ (b) $0.36T_s$ (c) $0.36T_s$.

automatic dc starters

4-1.
GENERAL

Automatic starters are designed to perform automatically the same functions as manual starters when controlled by one or more remote or locally operated manual momentary contact pushbuttons or starting switches. In general, it may be stated that automatic starters overcome the limitations of manual starters in that:

1. Automatic starters will not tire an operator during frequent start-stop cycles and, therefore, the operator is encouraged to stop the motor when it is not in use, thus reducing the electric load.

2. Automatic starters are not as limited in physical size as are manual starters by the human operating power required for a switching sequence.

3. The operator and the momentary contact control stations may be remote from the starter and the operator is thus protected by intervening distance.

4. Small pushbutton stations are more easily located in confined accessible spaces than drum or cam starters.

64

5. There is less danger of internal rupture and less need for internal ventilation in automatic starters.

6. Human error is eliminated and the motor may be started by an inexperienced operator in a minimum amount of time.

It should be pointed out, however, that the five advantages of manual over automatic starters covered in Section 3-11 are significant, and the choice between a manual and an automatic starter to start a given motor must take into account these considerations as well as the factors above.

As with all automatic devices, there are two general classes of operation: *open loop* and *closed loop*. Open-loop control governs the power to a motor in a predetermined manner *independently* of the operation of the motor. Closed-loop control governs the power to a motor in a predetermined manner which is *dependent* in part on the performance of the motor.

Automatic dc starters of the open-loop type are classed as *definite-time-acceleration starters*, and those of the closed-loop type as *current-limit-acceleration starters*. Space does not permit a complete discussion of all the various principles or techniques employed in each class of operation. For purposes of illustration, however, a few of each class will be presented, starting with the open-loop definite-time acceleration starters.

4-2.
DEFINITE-TIME-ACCELERATION DC MOTOR STARTER USING TIME-DELAY CONTACTORS

By way of illustration, Fig. 2-3a presented a dc starter using dc time-delay relays. As described in Section 1-5, these are inductive relays equipped with a dashpot or some other suitable time-delay escapement device so as to provide the time-delay sequence shown in Fig. 2-3b. Since the operation of the starter shown in Fig. 2-3a has been described in detail, no repetition of its description will be given here. Instead, a variation of this principle is shown in Fig. 4-1, in which time-delay *contactors* are used instead of time-delay relays.

A time-delay contactor is an ordinary inductive relay with a copper or brass sleeve inserted in its core. The effect of the nonmagnetic sleeve causes the relay to operate as an instantaneous magnetic relay under steady-state conditions, but as a time-delay relay whenever a transient current is flowing in the relay coil. Thus, for the normally closed relay contacts shown in Fig. 4-1, a rising current in the contactor coil will produce a delay in *clearing* contacts 1A, 2A, and 3A, respectively, and a decaying current will produce a delay in *closing* the same contacts. This delay is due to the opposing mmf's created by eddy currents induced in the sleeve in opposition to the rising or falling current, in accordance with Lenz's law.

The starter shown in Fig. 4-1 operates in the following manner:

1. The momentary contact START button is depressed and held in a depressed position long enough to permit contactors M, 1A, 2A, and 3A

65

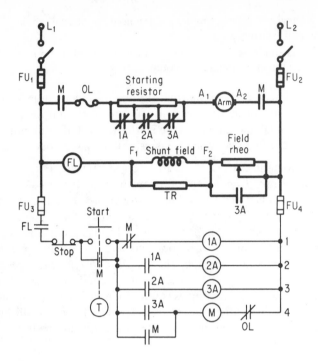

Figure 4-1

Definite time-acceleration dc starter
using time-delay contactors.

to operate their respective n.o. and n.c. contacts. A mechanical timer, *T*, coupled to the pushbutton, activates an alarm indicating when the button may be released. Releasing the START button resets the timer.

2. When the START button is released, all the relays are energized. Relay *M* in control line 4 remains energized through the auxiliary *M* contact, in parallel with the START button, with its interlock contact *M* shunting contact 3*A*. Relay *M* will be deenergized in the event of (a), undervoltage; (b), loss of field; (c), overload; (d), depressing the STOP button; (e), a blown fuse in the motor line circuit; or (f), a blown fuse in the control circuit.

3. The motor starts with its full starting resistance in series with the armature and its full field current, since relay 3*A* is energized.

4. Time-delay contact 1*A*, deenergized by relay contact *M* in control line 1, produces a specific time delay opposing decay of current in its coil, and it ultimately drops out, restoring its open contacts to their n.c. condition. The motor accelerates to a higher speed, as determined by the reduced series resistance with contacts 1*A* shorted.

5. Simultaneously with the closing of contacts 1*A*, time-delay contactor 2*A* is deenergized in control line 2 by the opening of contacts 1*A*. After a specific time delay, coil 2*A* "clears," shorting additional resistance in

series with the armature and opening control line 3. The motor accelerates to a higher speed, as determined by the reduced series resistance with both contacts $1A$ and $2A$ shorted.

6. Simultaneously with the closing of contacts $2A$ across the series resistor, contacts $2A$ are opened in control line 3, deenergizing time-delay coil $3A$. After a specific time delay, coil $3A$ "clears," shorting the remaining series armature resistance, opening interlock $3A$ in control line 4, and opening the field rheostat to permit the speed to rise to the desired speed setting. (This last step is actually a fifth step of motor acceleration.)

The circuit of Fig. 4-1 offers a number of specific advantages over the circuit of Fig. 2-3a. The only relay energized in the former during the running period is coil M, whereas in the latter case, all the relays are energized. This is not only wasteful, from a power point of view, but requires that the relay coils of the latter have continuous-duty ratings. A second, and perhaps more important advantage, is that the relay contacts $1A$, $2A$, and $3A$, respectively, are open and are not carrying current during the motor starting period. When they drop out, therefore, they do not interrupt current; nor are the relays energized at the instant that the contacts finally carry current and short out the series resistance. This starter requires smaller time-delay relays (using relatively little power to close the contacts not normally carrying current) having an intermittent rating. A disadvantage of this starter is that the time delays of the relays are *not* adjustable.

The automatic starter shown in Fig. 4-1 employs a single START and STOP station. Additional remote START stations may be added (to this and all other starters regardless of type) in *parallel* with the START pushbutton, and additional remote STOP stations may be added in *series* with the STOP pushbutton, in control line 1. The thyrite resistor TR absorbs the field energy whenever the motor is disconnected from the dc supply and the field is deenergized.

4-3.
DEFINITE-TIME-ACCELERATION DC MOTOR STARTER USING DASHPOT RELAYS

Another improvement over the definite-time-acceleration dc starter using dashpot relays, shown in Fig. 2-3a, is the type shown in Fig. 4-2. This starter also employs dashpot relays, but it attempts to overcome the disadvantages of a large number of energized relays during the running period and it provides adjustable time delays during the starting sequence.

The starter shown in Fig. 4-2 operates in the following manner:

1. Depressing the START button energizes coil M and closes all the n.o. M contacts. The motor starts across the line with full resistance in series with the armature and full field current, because n.c. contact $3A$ shunts the field rheostat.

67

2. The inductive dashpot time-delay relay, TD_1, is also simultaneously energized through the M contacts and the n.c. contacts of deenergized relay $3A$. After a suitable time delay, TD_1 closes its n.o. contacts, energizing instantaneous control relay $1A$ and time-delay relay TD_2, simultaneously.

3. When n.o. relay $1A$ closes, it shorts out one-third of the armature series resistance, causing the motor to accelerate toward a higher speed. After a suitable time delay, TD_2 closes its n.o. contacts, energizing instantaneous control relay $2A$ and time-delay relay TD_3, simultaneously.

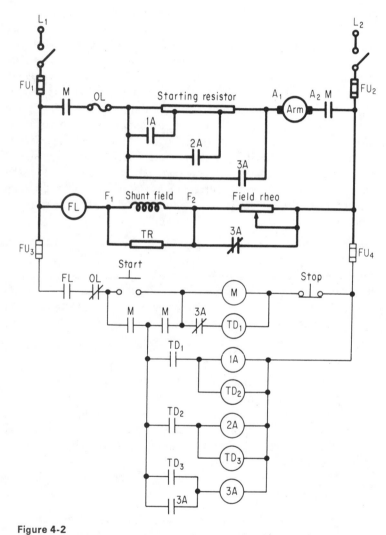

Figure 4-2

Definite time-acceleration dc starter
using dashpot relays.

4. When n.o. relay 2*A* closes, it shorts out two-thirds of the armature series resistance, causing the motor to accelerate toward a higher speed. After a suitable time delay, TD_3 closes its n.o. contacts, energizing instantaneous control relay 3*A* and shorting the remaining resistance in series with the armature.

5. Control relay 3*A* is the only relay to remain energized in addition to *M* because, when the former is energized, it opens n.c. contacts 3*A* in series with TD_1. When relay TD_1 is deenergized, it simultaneously clears relays 1*A* and TD_2. TD_2, when deenergized, simultaneously clears 2*A* and TD_3. Relay 3*A* remains energized through its electric interlock shunting the TD_3 contacts.

It should be noted that, although 1*A* and 2*A* are deenergized, the full series armature resistance is shunted by the contacts of 3*A*. At the same time, a fifth acceleration step is provided when n.c. contacts 3*A*, across the field rheostat, are opened, and the motor rises to its adjusted final speed. The dc starter of Fig. 4-2 has the advantage of: (1) the time delay is adjustable through the use of dashpot relays, (2) all the time-delay relays are deenergized during the RUN period and may be intermittently rated, and (3) a minimum number of relays are energized in the RUN period, thereby conserving energy and reducing heat within the starter. Starters using the same contact sequence as that shown in Fig. 4-2 are also available using relays having escapement mechanisms, toggle mechanisms, mechanical damper or drag mechanisms, pneumatic chambers, permanent magnets, etc., to provide the necessary time delay for definite time acceleration. (see Section 4-4).

4-4.
DEFINITE-TIME-ACCELERATION DC MOTOR STARTER USING MOTOR-DRIVEN TIMER OR TIMING MECHANISM

A single timing or time-delay relay having sequential multiple contacts may be used to provide the definite-time-acceleration sequence required to accelerate the motor. The starter of Fig. 4-3 employs a motor-driven timer (a slow speed, synchronous timing motor having timing contacts secured to an insulated drum) to close contacts TM_1, TM_2, and TM_3, respectively, in a predetermined sequence. These contacts energize n.o. control relays 1*A*, 2*A*, and 3*A*, respectively, in the same sequence with suitable time delays between each. Electrically actuated pendulums, solenoids, toggle and balance-wheel escapement time mechanisms, and pneumatic timers are also used to accomplish a predetermined sequence of 1*A*, 2*A*, and 3*A* in a similar manner.

All timing mechanisms, however, regardless of type, require some method of automatic reset should the sequence be interrupted temporarily by a power failure. In some mechanical timers, contact armatures are returned to their original START position by weights or springs. In other cases, as in the case of the motor-driven timer shown in Fig. 4-3, normally closed TM_5 and *M* contacts in series with the timer motor *TM* permit the

69

timer mechanism to be returned to its START position (when the power is restored), and to run until it opens its TM_5 contact. Thus, the timing motor is capable of running through its sequence and returning to START when power is reapplied independent of the n.o. START button. However, since the M coil is not energized, closing contacts 1A, 2A, and 3A during the period does not affect starting or produce any effect whatever. Because of the intervening reset time, however, rapid stops and starts (jogging) cannot be accomplished using this type of starter.

Timing-motor time sequences are adjustable, as are all timing mechanisms, and this is an outstanding advantage of this type of starter.

The starter shown in Fig. 4-3 operates in the following manner:

1. Start button closes n.o. M contacts when relay M is energized.
2. The timing motor, TM, begins to rotate when energized via n.c. TM_4 in line 1a and M contacts shunting start button, closing TM_5. TM_5 is thus closed and remains so until step 7.

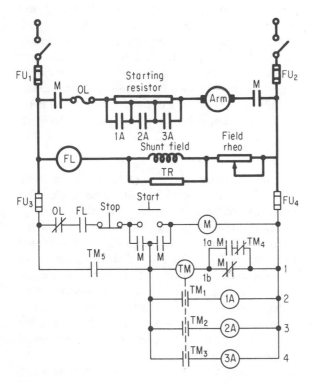

Figure 4-3

Motor-driven timer, definite time-acceleration dc starter (also pendulum or balance-wheel escapement).

3. Motor timer operates closing TM_1, TM_2, and TM_3, in sequence. These accelerate motor by energizing relays $1A$, $2A$, and $3A$, in turn, as predetermined by timer. When $3A$ closes, motor is running across the line.

4. Timing motor, TM, stops when n.c. TM_4 is activated, deenergizing line 1a.

5. With motor running across the line, all TM contacts of timing motor are in their activated state.

6. When STOP button is depressed, relay M is deenergized, and motor stops.

7. But timing motor is now energized via n.o. $TM5$ and n.c. M contacts in line 1b. The timing motor continues to run, opening simultaneously TM_1 through TM_3 and TM_5, and simultaneously closing TM_4.

8. When TM_5 is opened, the timing motor stops and the contacts are now completely recycled and ready to accelerate the motor, once again.

9. Note that in the event of overload, field loss, or opening the main switch at any time during and after the accelerating sequence, the timing motor continues to run via n.o. TM_5 (closed in step 2) and line 1b until all contacts are reset and TM_5 opens (steps 7 and 8). In the case of loss of power, this occurs immediately whenever power is restored.

4-5.
DEFINITE-TIME-ACCELERATION DC MOTOR STARTER USING INDUCTIVE TIME CONSTANT OF HOLDING COIL RELAYS

A starter using differential relays (see Table 2-2 and Fig. 1-6), consisting of two coils wound on a common relay core, is shown in Fig. 4-4. In addition to main contactor M, three relays are employed, each relay having an inductive HC (holding coil) winding acting in opposition to closing coil winding AC that tends to close contacts $1A$, $2A$, and $3A$, respectively. Thus, holding coils $1HC$, $2HC$, and $3HC$ tend to "hold" n.o. contacts $1A$, $2A$, and $3A$ in an open state as long as these coils are energized.

The starter shown in Fig. 4-4 operates in the following manner:

1. Depressing the START button energizes line contactor M and "closing" coils $1AC$, $2AC$, and $3AC$. But, since $1HC$ was energized (even before the START button was pressed), the n.o. contact $1A$ remains open. The starting current through the starting resistance also energizes $2HC$, maintaining n.o. contact $2A$ open as well. The motor starts, therefore, with its maximum series armature resistance across the line.

2. The START button and interlock contact n.o. M both short out holding coil $1HC$, and current decays in this relay in accordance with its own inductive time constant. When this current is sufficiently small, n.o. contact $1A$ closes as a result of the mmf of relay $1AC$. The motor accelerates to a higher speed with $1A$ closed.

3. Holding coil $2HC$, shorted by closed contact $1A$, undergoes an exponential decay, and ultimately contact $2A$ closes, accelerating the motor once again.

71

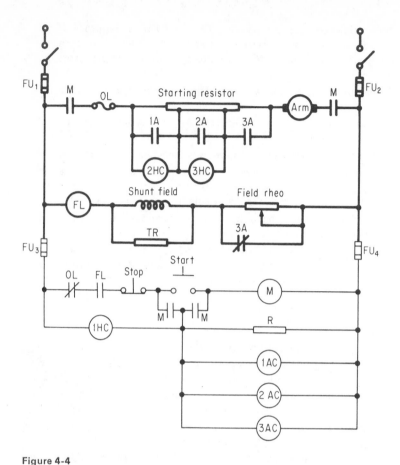

Figure 4-4

Definite time-acceleration dc starter using inductive time constant of holding coils.

4. Holding coil 3HC, shorted by closed contact 2A, undergoes exponential decay; and, after a period of time, contact 3A closes and the motor accelerates to its desired speed as determined by the field rheostat setting when n.c. contact 3A, shunting it, opens.

This starter has the disadvantage of requiring that relays M, 1AC, 2AC, and 3AC be continuously energized; in this respect it has the disadvantage of the starter of Fig. 2-3. On the other hand, holding-coil differential relays can be designed with extremely short air gaps that they develop powerful mmfs in comparison to dashpot time-delay relays, so that the coils of the differential relays of Fig. 4-4 do not require as much current as the coils of Fig. 2-3. A second disadvantage of this starter, however, is that the time of closure depends on the inductive time constant of the HC holding coils and is not readily adjustable.

Whether capable of adjustment or not, the greatest single disadvantage of all five definite-time-acceleration starters, discussed in Sections 4-1 through 4-5, is that the timing sequence is completely independent of the motor load current and speed. Thus, if the motor is heavily loaded, it may be accelerated too rapidly by means of a definite time starter. Similarly, if lightly loaded or unloaded, the motor may be accelerated too slowly, wasting precious production time. There is no feedback because of the open-loop nature of the control devices described. This disadvantage is overcome by the closed-loop current-limit-acceleration types described below.

4-6.
CURRENT-LIMIT-
ACCELERATION DC
STARTER USING
COUNTER-EMF
RELAYS (SPEED-
LIMIT METHOD)

The principle of closed-loop control is to provide a *feedback* from the power or line circuit to the control circuit. As shown in Fig. 4-5, this is done by means of "sensing" relays V_1, V_2, and V_3, connected in parallel across the armature, whose n.o. contacts V_1, V_2, V_3 control the sequence of relays in control lines 2, 3, and 4 shown in Fig. 4-5. The essential difference, then, between any open-loop and closed-loop starter method of control is that the former uses a *predetermined* control circuit sequence to initiate changes in the power circuit, whereas the latter uses "sensors" in the power circuit itself to initiate a response in the control circuit. This in turn produces or initiates a change in the *sequence* of the control circuit. This property of feedback in a closed-loop system will also be the major topic covered in Ch. 10.

The dc starter shown in Fig. 4-5 operates in the following manner:

1. Depressing the START button starts the motor with full field and full armature series resistance as contacts M, both main and auxiliary, close. The motor accelerates from standstill with a maximum voltage drop across the series starting resistance and practically zero voltage drop across the armature. As the armature accelerates, its counter emf and the voltage across the armature increase as a function of speed ($E = k\phi S$). Relay V_1, designed to close its n.o. contacts at about 50 per cent of the rated voltage, will not close and accelerate the motor until the speed of the motor with full field flux develops a counter emf of at least this value. In order for the counter emf to reach this value, moreover, the current must be limited sufficiently to reduce the voltage drop across the series resistance and, thereby, increase the drop across the armature.

2. When relay V_1 closes its n.o. V_1 contacts in control line 2, control relay $1A$ is energized, shorting out the first step of series armature resistance. The current surge increases, the motor accelerates, and more counter emf is developed. When the counter emf reaches 70 per cent of its rated value, V_2 is energized.

3. When relay V_2 closes it n.o. V_2 contacts in control line 3, relay $2A$ is

73

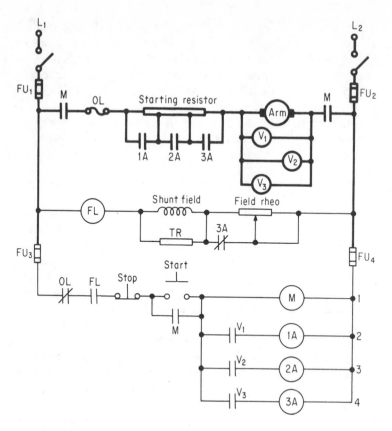

Figure 4-5

Current-limit acceleration dc starter using
counter emf relays (speed limit method).

energized, shorting out the second step of series resistance. The current
surge accelerates the motor once again, developing more counter emf.
When the counter emf reaches 85 per cent of the rated voltage, relay V_3
is energized.

4. When V_3 closes its contacts in control line 4, control relay $3A$ shorts
out the last step of series resistance and simultaneously opens the field
rheostat to its predetermined speed setting. The motor accelerates to
its load setting, with its armature across the line.

The advantages of this dc starter are that the motor is not accelerated
to a higher speed step until the current is sufficiently limited and the
necessary speed step (as measured by the counter emf) has been attained.
Thus, the time required for acceleration varies with the nature of the load
placed on the motor, and the motor is always accelerated automatically
at the proper time, rather than at a definite time. A second advantage of

this starter is the protection that it affords the motor in the event of a sustained heavy load which may cause a severe drop in speed; relay V_3 may drop out if the overload coil fails to function, and will not close contact $3A$ until the load condition is restored to a normal value.

There are several disadvantages to speed limiting using counter emf relays, however. The first is that the relays are hard to adjust so as to provide the desired action at the proper voltage and speed for all load conditions. Relays are inductive devices, and do not respond in the same manner to a rapidly rising transient current as they will to a slowly rising current. A second disadvantage is that seven relays are energized during the running period. In small starters up to about 5 hp, the control relays may be eliminated and relays V_1, V_2, and V_3 may operate contacts $1A$, $2A$, and $3A$ directly. A third disadvantage is the annoying possibility of a severe drop in speed as a result of a heavy load or undervoltage. A fourth disadvantage is that this device cannot be used when speed control below the basic speed is employed.

4-7.
CURRENT-LIMIT-
ACCELERATION DC
STARTER USING
HOLDING-COIL
RELAYS

Differential relays, similar to those described in Sec. 1-12, may be used to "sense" the armature current and to respond when the current has been sufficiently limited to produce the required acceleration. As previously shown, relay coils marked HC (holding coil) tend to hold the relay *open* whenever their currents and mmfs are large. Such a low-resistance series relay is also called a *series lockout* relay, and this type of starter is sometimes known as a dc series lockout starter. The starter, shown in Fig. 4-6, operates in the following manner:

1. Depressing the START button simultaneously energizes contactor M and $3AC$ in the control circuit and relays $3HC$, $1HC$, and $1A$ in the armature power circuit. The motor accelerates with full series resistance in the armature circuit, because coils $3HC$ and $1HC$ receive initial surges of current sufficient to lock out their relay contacts. Coil $3AC$, shunting M in the control circuit, is a closing relay which operates effectively in assisting $3A$ only when the relay armature has been pulled up to near its closure point.

2. As the armature current diminishes and the motor accelerates, current in $1HC$ in opposition to $1A$ decreases; and, since $1A$ has many more turns and is carrying the same current, under steady-state conditions it develops sufficient mmf to close contacts $1A$.

3. Armature current is now diverted through the low-resistance path of series relays $2HC$, $2A$, and $1A$ via n.o. contacts $1A$ closed by relay $1A$. When $1A$ closes, the current surges and the motor accelerates. The transient surge in $2HC$, in series with $2A$, is sufficient to permit the former to overcome the latter until the current returns to a sufficiently steady and low value. Under steady-state conditions, however, $2A$ closes its n.o. contacts.

75

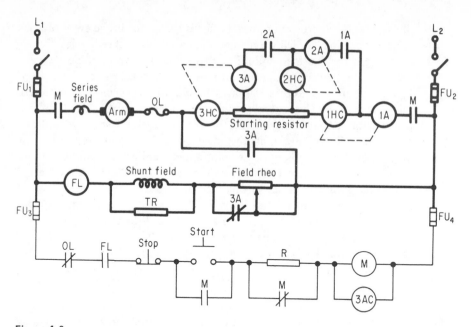

Figure 4-6

Current-limit acceleration dc starter using
holding coil relays.

4. Armature current is now diverted and surges through the low-resistance
   path of series relays $3HC$, $3A$, closed contacts $2A$, relay $2A$, closed
   contacts $1A$, and relay $1A$. During the transient current surge, $3HC$
   is sufficient to overcome $3A$. But when the motor acceleration and
   speed produce sufficient counter emf to reduce the armature current,
   relay $3A$ in combination with its aiding coil $3AC$, shunting M, close
   n.o. contacts $3A$, bypassing the entire network of starting resistance
   and relay coils across the line. Coil AC serves to hold n.o. contacts $3A$
   closed when relay $3A$ is shorted.

5. Simultaneously, n.c. contacts $3A$ (shunting the field rheostat) open,
   and the motor accelerates to its final speed setting. Resistor R, shunted
   by n.c. contacts M, serves both to provide undervoltage protection and
   to weaken $3AC$ sufficiently so that it will drop out rapidly when the
   motor is stopped, resetting all relays.

The chief advantage of the above dc starter is its relative simplicity.
Only three relays are required in addition to relay M. Its major disadvantage is that, because of its series coils, it is not easily adapted to speed
control at speeds below the basic speed, using lower voltages across the
armature. A second disadvantage is that the armature current rating
determines the design of the operating and lockout coils, so that one
starter cannot be readily adaptable for a range of motors of different
ratings. In the higher current ratings, furthermore, the armature current

CHAP. FOUR / *Automatic dc Starters*

is so large that the lockout coils must be wound with heavy copper bus bars of few turns, restricting the design severely and making it difficult to construct.

<table>
<tr>
<td>

4-8.
CURRENT-LIMIT-
ACCELERATION DC
STARTER USING
SERIES RELAYS

</td>
<td>

Fast-acting series relays, wound with a few turns of heavy wire or bus bar, may be used directly to "sense" the armature current surges and to accelerate the motor accordingly. The strong springs and air-gap stops of these relays may be mechanically adjusted to provide fast operating *closure* as well as fast contact *release*. A dc series relay starter used to start a dc series motor

</td>
</tr>
</table>

is shown in Fig. 4-7; it operates in the following manner:

1. When the START button is depressed, relay M energizes the armature power circuit and all the control lines. The heavy current surge through relay $1S$ opens n.c. contacts $1S$ in control line 2. Relay $1S$ is a fast-acting short-time constant relay compared to $1A$, and the motor starts with full armature series resistance. After the starting current surge decreases to a value where $1S$ can no longer remain closed against its spring, relay $1S$ opens releasing its contacts to a deenergized state.

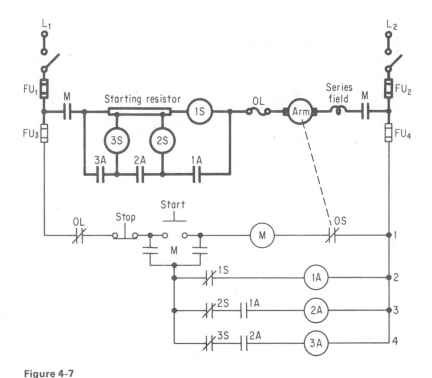

Figure 4-7

Current-limit acceleration dc starter using
series relays.

77

2. Control line 2 is energized by n.c. contacts $1S$ of deenergized relay $1S$. This energizes control relay $1A$, causing contacts $1A$ to close. Current now surges through low-resistance relay $2S$ and n.o. contact $1A$, shorting relay $1S$ and a portion of the starting series resistance. The motor accelerates until the starting current surge decreases to a value where $2S$ can no longer remain closed, and relay $2S$ opens.

3. Control line 3, energized by n.c. contacts $2S$ of deenergized relay $2S$, causes operation of relay $2A$, closing its n.o. contacts across the starting resistance. Current now surges through low-resistance relay $3S$, shorting relays $2S$ and $1S$ as well as the additional armature circuit series resistance. The motor accelerates until the starting current surge decreases to a value where $3S$ can no longer remain closed, and $3S$ opens.

4. Control line 4 is energized, operating relay $3A$ and shorting all series relays in the power line circuit as well as all armature series resistance. The motor accelerates to its normal rated speed at the rated load. Since the motor is a series motor, a centrifugal overspeed device in control line 1 protects the motor from running away in the event of loss of mechanical load.

When the series-relay starter shown in Fig. 4-7 is operating, four relays are always energized: the M relay and the three control relays $1A$, $2A$, and $3A$. This is the major disadvantage of this starter. Its coil design has the same disadvantages as those pointed out previously, however.

It should also be noted that the operation of the series relays is about 100 times *faster* than the operation of the control relays. There is no possibility, therefore, that, as the motor accelerates, a control relay will close its contacts before a series relay opens. The series-relay starter and its method of control are considered highly reliable, although the relatively large number of energized relays and the interlocks between series and control relays increases somewhat the cost of maintenance. This disadvantage is overcome in the holding-coil, voltage-drop starter considered in Section 4-9.

4-9.
CURRENT-LIMIT-
ACCELERATION DC
STARTER USING
HOLDING COILS
(VOLTAGE-DROP
ACCELERATION)

Figure 4-8 shows a dc holding-coil starter used to start a series motor. This starter "senses" the current surges across the protective starting resistance by means of HC holding coil relays. The relays are differential relays whose mmfs are in opposition. Operation of the starter shown in Fig. 4-8 is as follows:

1. Depressing the START button energizes contactor M, closing its main and auxiliary contacts. Armature current surges in the power line circuit, creating large voltage drops across all holding coils of the differ-

ential relays. Contacts 1*A*, 2*A*, and 3*A* remain open, therefore, as the motor accelerates.

2. As the surge current decreases, the voltage across the weakest relay *HC*1 decreases sufficiently, in time, to cause its countercoil 1*A* to close n.o. contacts 1*A*. When contacts 1*A* close, relay *HC*1 is shorted, as well as a portion of the starting resistance in series with the motor armature. The motor accelerates to a higher speed because of the armature current surge.

3. As the surge current decreases, the voltage across the weaker relay *HC*2 decreases sufficiently, in time, to cause its countercoil 2*A* to close n.o. contacts 2*A*. When contacts 2*A* close, relay *HC*2 is shorted, as well as a second portion of the starting resistance in series with the motor armature. The motor accelerates to a higher speed because of the armature current surge.

4. When the surge current decreases, the voltage across *HC*3 decreases sufficiently, in time, to cause its countercoil 3*A* to close n.o. contacts 3*A*. The motor accelerates to its normal speed with its armature circuit across the line. A n.c. overspeed contact *OS* in control line 1 prevents

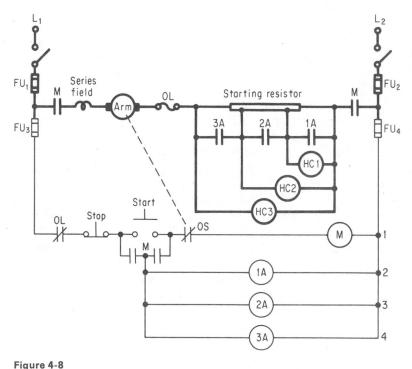

Figure 4-8

Current-limit acceleration using holding coils
(voltage drop method).

runaway in the event that the series motor is disconnected from the load, mechanically, or if the load decreases substantially.

The starter of Fig. 4-8 runs with four relays energized, but, in comparison to Fig. 4-7, it does not require any electric interlocks in its control circuits between individual relays. The voltage-drop method, moreover, lends itself to high-current starters more readily than devices using series relays whose coils must carry the full armature and surge currents.

It should be noted that all the current-limiting *closed-loop* starters shown in Figs. 4-5 through 4-8 have the advantage of accelerating the motor at the proper time, i.e., when the current surge has decreased to normal (see Fig. 3-2). Under extremely heavy loads, with high inertia, it may take quite a bit of time before the motor can accelerate, and the current will decrease slowly at each step. Unlike the definite-time, open-loop type of starter, these starters are capable of sensing the load and varying their time sequence to suit the load condition by using the surge of armature current as an indication of load.

4-10.
CURRENT-LIMIT-ACCELERATION DC STARTER USING HOLDING COILS AND MAGNETIC AMPLIFIER

*Magnetic amplifiers** have also been used in starting and control circuits to amplify the sensed surges of current and to provide the power to operate differential relays. Figure 4-9 shows an output voltage across the compensating winding or interpole of a shunt motor, x–x'. In the case of a series or compound motor, this output may be taken across the series field.

When the current surge is high, the output voltage across the series element is high. This sensed output is connected to the dc excitation or *control winding* of a magnetic amplifier (or saturable reactor), which is a connected to an ac supply. When the control winding saturates the core, the incremental inductance of the magnetic amplifier is small, and the ac output voltage E_{ac} across the magnetic amplifier is large. Conversely, when the dc control excitation is low or zero, the self-inductance and reactance of the series windings are high, and the ac output is low.

A full-wave silicon or germanium semiconductor rectifier converts the ac output to direct current, to excite the parallel-connected holding coils of the differential relays shown in Fig. 4-9. The holding coils shown in Fig. 4-9 are similar to those shown in Fig. 4-8 and operate in a sequence created by different values of dc voltage. The control circuit, therefore, is identical to that shown in Fig. 4-8. The starter of Fig. 4-9 operates in the following manner:

1. The motor starts with maximum series armature resistance and maximum surge current through the armature circuit. The high dc input

* See Section 8-4 for a more complete description of magnetic amplifier theory and operation.

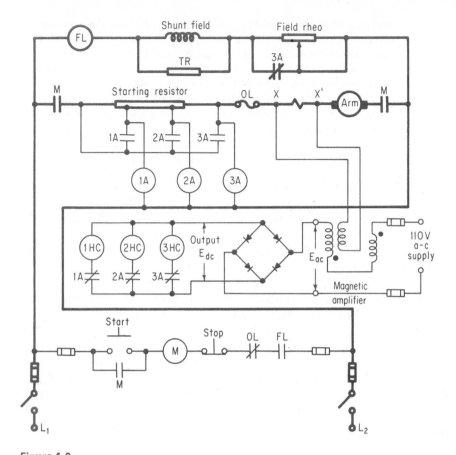

Figure 4-9

Current-limit acceleration using magnetic
amplifier and holding coils to start a dc shunt
motor.

to the magnetic amplifier produces a high dc output voltage across all
holding coils, maintaining contacts 1*A*, 2*A*, and 3*A* open. When the
current surge decreases, the weakest holding coil, 1*HC*, can no longer
restrain its armature, and differential relay 1*A* closes contacts 1*A*.
Differential relay 1*A* is the strongest relay because it has the largest
voltage across it.

2. When relay 1*A* operates, it simultaneously shorts out one step of the
series armature resistance and opens a n.c. 1*A* contact in series with
1*HC*, thereby preventing 1*HC* from responding to current surges.
When 1*A* shorts out the series armature resistance, the voltage differential across relay 1*A* is the same as the armature voltage; but, since its
opposing coil is no longer in the circuit, relay 1*A* remains energized.

3. The current surge created by shorting out one step of series resistance
accelerates the motor and provides a high dc output voltage from the

81

magnetic amplifier to holding coils $2HC$ and $3HC$. As the armature current surge decreases, since the voltage across $2A$ exceeds that across $3A$, differential relay $2A$ will close before $3A$ whenever the voltage $x - x'$ drops to a low value. When $2A$ closes, it shorts out additional armature circuit resistance and opens its n.c. contacts in series with $2HC$. Relay coils $1A$ and $2A$ are now in parallel across the armature.

4. When relay $2A$ operates, the current surges once again, locking out relay $3A$ by the action of $3HC$. Ultimately, however, the armature current drops to a normal value, and $3A$ prevails over $3HC$. The motor accelerates to its normal speed with its armature across the line.

The magnetic amplifier starter shown in Fig. 4-9 operates in the RUN position with all its closing relay coils, $1A$, $2A$, and $3A$, connected in parallel across the armature, and with the restraining differential (HC) coils across the magnetic amplifier output E_{dc} open-circuited. Any load surges or speed control changes, therefore, will not affect the motor operation except during the starting period.

The magnetic amplifier circuit, shown in Fig. 4-9, also has the advantage of sensing relatively small armature current changes and converting them into voltages which may be used to operate differential relays. The starter may be used with minor relay modifications on a wide range of motors. This is so because both coils of each differential relay are potential coils rather than current coils. A second advantage of the starter is the relatively fast response of the magnetic amplifier to changes in saturation (it responds in about five cycles of a 60-Hz circuit, or in $\frac{1}{12}$ sec). The amplifier itself, compared to electronic ac amplifiers having dc input and output modulators and demodulators, is relatively simple, insensitive to shock or vibration, and electrically rugged; and, in addition, it requires no more maintenance than an ordinary transformer. It can be used for ac starting and for dc and ac motor speed control, as well as in servomechanism applications, as shown in subsequent chapters.

The foregoing discussion of automatic dc starters is by no means a complete exposition of the various principles used in definite-time or current-limit acceleration. Only a few representative types of dc starters were presented to convey the nature of the techniques used to achieve automatic sequencing of the manual switching discussed in the preceding chapter.

BIBLIOGRAPHY

"Electronic Standards for Industrial Equipment" (General Motors Technical Center), *Electrical Manufacturing* (August 1958).

Fitzpatrick, D. "Reduced-Voltage Starting for Squirrel-Cage Motors," *Electrical Manufacturing* (March 1960).

Graphical Symbols for Electrical Diagrams (ASA Y32.2) (New York: American Standards Association) as shown in IEEE Std. No. 315-1970.

Harwood, P. B. *Control of Electric Motors*, 4th ed. (New York: John Wiley & Sons, Inc., 1970).

Herrmann, P. A. "Electronic Controls for Timed Acceleration of D-C Motors," *Electrical Manufacturing* (March 1957).

Heumann, G. W. *Magnetic Control of Industrial Motors*, 3 vols. (New York: John Wiley & Sons, Inc., 1961).

Industrial Control Equipment (Group 25) (ASA C42.25) (New York: American Standards Association).

Industrial Control Equipment (UL508) (Chicago: Underwriters Laboratories, Inc.).

James, H. D., and L. E. Markle. *Controllers for Electric Motors* (New York: McGraw-Hill Book Company, 1952).

Jones, R. W. *Electric Control Systems* (New York: John Wiley & Sons, Inc., 1953).

Kosow, I. L. *Electric Machinery and Transformers* (Englewood Cliffs, N.J.: Prentice-Hall, Inc., 1972).

Motor and General Standards (MG1) (New York: National Electrical Manufacturers Association).

Ogle, H. M. "The Amplistat and Its Application," *General Electric Review* (February, August, and October 1950).

Storm, H. F. *Magnetic Amplifiers* (New York: John Wiley & Sons, Inc., 1955).

QUESTIONS AND PROBLEMS

4-1. List six advantages of an automatic over a manual starter. Compare these to the answers given in Question 3-8. Summarize advantages and disadvantages in tabular form

4-2. a. Compare differences and relative merits of closed-loop vs. open-loop dc motor starting, including the necessity for starting under varying load conditions
b. List several types of open- and closed-loop dc starters described in this chapter
c. Would you consider a manual starter to be a closed-loop starter? Explain.

4-3. Explain the purpose of the copper sleeve used in the time-delay contactors of the starter shown in Fig. 4-1. Is it possible to vary the timing of such a contactor? How?

4-4. a. Discuss advantages of the starter of Fig. 4-1 over that shown in Fig. 2-3a. What are some of the disadvantages of the former in comparison to the latter?

b. Redesign the starter of Fig. 2-3 so that *one* relay (in addition to *M*) is energized during the running period instead of all and the motor is started and stopped from two remote locations.

4-5.   What specific relay characteristics are required so that the automatic starter shown in Fig. 4-2 will properly accelerate a dc motor?

4-6.   a. Draw a development of the cam of the timing motor used in Fig. 4-3 including the reset contacts and stationary contacts, labeling all parts

b. With reference to the diagram drawn in part a, describe the operation of the timing contacts in sequencing a motor through its acceleration to rated speed

c. Explain how the timing motor is reset when the motor is stopped at any time during its acceleration cycle or in the event of power loss or overload during this period

d. If the timing motor rotates at a speed of 2 rpm, explain why the starter of Fig. 4-3 cannot be used for jogging, reversing, or frequent start-stop operation.

4-7.   Discuss the relative advantages and disadvantages of the current-limit-acceleration dc starter shown in Fig. 4-5 in comparison to the open-loop starters shown in Figs. 4-1 through 4-4.

4-8.   Repeat Problem 4-7 for the starter shown in Fig. 4-6.

4-9.   Repeat Problem 4-7 for the starter shown in Fig. 4-7.

4-10. Show how the starter given in Fig. 4-8 overcomes the disadvantages of the starters discussed in Problems 4-7, 4-8, and 4-9.

4-11. a. From a reading of Section 8-4, modify the circuit of Fig. 4-9 so that the desaturating saturable reactor shown is converted to a self-saturating magnetic amplifier with external feedback having separate bias and control windings, including rheostats to adjust bias and control excitation

b. Discuss the relative merits of your redrawn circuit over that shown in Fig. 4-9.

automatic ac starters

5-1.
AUTOMATIC AC
STARTERS

A discussion of automatic ac motor starters is presented below. In general, it may be said that the majority of these are of the open-loop (rather than closed-loop) type, employing definite-time acceleration. Most ac motor starters contain ac relays which operate on the same principles as dc relays. High-capacity motor starters, however, employ one or more sets of bridge rectifiers (silicon, germanium, or selenium) and use dc control and accelerating contactors which have the advantage of stronger and more positive magnetic closure with less tendency to chatter. The use of dc relays in the control circuit, moreover, permits the use of both inductive and capacitive time constants to provide suitable time delay, as well as such devices as ignitron tubes and silicon-controlled rectifiers.

In addition to the use of direct current in the control circuits, it is also customary to use reduced voltage for the control circuits of higher voltage ac machines. Starters for ac motors, both single-phase and polyphase, that have rated voltages of 220 V or more, generally use transformers to

85

obtain a lower voltage (usually 115 V) for ac or dc control circuits. Those that use a lower voltage are called "reduced control voltage starters"; those that use direct current are called "dc control circuit starters."

Various types of definite-time starters are available for starting single phase and polyphase motors in the three general classifications of full-voltage or across-the-line starting, reduced-voltage starting, and part-winding starting.

No discussion of polyphase-induction-motor starters is possible without consideration of the various classes of squirrel-cage induction motors (SCIMs) based on starting torque and starting current.* Table 5-1 classifies the differences in rotor construction according to the NEMA class letter and shows the effect of rotor-construction differences on starting torque, starting current, efficiency, etc.

Examination of the applications in Table 5-1 reveals that the rotor modifications were developed to achieve across-the-line starting. Only the NEMA Class A SCIM may require a reduced-voltage starter.

5-2.
AUTOMATIC
MAGNETIC
ACROSS-THE-LINE
STARTER FOR
POLYPHASE SCIMS

An automatic across-the-line starter using reduced *control* voltage is shown in Fig. 5-1. Starters of this type are available in sizes that have been standardized by NEMA specifications. For example, NEMA size 00 is for $\frac{3}{4}$-hp, 110-V, single-phase and three-phase motors, for 1.5-hp polyphase motors using 220 V, and for 2-hp motors (three-phase) at 440 to 550 V.

The NEMA sizes run up to size 9, intended for 800-hp, 220-V motors and for 1600-hp motors using 440 or 550 V, with a continuous current rating of 2250 A. The NEMA sizes are standard items, readily available from manufacturers of starter equipment; larger sizes can be obtained on special order. Table 5-2 gives the ratings of standard NEMA across-the-line starters.

As in the case of all tables, it is customary to use the next larger size available for a horsepower not specifically designated. For example, a 10-hp, 220-V three-phase motor would require a size 2 NEMA across-the-line starter. It is also customary to go to the next larger size for a motor which is frequently and repeatedly started and stopped as in reversing, jogging, and plugging (rapid service). *Rapid service* is defined as *more than five operation changes* per minute. Thus, a 30-hp, 220-V, three-phase motor which may be subjected to rapid service would require a size 4 NEMA across-the-line starter.

The NEMA starters listed in Table 5-2 are available in a variety of enclosures: drip-proof, waterproof, explosion-resisting, etc., and it is customary to specify the starter enclosure along with the NEMA size, and the motor line voltage and horsepower.

* For a more detailed theoretical discussion, see I. L. Kosow, *Electric Machinery and Transformers* (Englewood Cliffs, N.J.: Prentice-Hall, Inc., 1972), Sec. 9-21.

TABLE 5-1 NEMA CLASSIFICATIONS AND CHARACTERISTICS OF POLYPHASE SCIMS BASED ON ROTOR CONSTRUCTION

NEMA CLASS LETTER	CONSTRUCTION OF ROTOR AND TYPE	TORQUE		STARTING CURRENT (per cent full load)	SLIP (per cent full load)	POWER FACTOR (per cent full load)	EFFICIENCY (per cent full load)	AVAILABLE RANGE OF COMMERCIAL SIZES (hp)	APPLICATIONS BASED ON CHARACTERISTICS
		Starting (per cent full load)	Pullout (per cent full load)						
A	Normal rotor, normal torque and starting current	105–150	200–250	500–1000	3–5 low	85–90 high	88–92	0–200	Usually requires reduced voltage starter but may be started across the line where capacity permits. Used for constant speed loads not requiring high starting torque, such as machine tools, blowers, and fans.
B	General-purpose rotor, low starting current	105–158	190–250	500–550 low	3–5 low	82–87	87–89	Same as above	May be started across the line. Applications similar to Class A with characteristics of somewhat lower power factor and starting current.
C	Double cage, high torque, low current	200–250	200–230	500–550 low	3–7	82–84	82–84	1–200	High-starting-torque, constant-speed loads such as pumps, compressors, refrigeration equipment, rock and pulp crushers, conveyor belts, and belt-line assemblies. May be started across the line.
D	Single cage, high-resistance rotor, high torque, high slip	250–350	Same as starting torque	300–800	7–15	50–75	50–75	0–150	Highest starting and accelerating torque but highest slip of all classes. Punch presses, shearing and stamping of heavy plates and other high inertia loads, such as hoists, cranes, and elevators. May be started across the line.
F	Double cage, low torque, low starting current	50–80	150–190	350–500 very low	2–4 low	82–88	87–90	40–200	Low inertia and low starting torque loads such as centrifugal pumps, blowers, and fans. High efficiency and low slip. May be started across the line.

TABLE 5-2   RATINGS OF NEMA ACROSS-THE-LINE STARTERS

NEMA SIZE NUMBER	8-hour Contactor Current Rating (A)	HORSEPOWER AT LINE VOLTAGES AS INDICATED BELOW		
		110-V, three-phase (hp)	220-V, three-phase (hp)	440/550-V, three-phase (hp)
00	9	0	1.5	2
0	18	2	3.0	5
1	27	3	7.5	10
2	45	—	15	25
3	90	—	30	50
4	135	—	50	100
5	270	—	100	200
6	540	—	200	400
7	810	—	300	600
8	1215		450	900
9	2250		800	1600

In their simplest form, NEMA starters contain a three-pole magnetic contactor, M, with one auxiliary M contact and two overload relays and their respective contacts as shown in Fig. 5-1. The step-down transformer is optional and only used with higher voltage motors. The START and STOP pushbuttons are usually external to the enclosure but may be mounted in the enclosure cover. A manual reset is sometimes provided with certain types of overload relays (Sections 1-6 and 1-7), and is located in the enclosure cover.

It bears repeating that across-the-line magnetic starters are usually used with NEMA class letter SCIMs B through E, Table 5-1. Nevertheless, reduced-voltage starting may be used with all SCIMs where frequent starting results in annoying line-voltage fluctuations due to excessive starting current.* Reduced-voltage automatic starters are described in Section 5-3.

The operation of the across-the-line starter shown in Fig. 5-1 is as follows:

1. Depressing the START button in the reduced-control-voltage circuit energizes contactor M, an ac relay, which immediately places the ac motor across the line.
2. The motor will stop under any of the following conditions:
   a. The opening of the line-disconnect switch.
   b. A short circuit in the high-voltage power line to the induction motor.

* For a detailed discussion of induction-motor starting, see Kosow, *Electric Machinery and Transformers*, Sec. 9-14 ff.

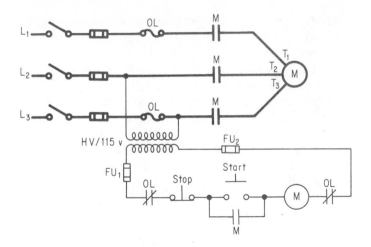

Figure 5-1

Automatic across-the-line ac starter.

    c. A short circuit in the 115-V, ac control circuit, clearing fuses *FU*1, *FU*2.

    d. An overload in the power line circuit that causes inductive or thermal relays to trip the n.c. *OL* relay contacts.

    e. The depressing of the n.c. STOP button.

    f. A temporary reduction or loss of line voltage sufficient to deenergize relay coil *M* in the control circuit.

It can be seen from the above that, despite the simplicity of its design, a great deal of protection is provided by the simple automatic across-the-line magnetic starter.

5-3.
AUTOMATIC
REDUCED-VOLTAGE
DEFINITE-TIME-
ACCELERATION AC
STARTERS

Where the capacity of supply permits, even extremely large induction motors may be started automatically across the line without damage to the motor. Where the inrush of starting current produces momentary line-voltage reductions affecting other electric or electronic devices, however, some method of *reduced-voltage* starting is required. Section 3-7 and its associated figures showed various closed-circuit and open-circuit transition techniques used in the manual starting of induction motors by primary resistance, primary reactance, wye-delta, autotransformer, and part-winding starting. All these circuits (Figs. 3-9 and 3-10) involved a *two-stage* START and RUN sequence in which the RUN contacts are closed after a suitable time delay. Thus, any control circuit which provides for (1) the closure of a set of START contacts, (2) a means of providing a time delay, and then (3) the closure of a set of RUN contacts may be used universally for primary resistance, primary reactance, wye-delta, autotransformer, and part-

89

winding starting. Section 3-7 also drew a distinction between *closed-circuit* and *open-circuit transition* in the switching sequence from START to RUN. A starter of each type will be discussed here.

Figure 5-2 shows a *closed-circuit transition reduced-voltage* or part-winding ac induction motor starter. In this instance, for purposes of illustration only, it is used as a *primary resistance* starter. As shown in the accompanying sequence diagram, three stages of definite time delay are provided before the RUN contacts are closed. The 115-V, ac control circuit is energized through a control circuit transformer, and, in the larger sizes, full-wave rectifiers are used to provide a dc control circuit using dc relays. The time-delay relays, whether they use alternating or direct current, are usually of the adjustable dashpot type. The starter shown in Fig. 5-2 operates in the following manner:

1. Depressing the START button energizes main and auxiliary contacts S in the power and control circuits, respectively. The motor starts with reduced primary voltage and current.

2. The ac dashpot time-delay relay, TD_1, energized through n.o. S contacts and n.c. TD_3 contacts, proceeds to pull its armature to a closed position. After a suitable time delay, relay TD_1 closes its n.o. TD_1 contacts, energizing time-delay relay TD_2.

3. TD_2 also provides a time delay before it, in turn, energizes time-delay relay TD_3 through n.o. contacts TD_2 and n.c. contacts R. When, after a suitable time delay, TD_3 closes, it energizes line contactor R.

4. When ac (or dc) line contactor R closes, it shorts out all primary starting resistance (in this instance), producing a second and smaller inrush of current as the motor is brought across the line. At the same time, auxiliary contacts R deenergize all the time-delay relays. Only relays S and R remain energized through their auxiliary interlock contacts.

The advantages of the starter shown in Fig. 5-2 are:

1. It provides closed-circuit transition from START to RUN where such transition is required, as shown in contact sequence table of Fig. 5-2.

2. Each of the time-delay relays is adjustable within limits to provide a fairly wide range over which time delay may be obtained to suit the starting requirements dictated by the nature of the load and the feeders supplying the load.

3. Only two relays are energized during the RUN period.

4. Additional time-delay relays may be added in the control circuit to provide longer delays if required.

5. The same starter may be used equally well with primary resistance, reactance, autotransformers, and part windings for a given horse-power rating, depending on the size of the line contacts, voltage and service, as noted.

6. The same starter may be used with induction motors of various voltage ratings by changing the control-circuit transformer.

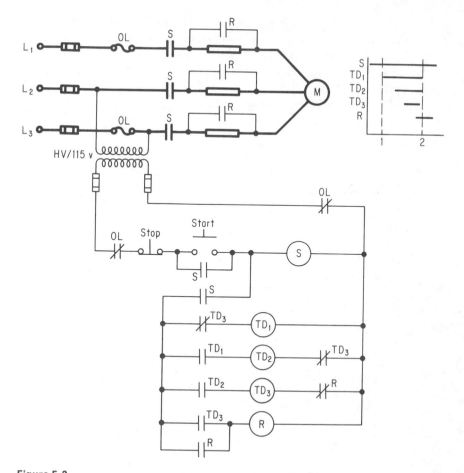

Figure 5-2

Definite-time acceleration using time-delay
relays in closed-circuit transition.

7. In some of the larger horsepower ratings, current transformers (Section 1-10) may also be used with overload coils and contacts of the same size since these operate in the control circuit only, greatly simplifying the parts required in manufacturing starters of various ratings.

Figure 5-3 shows only the control circuit and contact sequence for a basic *open-circuit transition* ac starter of the definite time-acceleration (open-loop) type. The circuit shown uses dc control relays (although ac relays may be used in the smaller ratings) energized through a full-wave rectifier from the reduced-voltage control-circuit transformer. The circuit also permits a separate set of *M* main and auxiliary contacts to be energized (in addition to a set of *S* start and *R* run contacts), should the switching transition require it (see Fig. 3-12). The dc timing relay *TR* may be any of the types described in Section 4-4, including a timing motor. The starter is provided with both mechanical and electric interlocks to ensure open-circuit transition (usually a necessity in simple wye-delta starting and in some types of autotransformer starting, although circuit schemes and protective resistors may be used to overcome disadvantages of this type of transition.

The operation of the starter shown in Fig. 5-3 is as follows:

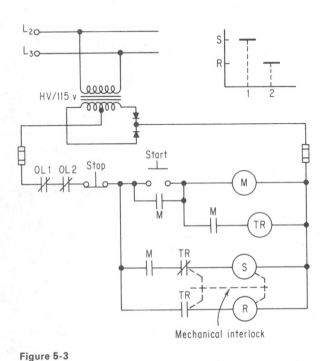

1. Depressing the START button energizes line contactor *M* and its auxiliary contacts in the control circuit, as well as the timing relay *TR*.

2. Energizing the *M* contacts also energizes START relay *S* through n.o. *M* and n.c. *TR* contacts.

3. After a suitable time delay, whenever the timing relay "times out," it opens n.c. *TR* contacts in series with *S*, and subsequently closes n.o. contacts *TR* in series with *R*. The latter contacts cannot close until the former contacts are opened.

Figure 5-3

Definite-time acceleration ac starter using dc latching and time-delay relays in open-circuit transition.

The starter shown in Fig. 5-3 is a basic definite-time-acceleration, open-loop type of ac induction-motor starter. It uses dc control elements around which many control circuit variations are developed for reduced-voltage, wye-delta, autotransformer, and part-winding starting, using various types of dc time-delay devices described in Sec. 4-4. Because most of these timing devices are mechanical or electromechanical, there is a waiting period until such a timing device is reset. This is a serious disadvantage where frequent starting, stopping, and reversing may occur. Electric timing devices, and relays using an inductive or capacitive time constant, are capable of faster resets and limited adjustment of the time constant to develop the necessary time delay.

Note that the use of *latching relays* S and R ensure the open-circuit transition. A latching relay has two coils: one closes a set of contacts, another opens them. When the relay closes, a latch operates to close the relay and maintain it in the closed position, even if the relay fails. Thus, when S is energized, all S contacts (not shown) are closed and all R contacts open. Similarly, when R is energized, all R contacts are closed and all S contacts are opened. This relay feature is simulated in static switching by RETENTIVE memory (see Table 9-1, Section 9-8).

5-4.
"NEO TIMER"
CIRCUIT AND
STARTER

An all-electric time-delay device capable of a wide range of time delays and *instant resetting* is the "Neo timer"* or neon-lamp timer starter. The basic timing circuit, shown in Fig. 5-4a, is essentially the circuit of the relaxation (gas-tube) oscillator.

A variable resistor R in series with a capacitor C across a dc supply will, when energized, develop an exponential rise of voltage across capacitor C at a rate determined by the RC time constant. When the voltage across C is sufficient to ignite the neon gas tube, the tube fires, sending sufficient current through timing relay TR to cause it to operate, reversing its normally closed and normally open contacts. Relay TR remains energized through its n.o. TR contacts, shorting the neon tube and opening and disconnecting the RC time constant circuit through n.c. TR contacts, for the period of the capacitor discharge. When TR is deenergized, the RC time constant circuit is restored and the cycle may be repeated again.

In the commercial (Square D) design, perhaps three or four timing periods are obtained to operate sequential relay contacts using the *same* RC timing circuit. The instant the circuit is deenergized, all the relays are returned to their normal condition.

A simplified version of Neo-timer acceleration is shown in Fig. 5-4b. The circuit combines both current-limit and definite-time acceleration through current and potential transformer circuits, respectively. Full-wave rectifiers are used in both circuits to provide a dc interlock between the load current circuit and the potential timing circuit. An ac control circuit, containing ac relays controlled by the dc control circuit, provides two stages of acceleration through accelerating relays $1A$ and $2A$, respectively, which are sequenced in the following manner:

1. When the START button (not shown) is pressed and the M contacts (not shown) are closed, the induction motor starts through full primary

---

* Developed by the Square D Company, Electric Controller and Manufacturing Division. The original device was proposed by H. L. Wilcox, in a paper presented before the M. I. & S. E. on Dec. 4, 1939.

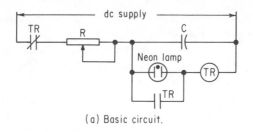

(a) Basic circuit.

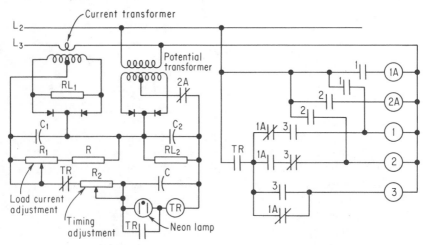

(b) Application to a-c motor reduced voltage starting.

Figure 5-4

Neon-lamp timer combining current-limit and
definite-time acceleration.

resistance or reactance in series with the stator (not shown). This energizes both the current and potential circuits of Fig. 5-4b.

2. The RC time constant controlling the firing of the neon lamp is a function of the resistance setting of R_1 and the timing adjustment R_2 in series with capacitor C. When the load current is high, the voltage drop across R_1 and R is high, tending to change the firing point of the neon lamp.

3. Capacitor C charges and ultimately fires through relay TR, opening the n.c. TR contacts that bridge the current and potential dc circuits, and closing the n.o. TR contacts in the ac relay circuit. TR remains energized for the full period of discharge of capacitor C through the neon lamp.

4. When TR closes in the ac relay circuit, it energizes ac control relay 3 through the n.c. $1A$ contacts. The latter holds itself closed through interlock contact 3 for the duration of time that TR is energized (the discharge time constant of C).

5. Energized ac control relay 3 energizes ac control relay 1 through the n.c.

$1A$ contacts and the n.o. 3 contacts. Energized control relay 1 now becomes self-energized across ac relay $1A$, via its n.o. contact 1.

6. When accelerating relay $1A$ is energized, it shorts out one step of primary resistance or reactance in series with the stator (not shown), or inserts a part winding in parallel (also not shown).

7. The ac control relay circuit operation ceases with relays $1A$ and 1 energized and control relays 2 and 3 deenergized. Furthermore, n.o. contacts TR ultimately return to their n.o. state when the discharge of C ceases and TR is deenergized, bridging the gap between current and potential circuits through normally closed TR.

8. The surge of current produced by the shorted $1A$ contacts in the power circuit recycles the RC time constant circuit once more and produces the time delay required before C charges. When the neon lamp fires a second time, timing relay TR is again energized, and the ac relay circuit is sequenced once again.

9. This time, relay 2 is energized through n.o. contacts $1A$ and n.c. contacts 3. Relay 2 energizes itself across the ac line, as well as accelerating relay $2A$. The accelerating relay shorts out the last step of primary resistance or reactance, placing the motor in the RUN position. At the same time, $2A$ deenergizes the secondary of the potential transformer circuit, thus ceasing all further RC time constant timing operations. Control relays 1 and 2 in the ac circuit, and line accelerating contactors $1A$ and $2A$, remain energized in the run position.

The advantages of the Neo timer circuit are: (1) closed-transition operation ($1A$ is energized during the time interval that $2A$ comes on), (2) instantaneous reset, (3) the circuit permits both definite-time and current-limit (relatively wide) adjustment within a single starter, (4) the circuit can be used for dc as well as ac operation, (5) no mechanical timing devices are employed, and (6) the basic components may be replaced at relatively low cost.

5-5.
WOUND-ROTOR
INDUCTION MOTOR
(WRIM) STARTING

The Neo timer (Section 5-4) may also be used to start, as well as to accelerate, wound-rotor motors, in addition to dc and polyphase squirrel-cage motors. The resistors of Fig. 3-11b are shorted out in three steps (rather than two) of acceleration, using suitable neon-timer accelerating relay sequencing, similar to that shown in Fig. 5-4, in closed transition, with a wound-rotor motor (Fig. 3-11b).

The relay circuit of Fig. 5-2, in which TD_1, TD_2, and TD_3 are energized in definite time-delay sequence, may also be used with some minor modification to short-circuit secondary rotor resistances effectively. Such a definite-time-delay starter, shown in Fig. 5-5, operates in the following manner:

1. When the START button is depressed, the stator of the wound-rotor motor is energized through contactors M. Auxiliary contact M, having

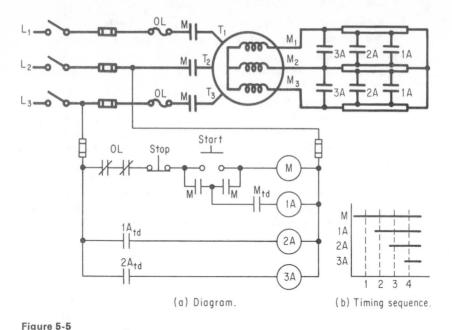

(a) Diagram.

(b) Timing sequence.

Figure 5-5

Wound-rotor secondary resistance starter using
definite-time acceleration relays.

a time delay contact, M_{td}, energizes ac time-delay relay $1A$. The motor
starts with full wound-rotor resistance and maximum slip in its secondary winding (thus developing maximum rotor torque and, consequently, reduced stator current to provide this torque).

2. After a preset time interval, as determined by M_{td} and time-delay relay
$1A$, the latter closes, shorting out one-third of the wye-conchected rotor
resistance. Relay $1A$ also closes its n.o. $1A$ contact, energizing relay $2A$.
Time-delay relay $2A$ closes after a preset time interval to short out an
additional one-third of the wye-connected rotor resistance.

3. The motor rises to a higher speed each time that the rotor resistance and
slip are reduced. After its suitable time delay, relay $2A$ also closes its
n.o. time-delay contact $2A_{td}$, energizing time-delay relay $3A$. The latter
closes its n.o. contacts after a suitable time delay, effectively shorting
out all rotor resistance.

The ac time-delay relays of Fig. 5-5 operate in closed-circuit transition
and may be used for sequencing any four-step closed-circuit transition for
any type of predetermined time-delay ac motor starting, as shown by
the timing sequence of Fig. 5-5b.

Another principle, previously mentioned but not yet shown, is the use
of RC time-constant discharge relays. This type of starting may be used
either with dc or ac motors to provide a closed-circuit transition sequence

similar to that shown in Fig. 5-5b. The starter whose control circuit (only) is shown in Fig. 5-6 operates as follows:

1. Depressing the START button energizes main contactor M in the stator circuit and control line 1. In the OFF or STOP position, however, the dc relay control circuit has been energized through a full-wave rectifier, and the contacts TD_1, TD_2, and TD_3 in lines 2, 3, and 4 are all open, since their dc relays are energized. At the instant of energizing M, n.c. auxiliary contact M (in series with TD_1 and its shunting R_1C_1 circuit) opens. TD_1 does not immediately deenergize, however, because the charged capacitor C_1 discharges through TD_1 and R_1, maintaining relay TD_1 energized for the greater portion of the dischange time-constant period.

2. During the period that M was energized, the motor started with full wound-rotor resistance. After a suitable time delay determined by the

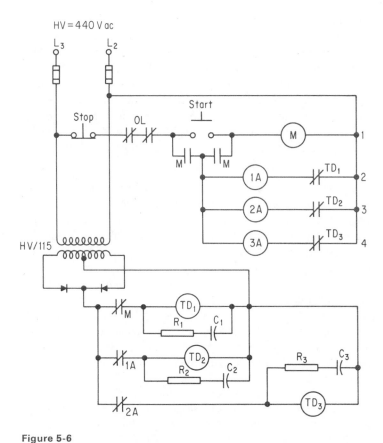

Figure 5-6

RC time constant relays for definite-time acceleration (wound-rotor motor).

97

discharge time-constant R_1–C_1, relay TD_1 is deenergized, closing the ac relay path to energize ac contactor $1A$ in control line 2. Relay $1A$ shorts out one-third of the rotor wye-connected resistance and simultaneously opens the circuit to relay TD_2.

3. Capacitor C_2 discharges through relay TD_2 and through R_2, maintaining relay TD_2 in an energized state for the major portion of its discharge time period. TD_2 is deenergized, ultimately, closing its n.c. contacts in series with ac relay $2A$. Relay $2A$ shorts out an additional one-third of secondary rotor resistance, causing the wound-rotor motor to accelerate with reduced rotor resistance and slip. At the same time that relay $2A$ is energized, it opens n.c. contact $2A$ in series with relay TD_3.

4. Capacitor C_3 discharges through relay TD_3 and through R_3, maintaining relay TD_3 energized for a portion of its time-constant discharge period. When TD_3 finally is deenergized, ac relay $3A$ shorts out the last step of external rotor resistance.

The wound-rotor motor starter shown in Fig. 5-6 operates in the RUN position with only its ac control circuit energized. If the motor is stopped for any reason, all the contacts in the dc control circuit are closed, deenergizing the ac control circuit relays. A line switch disconnects the motor and starter from the ac supply so that, when not in use, the dc circuit is also deenergized. The time delay of the starter may be increased by increasing the *RC* time constant of the circuit that shunts the time-delay relays. Usually the resistance (R_1, R_2, and R_3) is increased. As previously mentioned, this principle works equally well for dc motor starting.

5-6.
CLOSED-LOOP
WRIM STARTERS

With the exception of the Neo-timer principle (which combines current-limiting with predetermined time delay), the ac starters previously discussed were of the open-loop type. Although the majority of such starters employed in ac polyphase squirrel-cage and wound-rotor starters are of the open-loop type, a number of starters have been devised using closed-loop or feedback principles. One such starter, using fast-acting ac series relays, is able to sense and respond to the ac surge currents developed in the rotor resistance during starting and acceleration periods. The series relay (as previously shown in Section 4-8) is a fast-acting type, capable of opening and closing in shorter time intervals than potential relays. Figure 5-7 shows a starter using ac *series relays* which operates in the following manner:

1. Depressing the START button energizes contactor M, starting the wound-rotor induction motor with full rotor resistance in its secondary circuit. The current surge on starting through relay $1S$ in the rotor circuit is sufficient to deenergize relay $1A$ in control line 2 before it has a chance to operate. The motor starts with maximum rotor resistance and slowly accelerates to that value of speed determined by its load and slip.

2. Relay 1S does not deenergize until the surge current is sufficiently low to cause its heavy springs (see Sec. 15-8) to open its contacts. Depending on the nature of the load, the time for operation of relay 1S is a variable, and the motor is accelerated only when the surge disappears and relay 1S is deenergized.

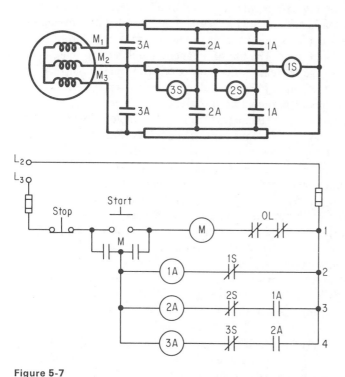

Figure 5-7

Current-limit acceleration wound-rotor starter using ac series relays.

3. The deenergized 1S contacts (n.c.) energize relay 1A, shorting out the first step of rotor resistance in the secondary circuit through n.o. contacts 1A and relay 2S. Fast-acting relay 2S opens control line 3 before 2A can act, and the motor accelerates to a new value of speed determined by its load and slip.

4. Relay 2S does not deenergize until the surge rotor currents are sufficiently reduced to cause its heavy springs to operate the contacts of relay 2S. At the proper time, relay 2S deenergizes, and control line 3 is energized through n.c. 2S and n.o. 1A.

5. The energized ac control relay 2A accelerates the wound-rotor motor once again, by shorting its n.o. contacts 2A in the secondary rotor circuit through low-impedance relay 3S. Fast-acting 3S opens control line 4 before ac relay 3A can act. The motor accelerates until its surge rotor currents decrease sufficiently to deenergize ac series relay 3S.

99

6. Deenergized 3S closes control line 4, energizing relay 3A and shorting the rotor circuit resistance completely. The motor accelerates as a squirrel-cage motor to its rated or normal slip, depending on the load.

Like the dc series relay current-limiting starter, the ac wound-rotor starter shown in Fig. 5-7 is virtually foolproof in respect to operation under various load conditions with frequent stops and starts. Because of the interlocking contacts, however, maintenance may be a problem, as the contacts pit and wear. The use of magnetic amplifiers, to sense either rotor or stator surge currents, in combination with differential relays in a principle similar to that of Fig. 4-9, eliminates contact wear and tear in a closed-loop magnetic amplifier ac starter.

Another closed-loop principle which may be employed with wound-rotor motors is the use of frequency-control relays (Sec. 1-15). The frequency-control relay, being inductive, is made part of a series-resonant circuit whose frequency response is "tuned" by means of a series capacitor and whose voltage response is adjusted by means of a series-connected resistor, as shown in Fig. 5-8. The relay circuit is connected to one slip ring and a tap on another line of the rotor resistance. Frequency-control relays 1F, 2F, and 3F are tuned so that they will drop out at a desired frequency below their series-resonant frequency. For example, relay 1F is resonant (minimum impedance) at 60 Hz but drops out at 45 Hz; relay 2F is resonant at 45 Hz but drops out at 30 Hz; relay 3F is resonant at 30 Hz but drops out at 15 Hz. The frequency-sensitive closed-loop starter, shown in Fig. 5-8, operates in the following manner:

1. When the START button is depressed in the primary control circuit (Fig. 5-8b), relay M energizes the stator and the motor starts with full secondary rotor resistance. Relay 1F, tuned to 60 Hz (the rotor frequency at standstill), opens n.c. contacts 1F in the primary control circuit. The motor accelerates with full rotor resistance from a slip of 100 per cent at standstill to a lower value of slip. As the motor accelerates, its current surges, and its rotor frequency decreases. At a rotor frequency of 45 Hz, relay 2F opens its n.c. contacts and relay 1F closes its n.c. contacts in the primary control circuit, the latter dropping out at this off-resonant frequency.

2. Energized relay 1A closes its contacts in the rotor secondary circuit, shorting out one step of rotor resistance. The rotor current increases, and torque is developed to accelerate the motor to a smaller slip and a lower rotor frequency.

3. As the motor gains speed, its current surges, and its frequency decreases to 30 Hz. At this frequency, relay 2F drops out, but relay 3F is energized. Control relay 2A is energized in the primary control circuit through n.c. relay 2F contacts and n.o. contacts of energized relay 1A. Energized relay 2A closes its contacts in the rotor secondary circuit, shorting out a second step of rotor resistance. The rotor current

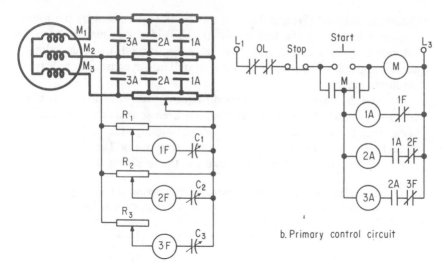

a. Secondary rotor circuit.

b. Primary control circuit

Figure 5-8

Acceleration of a wound-rotor motor based on
rotor frequency using frequency relays.

increases, and torque is developed to accelerate the motor to a smaller
slip and a lower rotor frequency.

4. The motor gains speed once more as its current surges, and its frequency decreases to 15 Hz. At this frequency, relay $3F$ drops out, energizing relay $3A$ in the primary control circuit. Relay $3A$ shorts out the last step of rotor resistance, and the motor accelerates to its rated normal slip (less than 5 per cent) and its rated rotor frequency (under 3 Hz).

It is a relatively simple matter to tune or adjust the frequency-control relays as well as the voltage at which they will drop out. The advantage of such closed-loop control, moreover, is achieved by sensing the frequency rather than the current. Thus, if the inertia of the load is great and the applied load heavy, the rotor frequency will be higher for longer periods and the motor will not be accelerated as rapidly. Conversely, if there is no load on the motor whatever, acceleration is relatively rapid because the frequency diminishes rapidly.

**5-7.
AUTOMATIC
SYNCHRONOUS
MOTOR STARTERS**

Since the stator of a synchronous motor is the same as that of an induction motor, the polyphase synchronous motor may be started as an induction motor on its damper windings by any of the automatic ac induction motor primary starting methods described in Sections 5-1 through 5-4.

In all these stator starting methods, however, it is customary to short-circuit the dc field during the starting period until the rotor is close to

101

synchronous speed (Section 3-9). At this speed (determined by the rotor frequency induced in the field winding), the field circuit may be opened automatically by a frequency-control relay (Section 1-15), and dc voltage may be applied to the field to permit it to pull into synchronism as a synchronous motor.

A typical starter using polarized-frequency relays is shown in Fig. 5-9. The primary stator circuit starting has been simplified (for purposes of emphasis) to show across-the-line starting (although the methods shown in Sections 5-1 through 5-4 are frequently employed). The synchronous motor starter shown in Fig. 5-9 operates in the following manner:

1. Depressing the START button energizes fast-acting control relay $1A$, which is fast enough to energize F_2 in the dc supply circuit (deenergizing relay $2A$) before the main ac contactor M and its auxiliary contact M can close relay $2A$. The motor thus starts as an induction motor across

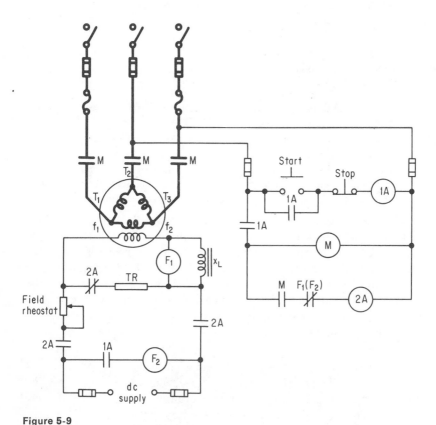

Figure 5-9

Synchronous motor starter using polarized
frequency relays.

the line with its field winding f_1–f_2 short-circuited through a low-resistance polarized-frequency relay coil F_1, a thyrite resistor TR, and n.c. relay coil $2A$.

2. Relay F_1–F_2 is a polarized-frequency relay (Section 1-15) having two windings in opposition. During the starting period, the frequency induced in the highly inductive field winding f_1–f_2 is high; and, since the reactance of coil X_L is high, most of the induced field current that flows in coil F_1 is polarized in opposition to F_2, thus holding n.c. contact F_2 open.

3. As the motor accelerates, the frequency induced in the field winding f_1–f_2 decreases, causing more and more of the ac induced current to be diverted to reactance coil X_L, which has a lower resistance than relay coil F_1. At a frequency near the synchronous frequency, coil X_L literally short-circuits coil F_1.

4. The polarized-frequency relay causes coil F_2 to act in such a manner as to close the contact (n.c.) in series with control relay $2A$. Energized control relay $2A$ connects the field to the dc supply and simultaneously opens the short circuit across the field winding to permit the field rheostat (and low dc resistance X_L) to be connected in series with the field. The field rheostat setting is usually adjusted to that value which is close to the unity power factor excitation of the synchronous motor.

The polarized-frequency synchronizing relay is designed in such a manner as to close the circuit to relay $2A$ at a point at which the ac wave induced in coil F_1 is most favorable to synchronism, i.e., when a south pole of the rotor is rotating almost in synchronism with the north pole of the rotating magnetic field. If the synchronous motor slips a pole, or if an overload occurs during the starting period, the induced frequency in coil F_1 is such as to maintain the F_1 (F_2) contact open in series with control relay A. The polarized-frequency or synchronizing relay acts as a *frequency transducer*, therefore, and synchronizes the synchronous motor at the proper time for synchronization rather than at a predetermined definite time interval.

Other methods of synchronization using power-factor relays or ac line current sensors are also employed for automatic synchronous motor starters, but space does not permit their inclusion.

**5-8.
MULTIPLE
STARTING**

No discussion of automatic starting, however, would be complete without a few examples illustrating the versatility of control circuits in starting and stopping more than one motor in a prescribed manner and under prescribed conditions. Like all design problems, various solutions are possible. The reader should attempt independent solution of each example before examining the solution described and illustrated.

103

EXAMPLE 5-1

Five wound-rotor motors in an automatic production setup, having separate secondary resistance starters, are to be started from a common pushbutton station with a time delay between each. Any overload on any motor will stop all of the motors driving their respective loads in the system. A light on the panel shall indicate when the fifth motor has started and the momentary pushbutton may be released. All motors are to be stopped from the same common station. Design (a) the common control station and (b) show a typical motor starter.

Solution

Shown in Fig. 5-10.

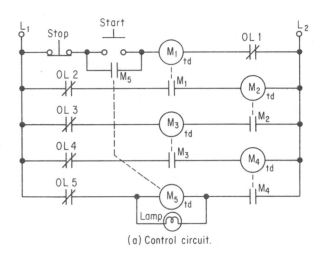

(a) Control circuit.

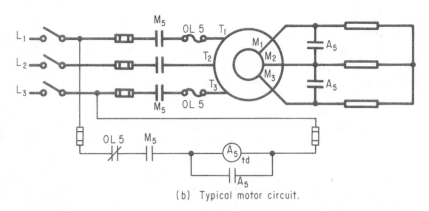

(b) Typical motor circuit.

Figure 5-10

Solution for Example 5-1.

EXAMPLE 5-2

Three motors are to operate in the following manner: when the START button is pressed, motor 1 is to start and is to run until it opens a normally closed microswitch and stops. Motor 2 is to start when motor 1 stops; motor 2 is to run until it trips a microswitch. When motor 2 stops, motor 3 is to start and run until it, too, reaches a mechanical stop, opening a microswitch, at which point motor 3 stops. All microswitches are reset to a normally closed position by the operation of the next motor. An overload in any motor shall instantly stop operation, requiring manual reset to resume operation from that motor, without recycling from motor 1.

Solution

Shown in Fig. 5-11.

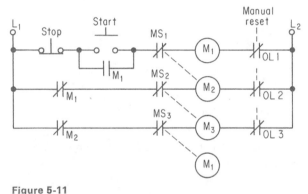

Figure 5-11

Solution for Example 5-2.

EXAMPLE 5-3

Three induction motors are to operate from a single master station in the following manner: when the START button is pressed, motor 1 starts. Motor 2 starts after a short time delay and runs for 40 s. When motor 2 stops, it starts motor 3; but motor 1 continues to run. Motor 3 shall stop after 20 s, simultaneously stopping motor 1. An overload in motor 2 or motor 3 shall individually cause these motors, respectively, to stop without affecting each other or motor 1. An overload in motor 1 shall cease all operation, as will the master STOP button.

Solution

Shown in Fig. 5-12.

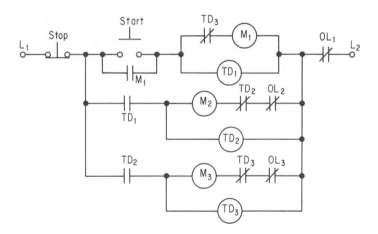

Figure 5-12

Solution for Example 5-3.

BIBLIOGRAPHY

"Electronic Standards for Industrial Equipment" (General Motors Technical Center), *Electrical Manufacturing* (August 1958).

Fitzpatrick, D. "Reduced-Voltage Starting for Squirrel-Cage Motors," *Electrical Manufacturing* (March 1960).

Graphical Symbols for Electrical Diagrams (ASA Y32.2) (New York: American Standards Association, or IEEE Standard 315-1971.

Harwood, P. B. *Control of Electric Motors*, 4th ed. (New York: John Wiley & Sons, Inc., 1970).

Heumann, G. W. *Magnetic Control of Industrial Motors*, 3 vols. (New York: John Wiley & Sons, Inc., 1961).

Industrial Control Equipment (Group 25) (ASA C42.25) (New York: American Standards Association).

Industrial Control Equipment (UL508) (Chicago: Underwriters Laboratories, Inc.).

James, H. D., and L. E. Markle. *Controllers for Electric Motors* (New York: McGraw-Hill Book Company, 1952).

Jones, R. W. *Electric Control Systems* (New York: John Wiley & Sons, Inc., 1953).

Kosow, I. L. *Electric Machinery and Transformers* (Englewood Cliffs, N.J.: Prentice-Hall, Inc., 1972).

Motor and General Standards (MG1) (New York: National Electrical Manufacturers Association).

5-1.   Review Table 5-1 and indicate which NEMA induction-motor classes are most suited for across-the-line starting and which may require some means of reduced voltage starting.

5-2.   a. Explain why any automatic starter incorporating a two-stage START and RUN switching sequence, in which the RUN contacts are closed after a suitable time delay, may be used universally for primary resistance, primary reactance, wye-delta, autotransformer, and part-winding starting
       b. Which of the above starting methods (in simpler form) usually require only open-circuit transition between contacts? Why? (*Hint:* See Figs. 3-9 and 3-10)
       c. Which may be used with either open- or closed-circuit transition?
       d. Is it possible to use closed-circuit transition for wye-delta starting? Draw such a circuit.

5-3.   A No. 1 size NEMA three-phase magnetic starter may be used for a 3-hp 110-V motor, a $7\frac{1}{2}$-hp 220-V motor, and a 10-hp 440/550-V motor under average operating conditions requiring no plugging or jogging. Where jogging, plugging for reversing, or stopping is required, a starter of the same size may be used with reduced ratings: 2-hp ,110-V motor, 3-hp 220-V motor, and a 5-hp 440/550-V motor. Explain why the NEMA size (current rating) of an automatic starter is affected by horsepower, voltage, and speed control duty cycle, respectively.

5-4.   From manufacturer's literature and/or the references cited at the end of the chapter investigate
       a. The purpose of a control transformer used with ac control circuits
       b. Constructional differences between control and power transformers
       c. Differences as to operating power factor
       d. Differences as to voltage range
       e. The use of current transformers in association with overload coils to extend the range of the latter.

5-5.   Redesign the starter shown in Fig. 5-2 to provide
       a. Dc time-delay relays using a suitable full-wave bridge rectifier
       b. Open-circuit transition to permit wye-delta starting of a three-phase induction motor
       c. Electrical and mechanical interlocks between start and run contacts. Describe fully the operation of the starter.

5-6.   Redesign the wound-rotor shown in Fig. 5-5 to provide
       a. A starter with dc time-delay relays and line contactors using any suitable full-wave semiconductor rectifier
       b. A definite-time acceleration starter using a dc timing motor.

5-7.   Compare the starters of Figs. 5-6 and 5-7 and discuss relative merits of each. In your discussion, include starting under varying load conditions as well as the ability to adjust the time required to accelerate the motor.

5-8. Referring to the wound-rotor starter of Fig. 5-8, specify the adjustments to be made in the event of the following

a. Frequency relays $1F$, $2F$, and $3F$ all produce an excessive time delay in accelerating

b. Frequency relays $1F$, $2F$, and $3F$ cause their contacts to drop out too rapidly

c. The first and second accelerating steps occur within the specified and desired time intervals, but the last step occurs too slowly. A check of relay $3F$ shows that it is tuned to 30 Hz

d. It is desired to change the resonant frequency of relays $2F$ and $3F$ to 50 Hz and 40 Hz, respectively. Specify the ratio of change in the capacitor values

e. Would there be an advantage in converting the primary control circuit shown in Fig. 5-8b to a dc circuit using dc relays? Discuss fully.

5-9. For the synchronous motor starter shown in Fig. 5-9, explain

a. Why the motor will not start if dc is applied to the field before or during acceleration

b. Why the starting torque may be less if the field is short-circuited during starting rather than left open

c. Why it is dangerous to leave the field open, even if the starting torque is higher

d. Why it is preferable to apply dc to the field automatically rather than manually

e. Whether it is possible for motor to "slip a pole" using the starter shown. Explain.

5-10. Three induction motors are used in a conveyor system. The sequence of operating the motors must conform to the following requirements: (1) motor 1 must be energized and running before either 2 or 3 may be started; (2) motors 2 and 3 may be started simultaneously from a single pushbutton and both may be stopped by a single pushbutton; (3) motors 2 and 3 may be individually started and stopped without stopping motor 1; (4) stopping motor 1 will stop the other two motors. Draw a control diagram which will satisfy these conditions. All motors are started across the line.

5-11. Three induction motors, A, B, and C, are used in an automated plant. The sequence of operations requires the following relationship among the motors: (1) When A and C are energized, motor B cannot be energized; (2) when A and B are energized, motor C must be deenergized; (3) deenergizing A will deenergize all motors. Draw a control diagram showing individual start and stop buttons for all motors, satisfying the above conditions. Use as many relays as you require to perform the logic indicated.

5-12. Design a relay circuit in which motor 3 can only be energized if motors 1 and 2 are energized. Motor 2 can only be energized if motor 1 is energized. Show individual start-stop buttons for each motor and use as many relays as required to perform the indicated logic.

CHAP. FIVE / *Automatic ac Starters*

manual and automatic speed control of dc motors

6-1.
GENERAL

The *speed* of any dc motor can be altered by a change in any of the variables in the fundamental speed equation, $S = k\dfrac{V_a - I_a R_a}{\phi}$.

Four methods of controlling the speed of a dc motor will be discussed in the early part of this chapter. The four methods are:

1. Changing the field flux, ϕ, by means of a variable series or shunt rheostat. This method is known as "field control" (Fig. 6-1a).

2. Changing the voltage V_a across the armature by using a variable resistance in series with the armature. This method is called "armature resistance control" (Fig. 6-1a).

3. Changing the voltage V_a across the armature, and the current I_a in the armature, by a combination of two variable resistances in parallel and in series with the armature. This method is called "series and shunt armature resistance control" (Fig. 6-1b).

4. Using a controlled source of variable dc voltage to change the voltage

V_a across the armature of a separately excited motor. This method is known as "armature voltage control" (Fig. 6-1c).

After the discussion of these four methods, this chapter will deal with various designs of manual and automatic controllers that incorporate the functions of starting, reversing, controlling speed, and braking.

6-2.
FIELD CONTROL

When the rated or line voltage is applied to the armature of a dc motor ($V_a = V_l$) and the field flux is manually or automatically varied by means of a field rheostat in series or in parallel with the field excitation winding, the method of speed control is called "field control." As shown in Fig. 6-1a, when the motor is started and the variable armature resistance is shorted out (at point a) so that V_a equals V_l, control of the speed may be achieved by varying the field rheostat from point a' (no added field resistance or full field current) to point b' (maximum field resistance or minimum field current).

The speed achieved with the full armature voltage and full field current (no added field resistance) is called the *basic speed* of the motor. Increasing the field resistance, therefore, will decrease the field current and field flux in the fundamental speed equation, causing the speed to rise. It may be said, therefore, that *field control can produce only speeds above the basic speed.*

Field control as a method of speed control to obtain speeds *above* basic speed has the following advantages over other speed-control methods: (1) field control is relatively inexpensive and simple to accomplish, both manually and automatically; (2) it is relatively efficient in terms of motor performance, since the field circuit loss is only 3 to 5 per cent of the total power drawn by the motor; (3) within limits, field control does *not* affect speed regulation in the cases of shunt, compound, and series motors; and (4) it provides relatively smooth, stepless control of speed.

The third advantage, however, carries with it a warning that this method of speed control is achieved by weakening the field flux *within limits.* If the field is weakened considerably, dangerously high speeds are produced. Since an increase in speed (created by flux reduction) results in an increase in the load and armature current, the increase in torque ($T = k\phi I_a$) is produced by a considerable increase in the armature current.

With a weak field and a high armature current, the dc shunt motor is particularly susceptible to the effects of armature reaction instability and may run away in the same manner as a differential compound motor.* It is precisely for this reason, moreover, that dc motors are started with

* For a detailed discussion of runaway and armature reaction effects see Kosow, *Electric Machinery and Transformers* (Englewood Cliffs, N.J.: Prentice-Hall, Inc., 1972), Secs. 4-10 and 4-14.

full field current. With a high speed and high armature current, moreover, commutation difficulties are increased, as the high armature currents are reversed more rapidly and serious damage to the commutator may be produced in arcing.*

It is customary, therefore, to set a maximum permissible limit on the overspeed when using field control as a method of speed control. Usually this is 1.5 times the basic speed. The disadvantages of field control as a method of speed control are (1) inability to obtain speeds *below* the basic speed, (2) instability at high speeds because of armature reaction, and (3) commutation difficulties and possible commutator damage at high speeds.

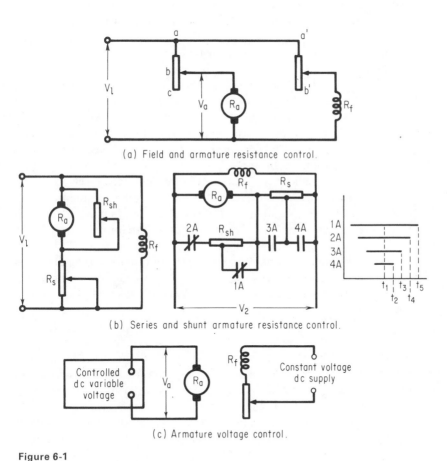

(a) Field and armature resistance control.

(b) Series and shunt armature resistance control.

(c) Armature voltage control.

Figure 6-1

Methods of dc motor speed control.

* *Ibid.*, Secs. 5-6 and 5-7.

6-3.

ARMATURE

RESISTANCE

CONTROL

When the field rheostat is set so that normal field excitation (in the saturation region) is produced, and the voltage across the armature is reduced, by means of a variable resistance in series with the armature, the method of speed control is called "armature resistance control." As shown in Fig. 6-1a, the field rheostat is adjusted to provide normal excitation, and the series armature resistance is adjusted so that the armature voltage, V_a, is varied below the line voltage, V_l. Control of the speed is obtained by varying the resistance in series with the armature. Increasing the series armature resistance reduces the voltage across the armature (at any given load) in the fundamental speed equation, $S = k(V_a - I_a R_a)/\phi$, causing the speed to drop. It may be said, therefore, that *armature resistance control can produce only speeds below the basic speed.*

The armature current in the fundamental speed equation is a function of the load. At any given setting of the series armature resistance, an increase in load will produce an increased voltage drop across the series-connected armature resistor, which produces a drop in speed. For any no-load speed setting below the basic speed, armature resistance control will produce a sharp drop in speed with the application of load, resulting in poor speed regulation. The greater the value of the series armature resistance, the poorer the speed regulation of the motor. Furthermore, the armature current flowing through the series-connected armature resistance will produce an appreciable power loss ($I_a^2 R_s$) which reduces the overall motor efficiency. While this power loss, fortunately, does not produce heat *within* the motor, it does require a larger, continuously rated, externally connected, variable resistor capable of carrying the rated armature current. This variable resistor may be used both for motor starting as well as for speed control by armature resistance control.

The advantages of armature resistance control are (1) the ability to achieve speeds *below* the basic speed, (2) simplicity and ease of connection, and (3) the possibility of combining the functions of motor starting with speed control.

The disadvantages of armature resistance control are (1) the relatively high cost of large, continuously rated, variable resistors capable of dissipating large amounts of power (particularly in higher horsepower ratings), (2) poor speed regulation for any given no-load speed setting, (3) low efficiency resulting in high operating cost, and (4) difficulty in obtaining stepless control of speed in higher power ratings.

The combination of armature and field resistance control of a shunt motor, shown in Fig. 6-1a, provides a reasonably effective and relatively inexpensive means of providing speeds both above and below the basic speed in the case of *smaller* dc motors. In the larger horsepower ratings, where extremely low speeds may be required for "inching" or "jogging" control, a fairly high resistance is required in the armature circuit of relatively high power rating. Such a resistance produces relatively inefficient operation and is relatively expensive. This difficulty is overcome by series and shunt armature resistance control.

6-4.
SERIES AND
SHUNT ARMATURE
RESISTANCE
CONTROL

Figure 6-1b shows the schematic diagram of a rheostatic speed control using combined resistances both in series and in parallel with the armature. R_{sh} is a variable resistor shunting the armature, and R_s is a variable resistor in series with the armature. The former acts as a diverter tending to reduce the armature current as its (R_{sh}) resistance is reduced. The latter, R_s, acts in the same manner as in the simple armature resistance control described in the preceding section. Thus, in the fundamental speed equation, $S = k\dfrac{V_a - I_a R_a}{\phi}$, for a constant field flux, at any given load, an increase in R_s will produce a decrease in V_a and a drop in speed. An increase in R_{sh} will produce an increase in the $I_a R_a$ drop and also a drop in speed. The speed may be raised, therefore, by decreasing *both* R_s and R_{sh} (the latter within limits). As in the case of field control, there is a maximum permissible limit to the shunting effect produced by R_{sh} at higher speeds. If R_{sh} approaches a short circuit across the armature, extreme torque instability occurs as a result of tendency toward high speeds and increased loads.*

The net effect of the shunt resistor, R_{sh}, is to make the operating speed less susceptible to changes in load torque and, as a result, improve the speed regulation of the motor over that which might be obtainable using only armature resistance control. It should be pointed out, however, that a reduction in shunting resistance produces a proportionate reduction in developed torque. Shunting resistors across the armature are used with armature resistance control, therefore, where it is desirable to maintain approximately the same operating speed and where the load torque may tend to vary. The shunting resistor may also be used to provide dynamic braking (as will later be shown).

Figure 6-1b also shows the basic switching circuit for starting and running with series and shunt armature resistance control, as well as the switching sequence used with these contactors. Note from the switching sequence that, on starting, only contactor 2A is closed. This provides the maximum protective resistance in series with the armature, as well as controlled current in the armature, in order to develop the necessary starting torque. The motor is accelerated in progressive steps by (1) opening n.c. 2A, (2) closing n.o. 3A, and (3) closing n.o. 4A. At time t_1 shown in the sequence diagram, therefore, all contacts are in their operating position and the motor is running as a shunt motor.

---

* This method of speed control verifies the basic principle of the doubly excited dynamo. Motor torque is produced as a result of interaction between the field and armature flux. A decrease in armature current and flux, ϕ_a, results in a decrease in torque. *More* armature current is required to produce the necessary armature flux and to provide the required motor torque to balance the applied torque.

113

Progressive decreases in speed with reasonably good speed regulation may now be obtained by opening (deenergizing) $4A$ at time t_2, followed by $3A$, at time t_3, deenergizing $2A$ ($1A$ is open and energized) to provide combined series-shunt armature resistance at time t_5. Thus, in the last case, running with all the contacts as shown in the figure provides the lowest running speed, with better speed regulation than would be provided by increasing the series armature resistance. Furthermore, since the applied torque of any given load is lower at reduced speeds, the motor does not suffer from the lower developed torque produced as a result of the diversion of armature current by R_{sh}.

The advantages of series-shunt armature resistance control are (1) improved speed regulation (better than with armature resistance control), and (2) possible use of the armature shunting resistor for dynamic braking.

The disadvantages are (1) reduced operating torque with increased diversion of armature current, and (2) reduced efficiency because of power losses in the series and shunt resistors.

**6-5.
ARMATURE
VOLTAGE CONTROL**

The relative efficiency of the smaller motors is not a serious consideration, whereas relative torque, speed regulation, and stepless control are of some importance in small motor applications. In the case of motors of higher horsepower, however, efficiency, torque, good speed regulation, and smooth, stepless speed control are all extremely important considerations. Heavy loads with high inertia require smooth acceleration over a wide speed range. All these criteria may be achieved by using a variable dc voltage from a source of sufficient capacity to supply the required armature voltage and current to a dc motor. The field is *always* separately excited from a constant-current or constant-voltage source, as shown in Fig. 6-1c. This method of speed control also eliminates the need for series armature starting resistance.

If the armature voltage supplied from the variable dc source is zero, the motor develops zero torque ($T = k\phi I_a$) and is at a standstill. If the armature voltage is increased slightly, in accordance with the fundamental speed equation, $S = k(V_a - I_a R_a)/\phi$, the motor starts and turns at a slow speed with a minimum of acceleration. The armature current is limited because of the low voltage across the armature. Reducing the armature voltage to zero, and reversing the polarity of the variable voltage source, will stop and reverse the motor in accordance with the left-hand motor rule.

For dc motors of fractional and relatively *low* horsepower rating, the variable dc voltage source may be a semiconductor (silicon-controlled rectifier) amplifier, operating from a three-phase or single-phase ac supply.

Motors of *moderate* rating, up to 100 hp, may be armature-voltage-controlled using rotary amplifiers such as Rototrol, Regulex, or smaller amplidynes.* In addition, static amplifiers such as magnetic amplifiers may be used as the adjustable dc voltage source (Section 4-10).

Larger dc motors, above 100 hp, are controlled in this manner by means of rotary amplifiers such as the amplidyne or the Ward-Leonard control system, shown in Fig. 6-2a.

The Ward-Leonard method is a rotary amplifier consisting of an ac (polyphase) motor driving a main generator and one or more exciters (smaller dc shunt generators). As shown in Fig. 6-2a, the (single) exciter supplies the motor and generator control field with a constant dc voltage. The amplification results from a small excitation power change in the field of the main generator, producing a larger power output from the armature of the main generator directly to the armature of the dc motor. For any given load, the motor armature current is constant. The motor field current and flux are usually held constant and (since $T = k\phi I_a$) the Ward-Leonard method thus represents in essence a constant-torque, variable-horsepower system (hp $= TS/5252$).

By means of reversing switches (not shown) in the armature circuit, it is possible to open and reverse the line connections to the motor armature, thus stopping and reversing the direction of motor rotation. Without doubt, the initial cost of the Ward-Leonard control system is much higher than any of the rheostatic control methods previously considered. Since the efficiency, neglecting the exciter efficiency, is essentially the product of the individual efficiencies of the two larger machines, the efficiency of this method is not as high as rheostat speed control by the field control method. But the advantages of armature voltage control using the Ward-Leonard method are:

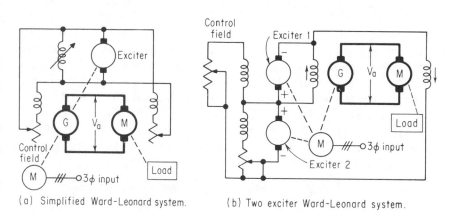

(a) Simplified Ward-Leonard system.   (b) Two exciter Ward-Leonard system.

Figure 6-2

Ward-Leonard methods of armature voltage control.

* For a complete discussion of these specialized rotary amplifiers, see Kosow, *Electric Machinery and Transformers*, Secs. 11-15 and 11-16.

1. A wide range of speed from standstill to high speeds in either direction.
2. Rapid and instant reversal without excessively high armature currents.
3. Starting without the necessity of series armature resistances.
4. Stepless control from standstill to maximum speed in either direction.
5. Larger units employing generator field reversal eliminate the need for heavy armature contactors for reversing, and at the same time prevent motor runaway since the motor field is always excited.
6. The method lends itself to adaptation of intermediate electronic, semiconductor, and magnetic amplifiers to provide stages of amplification for an extremely large motor. Thus, the power in the control circuit may extremely small.
7. Extremely good speed regulation at any speed.

These advantages offset, by far, the high initial cost and the somewhat intermediate overall motor-generator set efficiency of the Ward-Leonard method compared to other methods of speed control.

The response time of the Ward-Leonard control system may be improved by the two-exciter method shown in Fig. 6-2b. In this design, two exciters, in addition to the main generator, are coupled to the ac motor shaft. Exciter 2 (sometimes called the main exciter) provides excitation for the control field of Exciter 1 as well as for the fields of the main generator and the driven motor in series connection (across Exciters 1 and 2). The (main) voltage of Exciter 2 is higher (usually 230 V) than the rated voltage output of Exciter 1 (usually 115 V). The armature of Exciter 1 is always connected so that it opposes the emf of the main Exciter 2. Thus, an increase in control field excitation (Exciter 1) will raise the voltage across the field of the main generator and will also raise the armature voltage, V_a, applied to the motor armature, thus increasing the motor speed. At the same time, the increased bucking action produced by Exciter 1 against Exciter 2 reduces the field current of the driven motor, also increasing the motor speed. Increased control-field excitation, therefore, has produced two effects, both of which tend to increase the motor speed. The response of the motor to slight changes in control excitation is improved, both in response time and in sensitivity. This system is, of course, no longer a constant-torque system, since the field flux is subject to change; it might now be called a variable-torque, variable-horsepower system, the horsepower increasing with increases in speed.

6-6.
REVERSING dc
MOTORS

It is possible to reverse the direction of rotation of any dc motor by reversing the direction of *either* its field flux *or* its armature current, in accordance with the left-hand motor rule. Reversing *both* the field flux and the armature current simultaneously, by reversing the line connections, produces torque in the same direction, and the direction of rotation is unchanged. Thus, a dc series, shunt, or compound motor may be reversed by (1) reversing the direction of current in its armature circuit (including the interpole and compensating winding), or (2) reversing the direction of its shunt and/or series fields only, the direction of armature current, compensating winding, and interpole flux remaining the same.

The most popular method of reversal is to reverse the armature connections, despite the fact that heavier currents are to be interrupted. There are two disadvantages associated with field reversal: (1) Opening the field for reversal purposes may cause dangerous runaway, instability, and exceedingly high armature currents; and (2) the field is more highly inductive than the armature—opening a highly inductive circuit will produce more severe arcing and voltage breakdown than an interruption of the armature circuit. Furthermore, it is usually customary to "brake" a motor prior to reversing so that (as we shall see) during the period that a motor slows down (prior to reversal) its armature is drawing little or no current from the line. The wear and tear on contacts opening and closing the armature circuit, therefore, is not as serious as one might expect.

An electrically *reversible motor** is defined by the ASA as one which may be reversed by changing the external connections, even when the motor is running. All dc motors fall into this category. Thus, if the motor armature connections to the supply are reversed, a form of braking known as "plugging" occurs automatically. *Plugging* is the principle of applying power to a motor in such a direction that it *attempts* to reverse. Since it obviously must stop, or pass through a standstill condition, before it can reverse, it is possible to stop or "brake" a motor by plugging. Regardless of motor size, therefore, no additional or special braking devices are required to reverse the direction of a motor when its armature is "plugged" into connections of reverse polarity.

Motors may be reversed *manually*, using cam-operated or drum switches, or *automatically*, by means of relays and contacts. The basic circuit for shunt-motor reversal is shown in Fig. 6-3a. A set of M contacts is used to energize the field circuit and (if automatic starting is used) a control circuit. Closing the F of forward contacts will produce one *direction* of rotation.

Closing the R contacts will produce the reverse direction of rotation and armature current. As may be seen from Fig. 6-3, closing *both* the F and R contacts simultaneously will produce a *short circuit* directly across the line. In the case of *manual* switching, this is hardly likely because, in moving from forward to reverse, a transition (open) occurs in which the F contacts are opened and the R contacts are closed. In the case of *automatic* reversal, it is necessary to use electric and mechanical relay interlocks to prevent both sets of contacts from being energized simultaneously, as will be shown subsequently.

*Section 8-1.

117

The circuit for reversal of a compound motor is shown in Fig. 6-3b. The shunt-field circuit is energized by the *M* contacts, and the direction of current through the armature determines the direction of rotation. Note that the direction of current through the shunt and series fields *always remains the same*, maintaining the cumulative compounding, while the reversal of armature current reverses direction of rotation.

The series motor may be reversed by reversing either the armature connections, as shown in Fig. 6-3c, or the series-field connections, as shown in Fig. 6-3d. The former is preferable because of its plugging braking

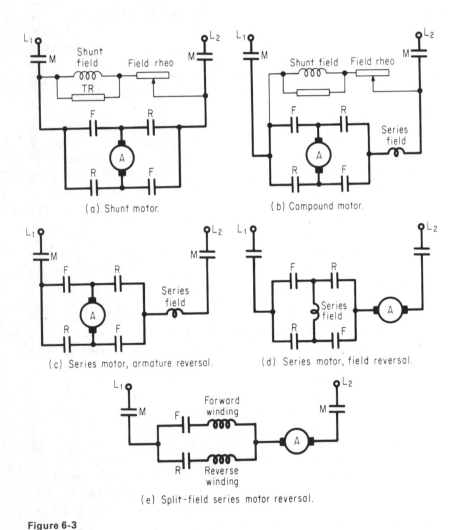

(a) Shunt motor.

(b) Compound motor.

(c) Series motor, armature reversal.

(d) Series motor, field reversal.

(e) Split-field series motor reversal.

Figure 6-3

Basic circuits for manual or automatic reversing
of dc motors (protective armature resistance not shown).

action as well as the fact that the field circuit is continuously energized. The specialized split-field series motor* may also be reversed by energizing either the forward or the reverse winding, as shown in Fig. 6-3e.

A controller for the automatic starting and reversal of a shunt motor using electric and mechanical interlocks of contacts and relays is shown in Fig. 6-4a. The configuration of the armature and field circuits is essentially the same as that shown in Fig. 6-3a with the addition of dashpot (or other suitable) time-delay relays 1A, 2A, and 3A. The unique features of this *controller* (see Section 3-1) are in the control circuit.

Pressing the FORWARD momentary contact will start the motor in a forward direction. The F relay, however, can be energized only through a n.c. R (reverse) contactor, i.e., when the REVERSE relay is deenergized. Similarly, depressing the REVERSE momentary contact will energize only the R relay through a n.c. F (forward) contactor when the FORWARD relay is deenergized. Furthermore, simultaneously depressing *both* contacts will result in neither relay being energized since (1) both R and F relays are disconnected from the line L_1, and (2) there is a mechanical interlock between the F and R relays (when R is energized it also pulls F into a de-energized position, and vice versa). The controller shown in Fig. 6-4a operates in the following manner:

1. Depressing the FORWARD contact closes all n.o. F contacts in series with the armature and in control line 1, energizing main contactor M. Simultaneously, F closes its interlocking self-energizing auxiliary contact in the forward relay circuit and opens a n.c. interlock in the reverse relay circuit.

2. The motor starts in the forward direction with full resistance in series with the armature. Relay M also energizes time-delay relay 1A in control line 2. The time-delay relays 1A, 2A, and 3A sequence the motor acceleration until the motor is operating as a shunt motor across the line, with all armature resistance effectively shorted by contacts 1A, 2A, and 3A.

3. Depressing the REVERSE momentary contact causes relay F to lose its line power to L_1 through the pushbutton interlock, and causes the F contact (n.o.) in control line 1 to deenergize relay M and, in turn, 1A, 2A, and 3A.

4. With its armature in series with the starting resistance across the line, and with the R relays in the armature circuit energized, the motor is plugged and rapidly loses speed. It accelerates in the reverse direction with its series starting resistors effectively shorted out in the same sequence as in the forward direction.

It should be noted when the motor is running in the FORWARD direction that, on pressing the REVERSE button, the series starting resistance is immediately inserted in the armature circuit so that the motor decelerates

---

* For a discussion of this and other types of specialized dc motors, see Kosow, *Electric Machinery and Transformers*, Sec. 11-13.

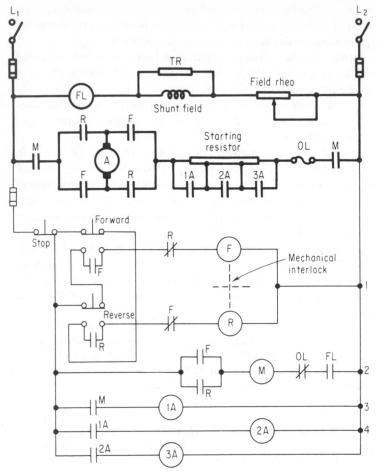

(a) Complete schematic using electrical and mechanical interlocks

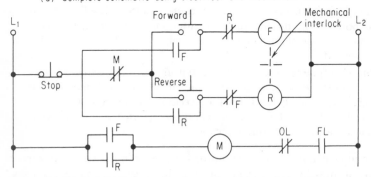

(b) Alternate control circuit using mechanical interlocks only.

Figure 6-4

Controller for automatic reversal of a dc shunt motor.

and starts in the opposite direction with protective armature series resistance. There is good reason for this, as will be explained in Section 6-7.

Figure 6-4b shows an alternative and somewhat simpler control circuit which uses mechanical interlocks only. If both forward and reverse buttons are simultaneously depressed, the L_1 line connection would tend to energize both the F and the R relay coils simultaneously. Since these relays are mechanically interlocked, however, neither relay is energized sufficiently to close its normally open contacts, and the M contactor in control line 2 remains deenergized.

There are circumstances in which it is absolutely necessary to bring a motor to a quick stop. Motors of large horsepower have excellent bearings and sufficient inertia so that they will continue to rotate for several minutes after the power has been disconnected. When such motors are to be reversed, jog-controlled, or operated at slow speeds, braking is necessary in order to avoid a time delay while the motor loses speed. One form of braking, "plugging," occurs automatically on the reversal of the armature voltage polarity. There are two other types of electric braking, called "dynamic" and "regenerative" braking. Each of these three types of electric braking will be considered, in turn, followed by a discussion of mechanical braking.

6-7. PLUGGING BRAKING

Although the reversing controller shown in Fig. 6-4a provides a plugging resistor in series with the armature during the plugging and reversing cycle, it can hardly be considered a plugging controller. When a motor armature is running in a specific direction and its armature applied voltage polarity is reversed, *at that instant* the counter emf is in phase (of the same polarity) with the applied voltage. The total applied voltage across the unprotected armature then is practically twice the voltage that may occur at the moment of starting, without any protective resistance in series with the armature. The maximum permissible current (usually 1.5 times rated) on starting is the same as that which should flow at the instant that plugging is initiated. *Additional* series armature resistance, over and *above* the starting series armature resistance, should be introduced, therefore, for proper plugging, in order to limit the armature current to a safe value. Thus, while the normal starting resistance in Fig. 6-4a may serve to limit the current as the speed decreases and reverses, it does not provide sufficient resistance for plugging at the time when such resistance is most required, i.e., at the moment when the polarity reversal occurs (since it takes time for the contacts to drop out).

The circuit of Fig. 6-5 shows a plug-reverse controller for a dc compound motor, in which a section of the starting resistance is shunted by a n.o. plugging contact, P. Control relays, P_f and P_r, through n.o. contacts F and R, are connected across a tap on the starting resistor and the armature of L_2 (since either F or R as well as M, are closed during running).

121

The starter also shows (in addition to Figs. 6-4a and b) a third method of interlocking the F and R contacts to eliminate a possible short circuit when both push buttons (F and R) are simultaneously depressed.

The operation of the starter shown in Fig. 6-5 is as follows:

1. Depressing the FORWARD button energizes relay F, closing all n.o. F contacts. Relay M in control line 3 is energized, and the motor starts across the line with full series resistance in its armature circuit. The voltage drop across the armature and across the series resistance is sufficient to energize coil P_f, through n.o. contact F, and close relay P in control line 6 through n.o. coil P_f. After a short time delay, therefore, when M closes and the voltage rises across P_f, relay P is energized in control line 6 and shorts out one step of series resistance. The motor accelerates as the armature current increases.

2. Relay M also energizes time-delay relay $1A$ in control line 4. After a specific time delay, $1A$ closes its n.o. contact across the series resistance, and the motor accelerates to a higher speed. Relay $1A$ also energizes relay $2A$ in control line 5. After a specific time delay, relay $2A$ closes its n.o. contact and shorts out the last step of starting resistance.

3. When the REVERSE button is depressed, the armature is plugged to a source of reverse polarity. Under these conditions, the induced armature voltage aids the line voltage and thus opposes the voltage drop across the plugging resistor. The voltage drop across coil P_f is now insufficient to hold n.o. contact P_f closed, and relay P_f drops out, deenergizing relay P in control line 6.

4. Depressing the REVERSE button also deenergizes relay F in control line 1, relay M in control line 3, relay $1A$ in control line 4, and relay $2A$ in control line 5. Simultaneously, all n.o. R contacts are energized by relay R.

5. As the motor approaches zero speed, the generated counter emf is negligible and relay P_r starts to close.

6. The motor reverses direction, with n.o. relay contacts P, $1A$, and $2A$ closing as the motor accelerates toward rated speed, in the same manner as described above.

It is often desired to stop the motor rapidly, without reversing the direction of rotation, by using electric braking (plug-stop braking). The control circuit of Fig. 6-5 may easily be modified to use the principle of plugging braking to bring the motor rapidly to a standstill. The modification shown in Fig. 6-6 involves a centrifugal switch (CS) which opens when the motor is running at, say, 10 rpm or less. When the STOP button in control circuit 1 is pressed, relay F is deenergized. With the motor running in a forward direction, centrifugal switch CS is closed and reverse relay R is energized in control line 2 through CS and the n.c. relay contact F. Reversal of the motor armature voltage produces the same action as in steps 3 and 4, and the motor is plugged with full series armature resis-

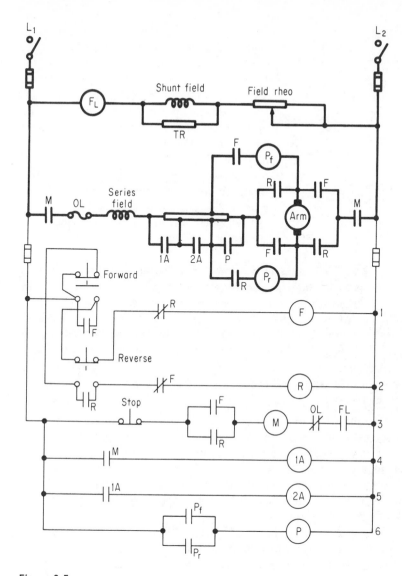

Figure 6-5

Plug-reverse controller for dc compound
motor.

tance. When the motor speed reaches 10 rpm or less, centrifugal switch
CS opens in control line 2 and reverse relay *R* is deenergized, bringing
the motor to a stop. A small, mechanical, solenoid-operated brake is
sometimes used to clamp the shaft when *CS* opens, and is deenergized
when the START (forward) button is again pressed (see Section 6-10).

SEC. 6-7 / *Plugging Braking*

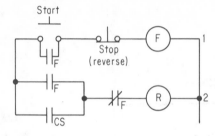

Figure 6-6

Control circuit modification for plug-stop nonreversing controller (plugging braking).

In contrasting the control-system operation of Figs. 6-5 and 6-6, the *former* may be considered a *three-stage* operation and the *latter* a *twostage* control. The sequence for the *plug-reverse* controller of Fig. 6-5 is (1) start and accelerate in the forward direction; (2) braking, by means of plugging, to a standstill; and (3) start and accelerate in the reverse direction. Using the control-circuit modification shown in Fig. 6-6, the control-system sequence for a *plug-stop* controller is (1) start and accelerate in a given direction, and (2) braking, by means of plugging, to a standstill.

6-8.
DYNAMIC BRAKING

When the armature of a motor is disconnected from its supply, it will come to a stop eventually, despite the inertia of its load, because energy is no longer provided to the armature and mechanical losses are present. If the field excitation of a dc motor is maintained at the time that the armature is disconnected from the supply, the moving armature conductors will have a voltage induced in them and the deenergized armature will act as a separately-excited generator. The prime mover of the armature is the inertia of the motor rotor and its connected load.

If an electric load in the form of a resistor is connected across the deenergized armature of a dc motor, the motor will be brought to a stop very rapidly, since the inertia of the motor armature (as prime mover) must supply both electric and rotational losses. This form of braking, in which the motor armature (only) is deenergized, connected across a resistor, and permitted to dissipate its rotational energy as a generator, is called "dynamic braking."

Because a resistor is required in series with the armature of a dc motor for purposes of motor starting, it is customary to use a *portion* of this resistor as a braking resistor to dissipate the generated energy of the motor when its armature is disconnected from the supply. Dynamic braking of a nonreversing shunt motor is shown in Fig. 6-7. A quick-acting braking relay, *B*, energized when the motor is started, maintains the necessary connection through its n.c. contact *B* from the starting resistor to the armature. The controller shown in Fig. 6-7 operates in the following manner:

1. Depressing the START button energizes quick-acting relay *B*, which simultaneously opens its n.c. contact *B* across the armature and closes its n.o. contact in control line 2. Relay *M* in control line 2 is energized and, through its n.o. interlock across the START contact, maintains relay

B in an energized state. When the main *M* contactor closes, the armature rotates with full series resistance starting.

2. Relay *M* also energizes time-delay relay 1*A* in control line 3. After a proper time delay, the accelerating relay 1*A* closes, shorting out one section of starting resistance in series with the armature. Relay 1*A* also closes its n.o. contact in control line 4, energizing time-delay relay 2*A*. After another suitable time delay, relay 2*A* closes, and the motor accelerates to normal speed with its armature across the line.

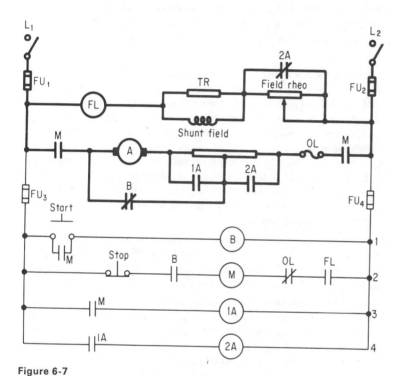

Figure 6-7

Dynamic braking of a nonreversing shunt motor using the starting resistance as a braking resistor.

3. When it is desired to stop the motor and bring it rapidly to standstill, the STOP button is depressed. Relay *M* is deenergized, disconnecting the armature of the motor from the line supply and deenergizing braking relay *B* in control line 1 through its n.o. interlock contact *M*. Deenergized relay *B* closes its n.c. armature shunting contact, placing half the starting resistance across the armature.

4. The motor rotates as a generator with full field excitation and a low resistance across the armature. The combined frictional, iron, rotational, and copper losses bring the motor quickly to a stop. Full field excitation is provided by the n.c. relay contact 2*A* across the field rheostat, required

125

both for starting and braking. Since relay M was deenergized when the STOP button was pressed, relay $2A$ is also deenergized, increasing the excitation and the losses, both electric and iron, to bring the motor to a quicker stop.

As in the case of plugging braking, it is sometimes desired to operate a motor using a sequence which calls first for braking and then for rapid reversal of the motor. Dynamic braking may be used to bring the motor to a standstill before the reverse polarity (plugging) is applied. A typical controller employing dynamic braking and reversing, or dynamic braking to a quick stop, is shown in Fig. 6-8; it operates in the following manner:

1. Depressing the FORWARD momentary contact energizes relay F_1 in control line 1 through n.c. contacts AP (antiplugging relay) and the REVERSE interlock. When F_1 is energized, it closes its auxiliary contacts, shunting n.c. AP contacts and its pushbutton to maintain F_1 energized in control line 1. Simultaneously, F_1 also energizes relays R and M in control line 3, through field loss and overload relays (both n.c.). Braking relay B energizes n.c. contacts B, shunting the armature, and also energizes relay F_2 in control line 1. With relay M energized in the power circuit, the motor starts with full armature series resistance in the forward direction through n.o. contacts F_1 and F_2 and full field current with n.c. $2A$ shunting the field resistance.

2. Acceleration is produced in the normal manner when relay M energizes time-delay relay $1A$ in control line 4, and $1A$ energizes time-delay relay $2A$ in control line 5. The latter also inserts the field rheostat to provide the desired speed of the motor in the forward direction.

3. Dynamic braking is applied when either the STOP or the REVERSE contact is depressed, since both of these deenergize relays F_1 and F_2. Deenergizing these contacts opens control line 3, deenergizing relays B and M, the brake relay and main contactor. The latter deenergizes relays $1A$ and $2A$ in succession. The former completes the armature shorting loop through a portion of the starting resistance via n.c. contact B. With full field excitation, the armature has been disconnected from the line and a braking resistance has been connected across it. The motor is rapidly stopped by means of dynamic braking.

4. In order to ensure that no plugging action occurs when the REVERSE contact is depressed, an antiplugging relay AP is connected across the armature. When control line 1 is opened by pressing the REVERSE button, the action described in 3 takes place. Relay R_1 cannot be energized during the time that dynamic braking is initiated, since it requires that relay AP be deenergized and contacts AP in the control line of R_1 be closed. Plugging cannot occur, moreover, unless R_2 is energized. Should R_2 close too soon (R_2 cannot close until braking relay B is energized in control line), n.c. contacts AP will open and prevent energization of R_1 in control line 2. This is equivalent to depressing the STOP button and permitting dynamic braking to occur as described in 3.

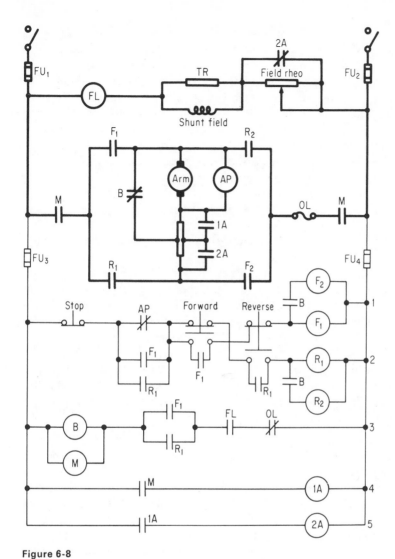

Figure 6-8

Dynamic braking of a reversing shunt
motor using starting resistance as a
braking resistor.

The sequence of operations of the controller shown in Fig. 6-8 permits the same three-step transition as that provided in Fig. 6-5 of (1) start and accelerate in the forward direction, (2) braking to a standstill by means of dynamic braking, and (3) start and accelerate in the reverse direction. Plugging or reversing may not occur until the voltage across the armature (during the dynamic braking process) is sufficiently small to deenergize antiplugging relay AP so that relay R_1 may be energized.

127

6-9.
REGENERATIVE
BRAKING

A third type of electric braking, in addition to plugging and dynamic braking, is "regenerative braking." The term "regeneration" implies energy return (or generating energy back) to the supply. Regenerative braking stems quite naturally from dynamic braking, since it would appear quite logical not to waste or dissipate the rotational energy of a large motor (operating as a generator during dynamic braking) by dissipation in a resistor, but to return that energy to the supply instead.

Any generator is loaded when it is paralleled and supplying energy to a bus. The greater the amount of energy supplied to the bus, the greater the prime-mover energy required to sustain the generator. In the case of a motor rotating at a fairly high speed, whenever the load tends to drive the motor in the same direction and full field excitation is applied, there is a strong possibility that the induced voltage may exceed the line voltage. If the nature of the motor load is such that it tends to drive the motor (as in a descending elevator, for example), the motor speed will increase and the generated voltage will tend to exceed the line voltage considerably. Motor loads for electric locomotives, trolleys, buses, elevators, cranes, and hoists will have sufficient potential energy (at the top of a hill in the case of traction, or a heavy load about to descend in the case of lifting devices) to drive their motor shafts at extremely high speeds. The speed of these motors may be reduced considerably, with practically little waste of energy requiring no mechanical or friction braking, by using regenerative braking. The power returned to the lines may be used for other motors, devices, or equipment served by the bus.*

The principle of regenerative braking is so simple as to be almost unbelievable. A dc dynamo acts as a motor when its armature is connected to a supply and when the dynamo develops a counter emf that is less than the terminal voltage. The motor will run at as speed determined by its excitation and counter emf ($S = kE/\phi$) as long as it is drawing current from the line (motor action). If, however, the dynamo excitation is increased and it is driven by a prime mover at a *higher* speed, the dynamo will send current *into* the line and act as a *generator*. The load on its prime mover as a result of generator action will reduce the prime-mover speed and will cause the dynamo to resume motor action when its counter emf and speed return to a value such that current is delivered *by* the line *to* the dynamo. In regenerative braking, therefore, it is not even necessary to disconnect the motor from the line. All that is required is that the motor

---

* In contrast to the internal combustion engine in an automobile, which dissipates its braking energy in brake linings, the electric car offers high energy-conservation possibilities. On fairly flat surfaces, energy is theoretically required only during acceleration. Similarly, energy is only required in uphill travel. During periods of deceleration and downhill travel, dynamic braking maintains safe speeds and *returns* energy to the electric-car prime mover. Thus, the only energy required for travel is that consumed by the losses inherent in the mechanical and electrical design.

speed and the field excitation be increased sufficiently to *reverse the armature current* and produce *generator* action.

Schematic diagrams of the switching circuits for plugging, dynamic, and regenerative braking are shown in Fig. 6-9 for purposes of comparison as well as reference. In each of the figures, the motor is accelerated to its rated speed by closing acceleration contacts A. At the end of a specific time, contacts A_R are closed, shorting out the protective series armature resistance. When it is desired to brake the motor, all A contacts are de-energized and all B contacts are energized. In the case of braking by plugging, shown in Fig. 6-9a, the armature polarity is reversed through protective resistor R and the full field flux is applied. In the case of dynamic braking, shown in Fig. 6-9b, the armature is disconnected from the line

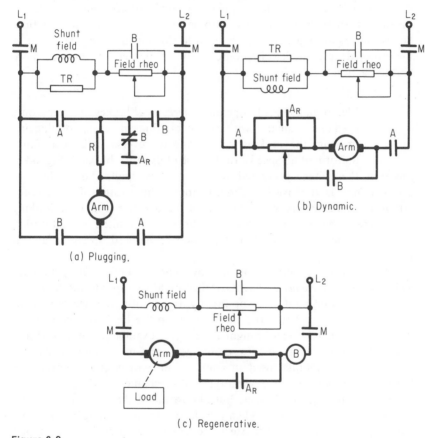

(a) Plugging.

(b) Dynamic.

(c) Regenerative.

Figure 6-9

Plugging, dynamic, and regenerative braking of a shunt motor; in schematic form for manual or automatic control (control circuits not shown).

129

when the *A* contacts open, and is connected across a portion of the starting resistance when the *B* contacts close. Both plugging and dynamic braking will bring the motor to a stop (at which point the *B* contacts should be opened).

In the case of regenerative braking, shown in Fig. 6-9c, braking occurs automatically whenever the speed of the motor armature is high and the load drives the motor as a generator. Contact *B*, a reverse current relay, closes to increase the field excitation and braking action during this period.

Regenerative braking will *not* bring the motor to a stop but serves only to reduce its speed (1) without the necessity for mechanical brakes, and (2) with little energy loss.

In some types of regenerative braking, the field (not the armature) is disconnected and is separately supplied (overexcitation) from an exciter on the same shaft as the motor, serving to provide more rapid braking action. Series motors are regeneratively braked in this manner.* As the speed of the motor drops, the field is progressively strengthened to provide more braking action.

6-10.
MECHANICAL
BRAKING
Electromagnetically operated *mechanical* brakes are commonly used in conjunction with electric braking. In the case of plugging braking, the braking action due to reverse polarity is fairly uniform at all speeds. In the case of dynamic braking (Fig. 6-9b) however, the voltage generated at low speed is small, and the braking action is not as effective for a motor approaching standstill as it is for a motor at its rated speed. In the case of regenerative (Fig. 6-9c) braking (even where the field is overexcited), braking is limited by saturation of the field poles, and the braking action at reduced speeds is almost negligible.

It is necessary, therefore, to employ some form of *auxiliary* mechanical braking, *electrically* actuated, to operate at low speeds and to ensure that the load is brought to a stop rapidly. In addition, the electromechanical brake serves to "pin" the shaft and keep the load from moving or drifting when it has been brought to rest (for example, as in electrically operated elevators, lifts, cranes, or bridges). The most common type of electromechanical brake used in conjunction with electrical braking is the *electromagnetic brake*. Several types of mechanical braking principles are used with magnetic brakes, but almost all of these are actuated by means of an *electromagnet*. When the electromagnet is *energized*, it operates as a powerful solenoid to exert a force on a set of brake shoes, brake

* For an extremely detailed discussion of the subject, see H. D. James and L. E. Markle, *Controllers for Electric Motors*, 2nd ed. (New York: McGraw-Hill Book Company, 1952), Chap. 10.

bands, or brake discs, which in turn tend to *release* a drum located on a motor shaft and *permit* rotation.

The fundamental advantage of an electromagnetic brake is that it may be operated electrically in a control circuit to (1) *release*, i.e., permit shaft rotation, whenever power is applied to a motor, and (2) *set*, i.e., lock the shaft, whenever power is removed from the motor. Energizing the brake coil, therefore, serves to actuate the brake solenoid to release the brake by lifting its shoes or bands away from the brake drum and to compress the brake spring. When the brake coil is deenergized, the compressed spring is no longer held by the solenoid and it sets the brake. For this reason, as well as the fact that a brake requires appreciable current for its operation, the brake coil is *usually* energized by and associated with the M contacts in the power circuit rather than in the control circuit.

Figure 6-10a shows a *series brake* used in connection with a hoist or crane. The solenoid, wound of a few turns of heavy wire, is connected in series with the armature of the series motor. When the M contacts are closed, the starting surge current instantly operates the short-time-constant quick-acting series brake solenoid, releasing the brake, and permitting the motor to operate during acceleration and under normal conditions of load. When the series motor is deenergized (or in the event of reduced current due to possible loss of load), the brake instantly sets and prevents the motor shaft from turning. Series brakes set at approximately 10 per cent and release at approximately 40 per cent of the motor load current.

Shunt brakes, shown in Fig. 6-10b, are generally used with shunt or compound motors. The solenoids of these brakes are wound with many turns of fine wire and are highly inductive. Since brakes are set in the absence of power, a time delay in brake operation when power is applied to a motor could be disastrous. The time delay is overcome and even eliminated by using fewer turns and by designing the coil to operate on approximately one-half of the rated voltage. A resistor, connected in series with the coil, prevents the coil from overheating, and simultaneously decreases the time constant ($t = L/R$). Reducing the turns by one-half, for example, decreases the inductance to one-quarter of its original value; and doubling the resistance in series reduces the time constant to one-eighth of its original value.

As shown in Fig. 6-10b when the motor circuit is energized, prior to starting, by closing the main safety switch, brake relay B is energized through the n.c. contact M shorting the series resistance to the brake coil. When the motor is started and contactor M is energized, the (part-voltage) brake coil has the full voltage impressed across it, and it quickly releases the motor shaft. As the motor accelerates, however, the n.c. time-delay M contact deenergizes the brake resistor relay B, opening the n.o. B contacts, and thus inserting protective series resistance to prevent the brake coil from overheating. A simple but unique feature shown in Fig. 6-10b is to

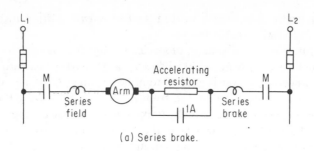

(a) Series brake.

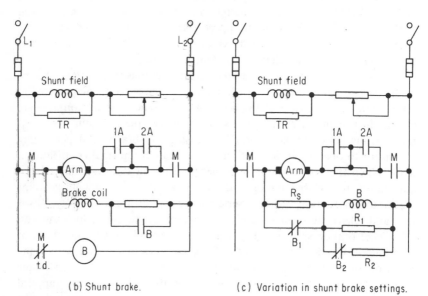

(b) Shunt brake.   (c) Variation in shunt brake settings.

Figure 6-10

Series and shunt brakes.

use the *M* contacts between the parallel-connected armature and the brake coil circuits. This prevents armature regeneration (when the *M* contacts are opened and the field is still excited) which may hold the brake released and prevent it from setting. The brake is released very rapidly when *M* is deenergized, because of the increased series resistance and the reduced time constant of the brake coil.

A combination of series and shunt resistance is shown in Fig. 6-10c. When the brake is released (energized), n.c. contact B_1 is closed to provide faster release. During the motor operation, contacts B_1 and B_2 open. When *M* is deenergized, the series resistor R_s reduces the setting time; and, as the energy of the inductive brake coil is discharged through shunt resistor R_1, contact B_2 closes, reducing the total shunt resistance and thus providing a softer and smoother brake setting action.

6-11.
JOGGING

Magnetic brakes are effectively and extensively used in conjunction with a type of speed control called *jogging*.* In this type of control, the momentary-contact JOG pushbutton is depressed, and the control circuit operates in such a manner that the armature rotates slowly only while the JOG button is depressed. The release of the JOG button immediately stops the motor by means of magnetic brake setting. Thus, it is possible to "inch" the motor shaft by depressing the JOG button momentarily and releasing it instantly.

Jogging is necessary wherever precise motor shaft position or alignment is required, as in some machine tool operations, on sewing and textile machines, for automatic belt-line processing, and in certain types of crane and elevator control. The control circuit for jogging is accomplished in a relatively simple manner, as shown in Fig. 6-11a for any dc motor. When the JOG button is momentarily depressed, it deenergizes the various motor acceleration relays which serve to eliminate protective series resistance. It energizes the main M relay only, and the motor is thus started with full protective series resistance (not shown). Energizing the M contactor, however, releases the magnetic brake (not shown). When the JOG button is released, M is deenergized and the magnetic brake is set simultaneously (see Section 6-10).

Occasionally, with forward JOG control as shown in Fig. 6-11a, the precise alignment position may be difficult to obtain. Repeated clockwise and counterclockwise "inching" may be necessary, or the machine tool processes (such as tapping and threading) may require both forward and reverse jog control. The circuit shown in Fig. 6-11b is the basic control circuit used for numerous control applications where forward as well as reverse jogging is required. Separate JOG controls are used in addition to the momentary-contact FORWARD and REVERSE controls. The circuit shown in Fig. 6-11b should be compared with the circuit shown in Fig. 6-4a. The former is a modification of the latter by the inclusion of forward and reverse jogging controls. The circuit of Fig. 6-11b operates in the following manner:

1. Depressing the JOG FORWARD contact energizes relay F_2. Main contactor M in control line 3 is temporarily energized to rotate the armature with full protective series resistance. As long as relay F_1 is deenergized, the accelerating relays (not shown) are deenergized. (This also is a modification of Fig. 6-4a.)
2. Main contactor M also energizes and releases the magnetic brake (not shown). When the JOG FORWARD contact is released, F_2 is deenergized, deenergizing M; and the motor stops instantly as the magnetic brake sets.

* Jogging as defined by NEMA is "the quickly repeated closing of a circuit in order to start a motor from rest for the purpose of accomplishing *small* movements of the driven machine."

133

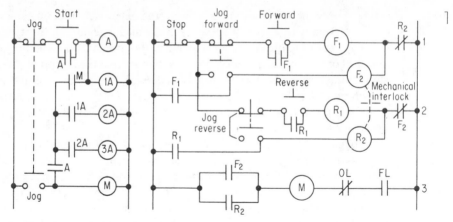

a. Basic jog control circuit.

b. Forward-reverse jog control circuit using separate jogging buttons with electrical and mechanical interlocks

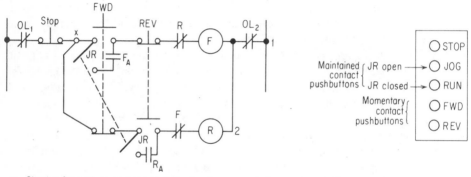

c. Simpler forward-reverse jog control using jog-run switch and same pushbuttons for forward-reverse jog and run.

d. Pushbutton station

Figure 6-11

Jog control circuits used with series or shunt brakes.

3. Depressing the JOG REVERSE relay energizes relay R_2, producing rotation in a reverse direction and energizing relay M in control line 3 in the same manner as described above.

The mechanical interlock between relays F_2 and R_2 prevents a short circuit across the line when both JOG buttons are simultaneously depressed. The n.c. R_2 and F_2 contacts in control lines 1 and 2 serve to prevent control relays F_1 and R_1 from energizing the accelerating relays, which cause the motor to accelerate when the JOG controls are operated.

A somewhat less complex forward-reverse jog control circuit is shown in Fig. 6-11c. This circuit uses the *same* pushbutton for forward jogging as for forward running and, conversely, the *same* pushbutton for reverse jogging and running. This circuit employs a DPST maintained-contact JOG-RUN pushbutton pair which selects the mode of operation, shown in Fig. 6-11d. The circuit operates in the following manner (see Fig. 6-11c):

1. In the RUN position, switch JR closed, the circuit is the conventional forward-reverse control circuit with electrical and mechanical interlocks similar to those shown in Figs. 6-4 and 6-5.
2. Energizing either relay F or R, by depressing the FWD or REV momentary contact pushbuttons, energizes either F or R contacts in the power circuits of Figs. 6-3a through e, respectively.
3. When jogging is desired, it is accomplished by depressing maintained-contact JOG switch, opening DPST switch JR.
4. In the JOG mode, with switch JR open, relays F or R are only energized, respectively, for the duration that the momentary FWD or REV contacts are depressed.

If a set of auxiliary M contacts is required in addition to the F and R contacts, a third control line may be added to Fig. 6-11c, similar to that shown in line 3 of Fig. 6-11b.

As previously noted, all JOG circuits require either series or shunt brakes so that the motor and its load are stopped, instantaneously, without coasting, whenever the momentary contacts are released. See Section 6-10 and Fig. 6-10.

6-12.
ELECTRONIC dc
MOTOR CONTROL

The history of motor control in the late nineteenth and early twentieth century began with the development of resistive dc motors manually controlled by armature and field control techniques. With the development of the ac motor and the extended distribution of ac power in the first three decades of the twentieth century, various types of ac motor control techniques were devised (See Chapter 7) and interest in the dc motor declined. The development of the electronic triode gas and thyraton tubes, coupled with the parallel development of the saturable reactor, the magnetic amplifier, and various rotary power amplifiers (Amplidyne, Regulex, Rototrol, and Ward-Leonard) in the next three decades (between 1930 and 1960), reawakened an interest in dc motor speed control.

The development of the *thyristor* or *silicon-controlled rectifier* (SCR) for low- and medium-power applications in the 1950s has created unlimited possibilities for dc motor control from an ac supply by electronic methods. The small size, high reliability, and relatively efficient SCR has begun to dominate the latter half of the twentieth century for controlling both dc and ac small-and medium-sized motors from an ac source. As of this writing, SCRs up to 400 A (rms) with voltage ratings (peak forward and reverse blocking) up to 1200 V (GE, type C 290 PB) are now available. Beyond this capacity, it is customary, at present, to use conversion systems

135

such as mercury-arc rectifiers, magnetic amplifiers, rotary converters, and motor generator sets to convert and supply the direct current required for extremely large dc motors (above 100 hp at 115-V dc and 200 hp at 230-V dc). Because the silicon-controlled rectifier may be used to control the speed of 115-V dc motors up to 50 hp and 230-V motors up to 100 hp from a single-phase or three-phase ac supply, its versatility and the reduced size of the control equipment needed indicate great promise for small- and medium-sized motor control by electronic methods.

The typical appearance of a silicon-controlled rectifier (SCR) is shown in Fig. 6-12a. A three-junction (alloyed silicon) wafer is contained within a ceramic housing from which a flexible, multistranded anode lead (or pigtail) is taken from a positive layer of semiconductor material. The cathode terminal, grounded to the metallic case, is brought out from a negative layer in the form of a screw thread attached to a hexagon nut, so that the silicon-controlled rectifier (SCR) may be bolted to a metallic heat sink for heat dissipation. A smaller pigtail control or gate lead, brought out of the housing, is taken from a positive layer of semiconductor material, separated from the anode and cathode, respectively, by suitable junctions as shown in Fig. 6-12b. The SCR may be considered as com- posed of two transistors: an *npn* and a *pnp* connected "back to back" as shown in Fig. 6-12b. The result is the production of a *pnpn* semiconductor consisting of three junctions: an anode, a control, and a cathode junction (J_1, J_2, and J_3, respectively) between the four regions produced in a single alloyed junction wafer.

The operation of the SCR may be analyzed using simple transistor theory in terms of the *pnp* and *npn* transistors, Q_1 and Q_2, respectively. As shown in Fig. 6-12b, the collector of Q_2 drives the base of Q_1, and simultaneously the collector of Q_1 drives the base of Q_2. If β_1 is the current gain of Q_1, and if β_2 is the current gain of Q_2, the product $\beta_1\beta_2$ is the gain

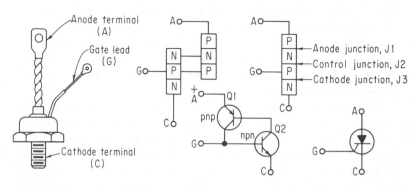

(a) Typical appearance.　　(b) Three junction transistor equivalent.

Figure 6-12

Silicon controlled rectifier (SCR).

of the positive feedback loop, or β_3. When β_3 is less than unity, the circuit is in a stable state and the SCR is "biased off"; i.e., the only current flowing between the anode and the cathode is the cutoff collector current between the two "transistor" sections, resulting in a very high impedance between anode and cathode (550 V, PFV and PIV). When a positive potential is applied to the control gate, G, "transistor Q_2" (the npn section) is biased to an "on" state, causing an increase in its collector current to a point where its current gain β_1 will cause loop gain β_3 to exceed unity. The circuit then becomes regenerative, causing the collector current of both sections to increase rapidly to a maximum value limited by the external circuit. As both npn and pnp sections are driven into saturation, the impedance between anode A and cathode C decreases to a very low value, and unidirectional current may flow in the direction from A to C.

When triggered by a positive potential at G from an "off" to an "on" state, it is no longer necessary to continue to apply a positive potential at the gate. The SCR will remain in the "on" state because the pnp section supplies sufficient current gain to drive the base of the npn section regeneratively. Thus, merely a positive pulse, of sufficient magnitude to increase β_1 so that β_3 exceeds unity, is all that is required to initiate self-generation and the "on" state.

The SCR may be turned off only by reducing the collector current of the pnp section so that β_1 is less than unity. This may only be done by lowering the emitter or anode A voltage of the pnp section, to reduce β_3 to less than unity and to restore the stable "off" state. Thus, while the gate initiates the conduction of the SCR, it has no control over conduction once it is initiated.*

Because it may be triggered from a state of high to low conduction (on-off) and vice versa (off-on) by reduction of anode potential and positive gate triggering, respectively, the SCR may be used as a switch for low and moderate currents (up to 400 A). Figure 6-13a shows the use of the SCR as a latching switch in which the armature of any dc motor is connected to a **dc** supply (fields and protective resistances not shown). In the absence of a positive pulse at G, the switch is open and the motor armature is not energized. The gating resistor, r_g, connected between the cathode and the gate, provides a negative gate bias current, ensuring a stable "off" condition. By increasing the value of r_g, the sensitivity may be increased, and any degree of reduced sensitivity may be obtained by proper selection of r_g. The motor is turned on by application of a low

---

* For this reason, the SCR is the semiconductor equivalent of the thyratron and the ignitron. In both of these electron tube devices, the positive potential applied to the control grid initiates firing, but conduction is cut off only when the plate potential is reduced to a point where deionization occurs. The SCR, however, requires no filament heating, no filament transformer, no time delay for ionization to occur, and no power waste while in the "off" state; in addition, it is much smaller in size for the same current rating. The circuits shown in Figs. 6-13 through 6-16 are essentially thyratron and ignitron circuits which have been simplified through the use of the SCR.

137

positive potential pulse (0.5 to 1.0 V at a current of approximately 25 mA for a duration of 1 μs). The operating motor can only be turned off once it is energized by opening n.c. contact M. Closing M once more will not cause the motor to turn on until a positive pulse reappears at G.

An alternative method of turnoff, using an RC time-constant principle, is shown in Fig. 6-13b also using a dc supply. The motor is switched on by a positive pulse at G. It will continue to operate until n.o. contact M is closed, placing the junction between R and C at ground potential. As capacitor C charges up, therefore, it places point A at a negative potential with respect to ground, and cutoff of the SCR occurs. Since M is a normally open switch, another SCR may be used to close the shunt capacitor circuit. Thus, the motor may be turned on by application of one positive pulse to the SCR shown in Fig. 6-13b, and turned off by application of another positive pulse to a second SCR, used in place of M. Transistor switching, or alternatively the method shown in Fig. 6-13e, may be used to reset the second SCR, with the result that a relay or a motor may be switched on and off by means of positive trigger pulses, as described below.

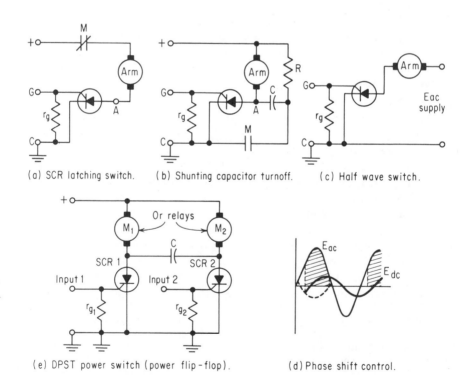

(a) SCR latching switch.   (b) Shunting capacitor turnoff.   (c) Half wave switch.

(e) DPST power switch (power flip-flop).   (d) Phase shift control.

Figure 6-13

SCR switching possibilities.

In the presence of a repetitive and positive gate potential, a dc motor may be operated from an ac supply, as shown by the half-wave switch of Fig. 6-13c. Whenever the anode voltage is negative at the SCR, it fails to conduct; but when the ac potential is positive, conduction is resumed, and the current through the motor armature is unidirectional.

If the phase of the voltage applied to the gate E_{gc} is alternating and *shifted* with respect to the ac supply, E_{ac}, the circuit may be used as a method of armature voltage control. The average value of dc voltage applied to the armature is a function of the phase relation between E_{gc} and E_{ac}. When E_{gc} lags E_{ac} by almost 180°, the voltage applied to the armature is almost negligible. Simple phase-shift control may be used (1) to start the motor, and (2) to serve as a method of armature voltage speed control, as shown in Fig. 6-13d.

The principle of the shunting capacitor switch shown in Fig. 6-13b, in which a second SCR is used to provide a negative bias cutoff by a charging capacitor, is shown in Fig. 6-13e, the so-called SCR power flip-flop circuit. Two dc motors or two dc relays may be alternately energized by application of positive pulses to inputs 1 and 2, respectively. A positive pulse at input 1 will turn on motor (or relay) M_1 and, simultaneously, charge C in such a manner that a negative potential is applied to the anode of SCR_2. Capacitor C must be sufficiently large to hold the anode of the SCR at a negative potential for that time in which the transition occurs from regeneration to a stable state. When a positive potential is applied to input 2, conduction of SCR_2 occurs, turning on M_2 and simultaneously charging C in such a manner that SCR_1 is turned off. This circuit may also be used to turn a single motor on and off by means of pulses by substituting an equivalent resistor in place of M_2.

In order to reduce ripple, or ac component in the wave which introduces increased motor iron losses, it is customary to use full-wave rather than half-wave rectification. Various methods of full-wave, unidirectional, armature voltage control techniques are shown in Fig. 6-14, to start and run a dc motor from a single-phase ac supply. The circuit of Fig. 6-14a is a bridge circuit in which two arms of the bridge are simple solid-state diodes, CR_1 and CR_2, respectively, and the other two arms of the bridge are SCRs. The phase of gate inputs 1 and 2, respectively, may be adjusted by means of either phase-shift circuits or trigger circuits* to provide identical conduction on each corresponding half-cycle of alternating current. For operation from higher ac voltages, the circuit shown in

* Space does not permit a discussion of the various methods and principles employed to obtain phase-shift and positive pulse circuits, as well as specific design problems in which these differ from conventional thyratron circuits. A fairly complete discussion of the subject, including the use of magnetic amplifiers, may be found in the following: W. J. Brown, "Firing Circuits for Silicon Controlled Rectifiers," *Electrotechnology* (July 1961), pp. 79–83; *Semiconductor Power Circuits Handbook* (Motorola Semiconductor Products Division, November 1968); F. W. Gutzwiller, ed., *G.E. SCR Manual*, 4th ed., (Syracuse, N.Y.: General Electric, 1967).

Fig. 6-14b may be used. This circuit places the dc motor armature across the dc output of a full-wave bridge. SCR_1 and SCR_2, respectively, are used merely as phase-sensitive switches that determine in which portion of each half-cycle conduction occurs. The pulses applied to inputs 1 and 2 may be adjusted to produce circuit completion for the entire cycle, for identical corresponding portions of each half-cycle, or for unequal durations of each half-cycle. In effect, this technique provides a range of control over the full 360° of the ac input wave. Additional *LC* filter circuits (similar to those used for power supplies) may be employed to improve the nature of the dc waveform supplied from the full-wave rectifiers to the motor armature. For higher voltages, a stepdown transformer (not shown) may be used in Fig. 6-14b.

A total of six SCRs may be used to drive a dc motor *unidirectionally* by armature voltage control from a three-phase supply as shown in Fig.

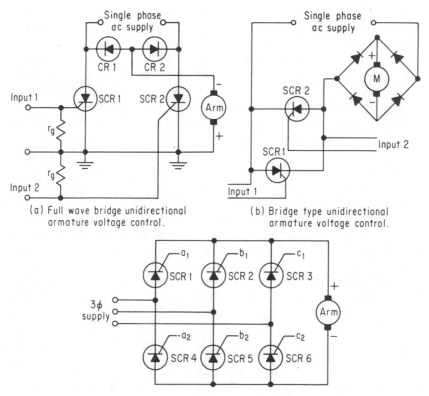

(a) Full wave bridge unidirectional armature voltage control.

(b) Bridge type unidirectional armature voltage control.

(c) Three-phase full wave unidirectional voltage control.

Figure 6-14

Unidirectional armature voltage control using SCRs.

CHAP. SIX / *Manual and Automatic Speed Control of dc Motors*

6-14c. Each pair of parallel-connected SCRs provides full-wave rectification of each phase. The resultant paralleled dc output voltage contains considerably less ripple. Control voltages a_1–a_2, b_1–b_2, and c_1–c_2 may be phase-shifted with respect to their phase voltages, or individually triggered to provide a smooth dc output voltage from zero to maximum, which may be used *both* for starting and armature voltage speed control.

The SCRs numbered 4, 5, 6 in Fig. 6-14c may be replaced by diodes of equivalent current and voltage rating. If the triggering signals a_1, b_1, c_1 are capable of being phase-shifted over 180°, the circuit still yields control from 0 to 100 per cent of full dc output voltage. If the triggering signals a_1, b_1, c_1 can be phase-shifted over 120°, the circuit produces full control from 25 to 100 per cent of full output dc voltage and power.

The circuits of Fig. 6-14 are adaptable primarily to starting and to the control of speed where it is desirable to rotate the motor in *one* direction only. Where motor *reversal* is desired, it may be accomplished either by switching or by using pairs of SCRs. Reversal by switching is shown in Fig. 6-15 while that by using full-wave pairs is shown in Fig. 6-16.

The switching method shown in Fig. 6-15a uses an *inverse parallel* arrangement of SCRs in series with the dc motor armature connected to a variable-voltage ac supply. R_1 and R_2 are equal-value *gating* resistors sufficient to maintain the SCRs in a stable state, i.e., when contact F and R are *both* open. Closing contact F, however, shunts resistor R_1 with an extremely low resistance, R_3, causing SCR_1 to conduct as a half-wave rectifier from anode to cathode every half-cycle.

Opening F and closing contact R will produce conduction of SCR_2 and simultaneously drive SCR_1 to cutoff, since it will not have sufficient current to produce regeneration. SCR_2 conducts as a half-wave rectifier from anode to cathode on the reverse half-cycle and in the reverse direction, thus reversing the armature polarity and the direction of rotation. A phase-shift circuit may also be connected between the two gates to provide OFF, FORWARD, and REVERSE operating point switching as well as armature voltage control. In the method shown in Fig. 6-15a, voltage control may be obtained by varying the ac single-phase input by means of a variac. The inductance shown in series with the armature reduces ac ripple and limits the ac current through the low-resistance armature and shunting resistors. It is designed and selected to saturate when a dc current flows through it, thus reducing its reactance under running conditions.

The full-wave dc output of a pair of SCRs may be connected in the usual FORWARD-REVERSE switching configuration to provide the necessary reversal of polarity across the armature, as shown in Fig. 6-15b. In the absence of a trigger input or alternating current applied to gating terminals 1 and 2, the armature is completely disconnected from the supply. By phase shifting the gating ac waveform (pulse or sinusoidal) with respect to the transformer secondary ac voltage, a variable dc voltage is produced of the polarity indicated in the figure. Closing either F or R will

141

produce armature current reversal as well as reversal of direction of rotation. The same switching circuit is used extensively to supply variable-voltage direct current from a three-phase source, in the manner shown in Fig. 6-14c, capable of reversing the motor armature.

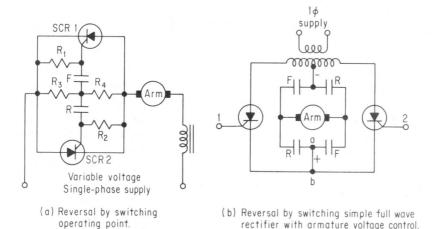

(a) Reversal by switching operating point.

(b) Reversal by switching simple full wave rectifier with armature voltage control.

Figure 6-15

Reversal and armature voltage control using SCRs (full-wave rectification).

The circuits shown in Figs. 6-13 through 6-15 show only the dc motor armature, since these circuits employ armature voltage control. The field circuit is usually *separately* excited from a half- or full-wave rectifier supply (Fig. 6-1). In the case of series or compound motors, where reversal of current in the armature is produced, the field must be connected in such a manner as to maintain the same direction whenever the armature current is reversed (Fig. 6-3d). In Fig. 6-15b, for example, the series field would be connected between and in lieu of the lead from points a to b.

Split-field series motors (Fig. 6-3e) may be controlled by full-wave conduction of SCR pair A or B, shown in Fig. 6-16a. When the gating polarity and phase are such that both a_1–a_2 cause respective full-wave conduction of SCR pair A, one particular direction of rotation is produced by the current in winding S_A and the armature. If SCR pair A is cutoff and pair B driven into full-wave conduction, the same direction of current flows in the armature, but the current in series field S_B will produce the opposite rotation. The gating circuits (not shown) are arranged with suitable interlocks to prevent simultaneous conduction of both pairs, which might result in motor runaway because of a field weakened by differential flux.

The circuit shown in Fig. 6-16b may be used with any motor, shunt, series, or compound; this circuit employs SCR pairs in an *inverse parallel* arrangement, each pair designed to produce full-wave direct current and to

CHAP. SIX / *Manual and Automatic Speed Control of dc Motors*

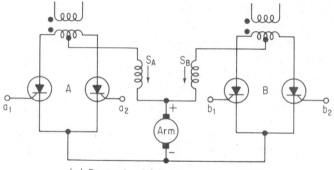

(a) Reversal and full-wave voltage control
using split-field dc series motor.

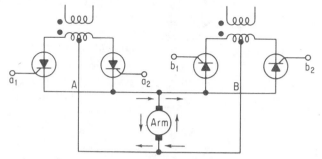

(b) Reversal and full-wave voltage control of any dc motor.

Figure 6-16

Reversal and armature voltage control
using pairs of full-wave rectifiers.

conduct in an opposite direction. Pair *A*, when conducting, will send full-wave current through the armature in a downward direction; and pair *B*, when conducting, will send current through the armature in an upward direction. The voltage across the armature of either pair is determined by the phase of the control voltage or pulse with respect to the secondary ac transformer. This method also has the advantage of requiring *no contacts whatever* for reversing the direction of rotation or for increasing the magnitude of the voltage applied across the armature. It is, in effect, combined static switching and full-wave rectification using armature voltage control.

The three-phase, solid-state, full-wave, unidirectional voltage control circuit shown in Fig. 6-14c is extremely popular for high-horsepower-capacity drives featuring dc motors which are driven in one direction from zero to rated speed (and even higher using field control). It *is* possible to reverse direction of rotation of this motor by field reversal, but the disadvantages of possible runaway (due to an open field) and excessive inductive arcing during switching discourages such designs.

The fundamental design of Fig. 6-14c gives rise to the *bidirectional* full-wave circuit, shown in Fig. 6-17, employing *twice* the number (12) of SCRs. Phase shift or trigger inputs a_1-a_2, b_1-b_2, and c_1-c_2 will provide full-wave rectification of the three-phase input from zero to maximum dc output voltage, with positive polarity at the top brush of the dc armature in Fig. 6-17. Similarly, inputs x_1-x_2, y_1-y_2, and z_1-z_2 will provide full-wave rectification in the *opposite* direction producing positive polarity at the bottom brush.

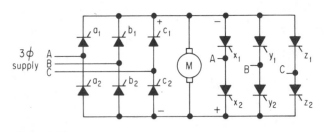

Figure 6-17

Bidirectional armature voltage control using SCRs, from 3ϕ supply.

While the circuit of Fig. 6-17 uses twice the number of SCRs compared to Fig. 6-14c, and despite its higher cost, it yields two important advantages. First, it provides control of speed smoothly and with good speed regulation in *either* direction of rotation from standstill, without any dead zones. Second, it should be noted that unidirectional SCR control is generally not suited for systems requiring power regeneration because operation is limited to only two quadrants in the torque-speed diagram, shown in Fig. 6-17. Thus, the bidirectional circuit of Fig. 6-17 enables "inverter" operation and power regeneration, as well as dynamic braking. In making cost comparisons, the advantages of the bidirectional SCR over the unidirectional SCR package should be kept in mind.

Figure 6-18 makes a comparison between the relative costs of rotary speed-control packages versus the stationary (static) unidirectional and bidirectional packages for dc motors. It should be noted that the bidirectional static SCR package is, in every way, the equivalent of the rotary amplifier (Ward-Leonard or amplidyne) with the added advantages of smaller size, silent operation, longer life, and higher efficiency. But below 100 hp, the static bidirectional dc drive (including phase-shifting circuitry) is more expensive than the rotary amplifier. Conversely, above 100 hp, the bidirectional dc drive is lower in cost than rotary amplifiers for the same horsepower range. This accounts for the data given in Table 6-1, showing the relative percentages of rotary and dc solid-state drives engineered and sold in the United States in a recent study.*

Table 6-1 shows the reduced manufacture of rotary MG sets in the 100-hp to 1000-hp range due to the new competition of SCR static dc

---

* P. G. Mesniaeff, "Solid State Sets New Trends in Motor Drives," *Control Engineering* (October 1970), pp. 73–78.

Deive	MG Set (Ward-Leonard) or Amplidyne	3-phase, full-wave unidirectional SCR	3-phase, full-wave bidirectional SCR
Detailed circuit	Fig 6-2b	Fig 6-14c	Fig 6-17
Schematic representation	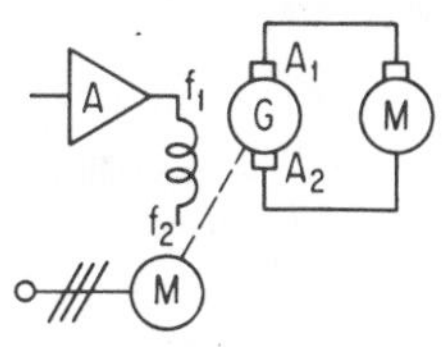	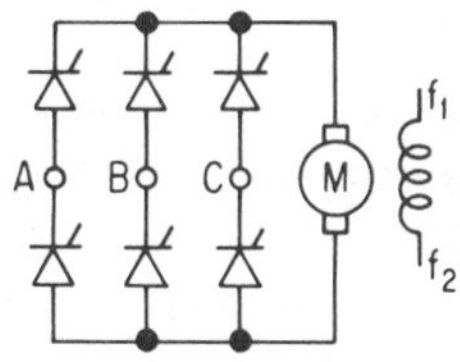	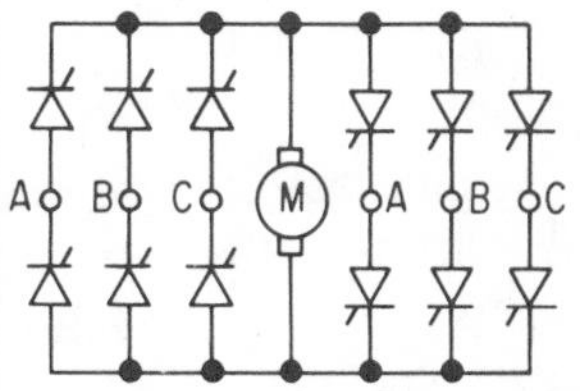
Torque-speed power flow	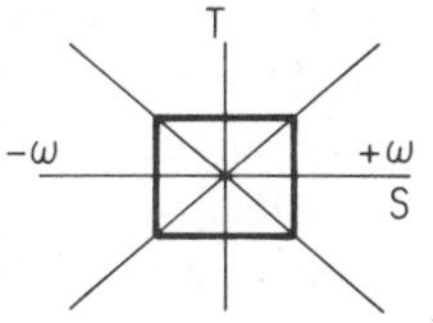	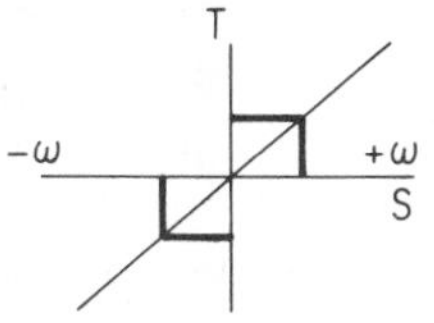	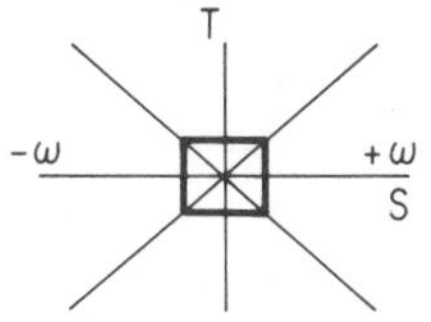
Speed and position control	Universal (zero to rated, in either direction)	2-quadrant control only. Switching of f_1-f_2 required for 4 quadrant control	Universal
Normalized cost factor	1.0	Less than 1.0 at any hp	Below 100 hp, more than 1.0 Above 100 hp, below 1.0

Figure 6-18

Comparisons between rotary and solid-state dc motor drives.

TABLE 6-1   PERCENTAGES OF THE VARIOUS TYPES OF DRIVES SOLD
            AS PARTS OF SPECIAL SYSTEMS BY HP RATING

TYPE OF DRIVE	HORSEPOWER RANGE					
	0–1 %	1–9 %	10–24 %	25–99 %	100–999 %	1000–10,000 %
Static, SCR, dc	4	17	18	18	25	18
Rotary, MG set	0	15	13	20	13	39

drives, where, correspondingly, the highest percentage (25 per cent) of all static drives are produced. Note that the rotary MG sets still dominate the very high power field (from 1000 hp to 10,000 hp) despite the availability of ignitrons and thyratrons, as well as mercury arc rectifiers, which could be used in preference to rotary amplifiers. Note also that the solid-state dc static drives currently cover a wider range of horsepower than

145

rotary amplifiers. The latter are currently preferred mainly in the higher power ranges, above 1000 hp.

It is expected that improvements in solid-state devices and technology will see further replacements of rotary amplifiers by solid-state static dc motor control packages.

6-13.
PERFORMANCE OF
dc MOTORS WITH
ELECTRONIC
RECTIFICATION

The ac component of current in a single-phase, half-wave rectifier exceeds that of a full-wave rectifier, which in turn exceeds that of a half- or full-wave polyphase rectifier. Thus, full-wave three-phase or six-phase rectification will produce a lower ac ripple component than single-phase rectification.*

Furthermore, as the trigger voltage is phase-shifted so that the saturable reactor,† magnetic amplifier,‡ thyratron, ignitron, or SCR, produces conduction over a smaller portion of the total ac cycle, the ripple is also increased. When used with polyphase full-wave ac control, it is customary to *derate* dc motors to about 90 per cent of their rated continuous-duty horsepower. Single-phase operated motors may be derated as much as 70 per cent. Thus, an equivalent horsepower dc motor used with an electronic speed control technique appears physically larger in size.

The reasons for the derating are (1) an increase in iron losses produced by the increased ac ripple; (2) commutation difficulties produced by the high ac component making the interpoles less effective (since they require dc), producing overheating and sparking at the commutator; and (3) poorer speed regulation, produced by the ac impedance drop across both the rectifier and the armature, resulting in lower speed and less efficient motor cooling. This disadvantage may be eliminated, however, by closed-loop techniques that sense and maintain speed regulation (Chapters 10 and 11).

Despite the derating and the poorer open-loop speed regulation that characterize some electronic speed-control methods, the advantages of smaller and lighter controllers, requiring no relays or contactors, together with the manner in which they may be adapted easily to closed-loop servo systems (Chapter 10) and the added possibilities of solid-state switching (Chapter 9) all have led to extensive use of this method of control.

Half-wave drives have the advantage of being relatively simple to design and maintain, as well as relatively low in cost compared to full-wave or three-phase drives. The disadvantages of half-wave drives are

1. Saturation of the supply transformer by the dc component of line current drawn by the motor,

* See Kosow, *Electric Machinery and Transformers*, Sec. 13-21.

† H. W. Weed and S. P. Jackson, "Saturable Reactor Control of dc Motors," *Electrotechnology* (July 1961), pp. 103–106.

‡ A. Kusko and J. G. Nelson, "Magnetic Amplifier Control of dc Motors," AIEE Paper 55–60.

2. At heavy loads and low speeds, the motor tends to produce uneven torque over a complete cycle, resulting in audible power pulsations as power is applied once per cycle and uneven motor speed.

3. The motor must be derated (as high as 70 per cent) because the current has a high rms to average ratio, due to the relatively small conduction angles, using half-wave rectification. Use of a Zener diode (called a free-wheeling diode) across the armature results in some improvement of operation.*

4. For the above reasons, half-wave drives are limited to dc motors of about 1 hp.

Single-phase, full-wave drives are available commercially for 115-V and 230-V operation, for motors of 1, 1.5, 2, 3, and 5 hp. While somewhat more expensive because of the greater number of rectifiers or SCRs employed, they do not suffer from the disadvantages described for half-wave drives above. Standard basic speeds for such drives are 1150, 1750, 2500, and 3500 rpm with speed variations of up to 20/1 at rated torque and almost 100/1 at light, reduced torques.

Packaged commercial "adjustable speed/torque drive systems" employing some type of armature voltage feedback sensing system (see Section 8-8) as a measure of output speed are capable of speed regulations of 3 per cent at basic speed. Using tachometer voltage speed-sensing techniques, speed regulations of up to 0.1 per cent are available. In addition to speed control, packaged systems also include such options as dynamic braking, reversing, speed setting, acceleration limiting, and extension of speed range by weakening the field.

Three-phase drives are currently available for dc motors from 1 to 150 hp. The most recent designs operate from either 230 or 460 V, three-phase supplies, eliminating transformers and matching the dc motor moltage to the rectified maximum voltage value. Double half-wave circuits may be used for reversing, as shown in Fig. 6-19. The armature choke, L, in series with the armature, tends to reduce the ripple and eliminates the necessity for derating the motor due to excessive

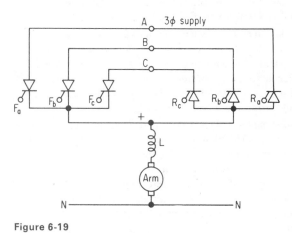

Figure 6-19

Bidirectional armature voltage control using 3ϕ, half-wave bridge and series choke.

* Gaudet drive circuit as described in U.S. Patent 3,123,757, issued March 3, 1964, to J. M. Gaudet for "motor speed control system."

heating. The choke may be eliminated and no derating is necessary using the bidirectional full-wave bridge shown in Fig. 6-17. The latter has the disadvantage of using twice the number of SCRs, however.

In the higher power ratings (100 hp and above), SCR parallel-type bridges are used to provide the required current rating. Commercially packaged drives provide approximately the same (or better) speed regulation and optional features as those described for single-phase full-wave drives above. Standard basic speeds include 850 rpm in addition to those described for single-phase full-wave drives above. Finally, as shown in Fig. 6-18, 100-hp, three-phase full-wave drives are currently less expensive, lighter in weight, and require less space than rotary amplifiers. In addition, their relatively quiet operation provides less restriction and greater versatility in selecting locations for their installation.

6-14.
BRUSHLESS dc
MOTOR

The function of the commutator in a dc motor is to convert the dc (applied to the brushes) to ac in the armature winding to produce continuous rotation of the armature in the same direction. Recent developments have seen at least three distinct designs for brushless dc motors (having no commutator or brushes) which are used to produce output torque in small servo packages. These are*

1. **The electronically commutated motor.** The rotor of this motor is a permanent magnet. The wound stator consists of several windings, each containing a rotor position sensor (optical, mechanical, or Hall effect) and solid-state switching circuitry. In effect, electronic switching (using transistors or SCRs) substitutes for brushes and commutator bars. The motor has the general characteristics of a small dc series motor.

2. **The ac motor/inverter.** This design combines a hysteresis synchronous (single-phase) motor with an electronic inverter (converting dc to ac). Some packages use a small single-phase induction motor. This brushless motor package, designed for dc operation, is generally larger (because of the inverter package) and somewhat less efficient but has better speed regulation than the electronically commutated motor.

3. **Limited rotation motors.** The motors described in 1 and 2 are continuous rotation motors. This motor is a variation of the step motor and provides torque over a range of $\pm 90°$, requiring no electronics for commutation. Limited rotation torquers have toroidally wound stators and pancake-shaped rotors with printed-circuit conductors. Extremely flat and requiring little space, these brushless motors are used to position mechanisms, drive pens of strip recorders, and position gyro gimbals on gyroscopic stable elements and platforms.

Brushless dc motors are more expensive for the same horsepower rating than conventional dc motors, but they possess certain advantages over dc commutator-brush motors:

* For a more complete discussion of the various designs and characteristics of brushless dc motors, see Kosow, *Electric Machinery and Transformers*, Sec. 11-18.

1. Little or no maintenance required.
2. Much longer operating life.
3. No arcing is produced, thereby eliminating the possibility of RF radiation or the hazard of explosion in areas containing combustible gases.
4. May be hermetically sealed and are even capable of operation in fluids or combustible vapors.
5. Generally higher efficiency.
6. No commutator or brush particles produced as a result of operation.
7. Capable of more rapid response (lower servo time constant) and fairly constant output torque vs. input current characteristics.

The last advantage, coupled with those previously cited, indicates a fairly bright future for the general class of brushless dc motors as new designs emerge. It should be noted, however, that current designs possess the following general disadvantages:

1. Greater overall total size of the motor package because of the additional space required for the associated electronic devices (although some brushless dc motors are themselves usually smaller than conventional dc motors of the same horsepower).
2. Higher initial costs compared to conventional commutator types of the same horsepower.
3. Somewhat limited choice (at present) in "stock" sizes and horsepower ratings, necessitating "special" orders for particular applications.

Still, despite the above-cited disadvantages, the brushless dc motor will continue to encourage new and wider dc motor applications. Originally developed for use in space-vehicle systems and military hardware (where cost is a secondary factor compared to the above-cited advantages), the brushless motor is now available in a wide variety of sizes (from small pump motors to large traction motors for electric cars). Increased use of semiconductors, integrated circuit packages, and servo-control systems have brought about reduced costs of these assemblies. The overall effect is to make brushless dc motor packages more competitive in price compared to conventional commutator-type dc motor and associated electronic control packages.

BIBLIOGRAPHY

Dailey, J. J. "Dual Circuit Electronic Motor Control," *Electrical Manufacturing* (July 1957).

Dinger, E. H. "Current-Controlled DC Motor Drive," *Electrical Manufacturing* (August 1959).

Electric Clutch and Brake Engineering Handbook (Warner Electric Brake and Clutch Co., Beloit, Wis., 1965).

"Electronic Standards for Industrial Equipment" (General Motors Technical Center), *Electrical Manufacturing* (August 1958).

Graphical Symbols for Electrical Diagrams (ASA Y32.2) (New York: American Standards Association), or IEEE Standard No. 315-1971.

Gutzwiller, F. W. "The Silicon Controlled Rectifier," *Electrical Manufacturing* (December 1958).

Harwood, P. B. *Control of Electric Motors*, 4*th* ed. (New York: John Wiley & Sons, Inc., 1970).

Heumann, G. W. *Magnetic Control of Industrial Motors*, 3 vols. (New York: John Wiley & Sons, Inc., 1961).

Industrial Control Equipment (Group 25) (ASA C42.25) (New York: American Standards Association).

Industrial Control Equipment (UL508) (Chicago: Underwriters Laboratories, Inc.).

James, H. D., and L. E. Markle. *Controllers for Electric Motors* (New York: McGraw-Hill Book Company, 1952).

Jones, R. W. *Electric Control Systems* (New York: John Wiley & Sons, Inc., 1953).

Kosow, I. L. *Electric Machinery and Transformers* (Englewood Cliffs, N.J.; Prentice-Hall, Inc., 1972).

Kusko, A. *Solid State DC Motor Drives* (Cambridge, Mass.: The MIT Press, 1969).

Kusko, A., and J. G. Nelson. "Magnetic Amplifier Control of DC Motors," AIEE Paper 55-60.

Motor and General Standards (MG1) (New York: National Electrical Manufacturers Association).

O'Brien, D. G. "D-C Torque Motors for Servo Applications," *Electrical Manufacturing* (July 1959).

Storm, H. F. *Magnetic Amplifiers* (New York: John Wiley & Sons, Inc., 1955).

Trofimov, L. A. "Wide Speed Range Electrical Drives," *Electrical Manufacturing* (March 1954).

Uri, J. Ben. "Variable Speed Control Systems," *Electro-Technology* (March 1961).

Variable Speed Systems, Minarik Electric Co., Catalog S-300A, 232E. 4th St., Los Angeles, Calif., 90013.

Weed, H. R., and S. P. Jackson. "Saturable Reactor Control of dc Motors," *Electro-Technology* (July 1961).

Weissmann, S., and M. Bracutt. "Transistor Magnetic Amplifier for Power Motor Control," *Electrical Manufacturing* (June 1957).

Winsor, L. P., and E. E. Moyer. "Adjustable Speed Drives," *Electrical Manufacturing* (November 1952).

Wulfken, H. J. "Magnetic-Amplifier Regulators in Adjustable-Speed Drives," *Electrical Manufacturing* (May 1955).

PROBLEMS AND QUESTIONS

6-1. Give four advantages and three disadvantages of field control as a method of speed control of dc motors in comparison to alternative methods.

6-2. Give three advantages and three disadvantages of armature resistance control compared to alternative methods of speed control.

6-3. Give two advantages and two disadvantages of series-shunt armature resistance control compared to alternative methods of speed control.

6-4. List seven advantages and two disadvantages of armature voltage control using rotary amplifiers for speed control of dc motors.

6-5. a. Show by switching diagrams (using **DPDT** switches) two methods of reversing a dc compound motor
   b. Give two disadvantages of the field-reversal method
   c. Give three advantages of the armature-reversal method and one disadvantage.

6-6. Define
   a. Plugging
   b. An electrically reversible motor (see Section 7-18)
   c. An externally reversible motor
   d. A nonreversible motor
   e. A reversing motor
   f. Jogging (see Section 6-11)
   g. Braking
   h. Plug to stop
   i. Plug reverse.

6-7. a. Are *all* dc motors electrically reversible? Explain
   b. Are *all* dc motors capable of being jogged or plugged? Explain fully
   c. How would you classify a commercial universal motor having only two line leads brought out from a sealed housing in view of your answers to parts a and b?

6-8. a. Using the control circuit shown in Fig. 6-6 for plugging braking and the basic reversing circuit shown in Fig. 6-3b, design a controller for automatic starting, accelerating, and braking of a compound motor in both forward and reverse directions. Describe the action of the controller using the technique employed in the text
   b. Modify part a by addition of jogging in both forward and reverse direction using method of Fig. 6-11c.

151

6-9. Repeat Problem 6-8 for a split-field series motor. Include mechanical and electrical interlocks in both the forward and reverse control circuits. Describe operation.

6-10. a. Explain the advantages of a controller using both mechanical and electrical interlocks for forward and reverse contacts over one using either mechanical or electrical interlocking, only
b. Consider and discuss the action of interlocking provided by the control circuit given in Fig. 7-10b as compared to Fig. 6-4a. Which is preferable and why?

6-11. Design a controller capable of starting, accelerating, reversing, and stopping a dc compound motor automatically, using *dynamic* braking (see basic circuit Fig. 6-9b) with provision for jogging in both the forward and reverse direction. Include mechanical and electrical interlocks in both the forward and reverse control circuits. Describe operation in detail, using the descriptive technique of the text.

6-12. Repeat Problem 6-11, using regenerative braking (see Fig. 6-9c) instead of dynamic braking. Describe operation in detail.

6-13. Using the basic power circuit for reversal of the shunt motor shown in Fig. 6-3a, design a controller using silicon-controlled rectifiers as static switches for providing start, stop, and reversal of a shunt motor. The forward and reverse accelerations are to be initiated by 1-V pulses obtained from a separate power supply to be shown.

6-14. a. Using the circuit shown in Fig. 6-15a as a model, design a controller using silicon-controlled rectifiers for armature voltage control of a dc shunt motor by full-wave rectification from a single-phase ac supply. Also use an additional pair of SCRs for static switching the forward and reverse accelerations from a separate power supply. Indicate the phase-shifting power supplies for armature voltage control and the positive pulse generator for static switching in block-diagram form
b. Why are protective armature resistors not required in this controller?
c. Discuss the relative merits of this controller over the one shown in Fig. 6-8.

6-15. Repeat Problem 6-14 using the circuit shown in Fig. 6-16b as a model.

6-16. Explain why universal (four-quadrant) control is obtained using
a. Ward-Leonard, amplidyne, or any rotary amplifier
b. Three-phase, full-wave bidirectional SCRs (see Fig. 6-17)
c. Three-phase, half-wave bidirectional SCRs (see Fig. 6-19).

6-17. Compare rotary amplifiers vs. solid-state bidirectional packages for use with dc motors with respect to the following:
a. Inverter operation and power regeneration (for regenerative braking)
b. Size
c. Efficiency
d. Life
e. Cost
f. Necessity for derating
g. Speed regulation.

6-18. Compare half-wave unidirectional vs. full-wave polyphase bidirectional solid-state packages with respect to each of the criteria a through g listed in Problem 6-17.

6-19. List seven advantages and three disadvantages of brushless dc motors.

manual and automatic speed control of ac polyphase motors

7-1.
GENERAL

The reader should be familiar with the torque relations of the two main types of ac polyphase dynamos, the *synchronous* polyphase motor and the *asynchronous* polyphase induction motor.* The former is a constant-speed (variable-torque), doubly excited dynamo, whose stator armature is excited with a polyphase alternating current of a given frequency and whose rotor field is a dc electromagnet or permanent magnet. The speed of the rotating field of the synchronous (and asynchronous induction) motor is a direct function of the number of poles and the frequency of the polyphase ac supply ($S = 120f/P$). The only ways to change the speed of a synchronous polyphase motor are (1) to change the applied frequency of the polyphase voltage applied to its stator, or (2) to change the number of poles of its rotor and its stator.

The asynchronous polyphase induction motor is also a doubly excited ac dynamo, whose stator armature is excited with a polyphase alternating

---

* For a complete discussion of these dynamos, see I. L. Kosow, *Electric Machinery and Transformers* (Englewood Cliffs, N.J., Prentice-Hall, Inc., 1972), Chaps. 8 and 9.

154

current of a given frequency and whose rotor winding is excited by an *induced* ac voltage of variable frequency. The speed of the polyphase induction motor is an asynchronous speed which may be varied (1) by changing the applied frequency to the stator, (2) by changing the number of poles of both the stator and the rotor, (3) by introducing applied voltages of desired frequency to the rotor, (4) by controlling the rotor slip by means of rheostatic rotor control, or (5) by mounting the stator in bearings and driving it with an auxiliary motor.

Both classes of polyphase motors of all types are speed-controlled either by manual switching or automatic controllers. Braking, jogging, and reversing techniques are used for the same purposes and in manners similar to those covered in Chapter 6. The various methods, special devices, and/or machines using these methods will be described in this chapter.

7-2.
CHANGE IN
FREQUENCY AND
OF THE VOLTAGE
APPLIED TO A
POLYPHASE STATOR

The synchronous speed of a polyphase synchronous or asynchronous induction motor, having a *given* number of poles, varies directly as the frequency of the voltage applied to its *stator*. If the supply voltage is maintained constant when the frequency is reduced, for example, the permissible flux density is increased*. In *changing frequency*, therefore, *it is necessary to change the applied voltage* in the same manner and to the same extent in order to maintain the same degree of saturation and mutual air-gap flux density.

In order to change speed by a variation of voltage and frequency, therefore, it is necessary to have a separate prime mover to drive a separate alternator at a variable speed to vary the frequency and voltage applied to the polyphase (or single-phase) stator. This method is employed extensively in ship propulsion (the so-called "turboelectric drive") and has the advantage of stepless speed control over an extremely wide range of speed.†

It is also possible to employ a rotary frequency changer called an *induction frequency converter*,‡ particularly where speeds much higher than synchronous speed are desired. In both instances (the separate alternator or the "frequency changer"), the rating of a driven polyphase induction or synchronous motor governs the rating of the prime mover and its frequency-generating source. Although the permissible flux density remains the same at higher frequencies (because the voltage is raised proportionately), increasing the frequency and the motor speed produces

---

* cf. Eq. (2-15) and Sec. 13-2, Ex. 13-5 in Kosow, *Electric Machinery and Transformers.*

† Closely related to this method is the method of changing the frequency of the voltage applied to the *rotor* of a polyphase wound-rotor induction motor; see Section 7-5, this chapter.

‡ *op. cit.,* Sec. 9-23.

increased rotational losses* in the driven motor. This is offset, in part, by the increased cooling effect produced at higher speeds, as well as by the increase in developed horsepower afforded by the increase in speed, so that no increase in rating is required for higher speeds.

When a dc motor is used as a speed-controlled prime mover for an alternator, the methods of speed control described in Chapter 6 apply. When an ac induction motor is used as a prime mover for a rotary alternator or frequency changer, the methods of starting described in Chapter 5 and the methods of speed control to be described in this chapter apply.

In addition to the *rotary* adjustable frequency drives described above and in subsequent sections, the emergence of the high current SCR (Section 6-12) and its use in the *static* (stationary) adjustable-frequency ac motor drives offers the promise of widespread future application similar to the dc motor drives described in Section 6-13. Two general classes of static adjustable-frequency ac drives have emerged thus far (for use with polyphase synchronous and induction motors) and the past few years has witnessed a dramatic rise in their use†:

a. The *cycloconverter*—a static frequency converter for transforming a higher frequency to a lower frequency without a dc link. This device, shown in Fig. 7-1, may be used with polyphase squirrel-cage induction motors (SCIMs), wound-rotor induction motors (WRIMs), and synchronous motors.

b. The *rectifier inverter*—a static frequency device which combines an *inverter* (for transforming dc to ac) with a polyphase rectifier. This solid-state device may also be used with synchronous motors as well as SCIMs and WRIMs.

7-2.1
The Cycloconverter

The *cycloconverter* is shown in schematic form in Fig. 7-1a. In the application shown in Fig. 7-1, it is used to supply a lower variable-frequency and voltage amplitude from a three-phase 60-Hz supply to the stator of a SCIM, WRIM, or synchronous motor. Proper triggering or phase shifting of inputs a_1–a_2, b_1–b_2, and c_1–c_2, respectively, at all three-phase banks permits simultaneous reduction of output frequency and voltage waveform. The polyphase motor, therefore, may be brought from synchronous speed down to lower speeds smoothly and with good speed regulation at any given desired speed below approximately one-third of the input frequency to zero speed.

A second possible application of the cycloconverter is found wherever an alternator is driven by a variable-speed prime mover, such as in an aircraft engine, turbine, or internal combustion engine.

The cycloconverter maintains a constant alternator output voltage at

* *Ibid.*, Sec. 12-2.
† P. G. Mesniaeff, "Solid State Sets New Trends in Motor Drives," *Control Engineering*, (October 1970), pp. 73–75.

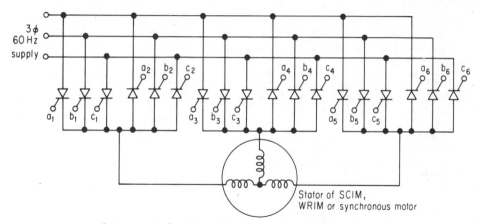

a. Cycloconverter (exclusive of triggering and interphase transformers)

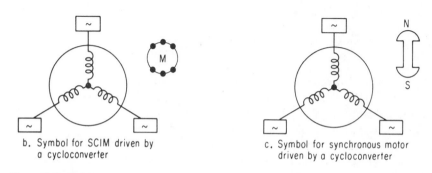

b. Symbol for SCIM driven by a cycloconverter

c. Symbol for synchronous motor driven by a cycloconverter

Figure 7-1

Basic 3-phase cycloconverter with symbols.

a constant frequency (either 400 or 60 Hz) regardless of variations in prime-mover speed. The SCRs are phase-shifted or delay-triggered in proportion to the change in prime-mover speed, thereby maintaining constant output voltage and frequency to the aircraft or vehicle power system.

Operating the cycloconverter from a constant frequency ac power supply as an adjustable-frequency ac motor drive, the configuration shown in Fig. 7-1a produces frequencies from 20 Hz down to zero, using a minimum of 18 SCRs. Because of the *half-wave* configuration, however, phase-sequence reversals are not obtained, nor are frequencies above 20 Hz.

These disadvantages are overcome using twice the number (36) of SCRs in the *full-wave* configuration shown in Fig. 7-2. The full-wave converter permits control of frequencies from +30 Hz down to zero and up to −30 Hz. This not only permits wider frequency variation but also makes possible (1) reversal of rotation, (2) power regeneration, and (3) dynamic braking.

157

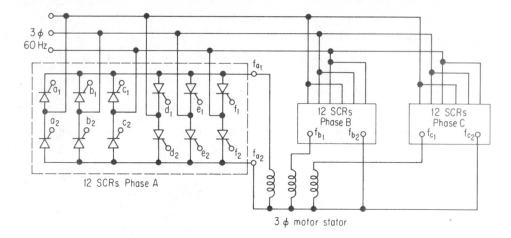

Figure 7-2

Cycloconverter with three separate single-phase outputs (interphase transformers and trigger circuits not shown).

The full-wave cycloconverter is not without disadvantages, however: (1) Twice the number of SCRs are required, increasing the cost appreciably, and (2) either interphase transformers (not shown in Fig. 7-2) *or* purchase of a special motor having both ends of each stator phase winding brought out is required to isolate the inverter outputs of each phase. Despite these disadvantages, the advantages of regeneration and full reversing capabilities, coupled with high efficiency and smooth stepless control, have witnessed the increased emergence of such packages, particularly in the range 1 to 25 hp. It should be noted, moreover, that compared to rotary frequency converters, the solid-state static converter features the following advantages: quiet operation, solid-state reliability, higher efficiency, reduced maintenance, smaller space requirements and longer life. Similar to the dc solid-state drives (Section 6-13), such ac solid-state packages are gaining in popularity over rotary frequency converters, particularly in the range from 1 to 10 hp and to some extent from 10 to 25 hp in driving polyphase SCIMs, WRIMs, and synchronous motors.

7-2.2
The Inverter

As opposed to a rectifier (equipment for transforming dc to ac) and a converter (equipment for transforming ac to ac), the inverter contains the devices and circuitry for transforming dc to ac. Two basic three-phase inverter configurations are shown in Fig. 7-3, both operating from a dc supply. The *half-wave inverter* of Fig. 7-3a uses only three SCRs but unfortunately produces a dc component in the waveform of the output load, or stator winding of a polyphase ac motor.

CHAP. SEVEN / *Manual and Automatic Speed Control of ac Polyphase Motors*

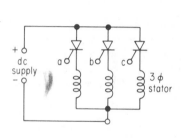

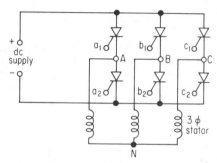

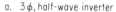

a. 3 φ, half-wave inverter

b. 3 φ, full-wave, bridge or multiple inverter

Figure 7-3

Inverter configurations, 3-phase ac output
from dc supply.

The three-phase, *full-wave bridge inverter*, shown in Fig. 7-3b, uses twice the number (6) of SCRs but has the advantage of containing no dc component for balanced three-phase loads. The inverter shown in Fig. 7-3b, moreover, lends itself to three fundamental methods of synthesizing the output waveform by controlling the pulse width and waveform applied to the gates of the SCRs.

*A. Control of phase relationship.** A transistor gating circuit is used to trigger the SCR pairs in sequences 120° apart, producing the phase voltages shown in Fig. 7-4a across each phase of a three-phase stator. The phasor sum of the phase voltages produces the corresponding line voltages shown in Fig. 7-4b. The output line voltages contain only higher-order odd

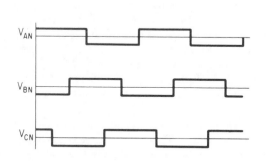

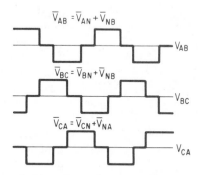

a. Phase voltages produced by control of SCR
switching use phase delay control

b. Line voltages of output waveform

Figure 7-4

Output waveforms of 3-phase inverter by control
of phase relationship.

* B. D. Bedford and R. G. Hoft, *Principles of Inverter Circuits* (New York: John Wiley & Sons, Inc., 1964), Chap. 9.

159

harmonics in phase with the fundamental and may easily be filtered to yield the pure sine wave, if required.

B. Control of multiple pulse-width* output waveform (MPW control). This method of inverter control of output waveform (sometimes called the *McMurray-Bedford* inverter) is fundamentally a method of controlling the pulse width by appropriate SCR triggering over each half-cycle. As shown in Fig. 7-5a for a single-phase inverter, it is possible to trigger

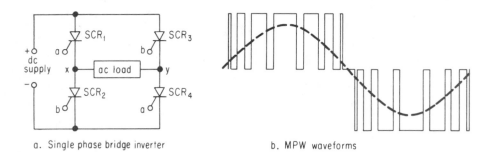

a. Single phase bridge inverter   b. MPW waveforms

Figure 7-5

Multiple width control of single-phase (bridge) inverter.

SCRs 1 and 4 for one half-cycle to produce current in the load in one direction (x to y) and trigger SCRs 3 and 2 for the second half-cycle to produce current in the load in the opposite direction. For each half-cycle the appropriate pairs of SCRs are triggered in such a way as to produce multiple pulse widths having effective current equivalent to a sine wave as shown in Fig. 7-5b.

The same type of triggering may be used for the three-phase full-wave bridge or multiple inverter shown in Fig. 7-3b to produce a three-phase McMurray-Bedford inverter using MPW control. Each phase voltage (V_{AN}, V_{BN}, and V_{CN}, respectively) is effectively sinusoidal and displaced 120°, respectively. A typical commercial McMurray-Bedford MPW inverter is shown in Fig. 7-6. Like all McMurray-Bedford circuits, class C inversion is used in which each LC circuit pair is switched by a load-carrying SCR.

C. Selected harmonic reduction for control of output waveform.† If the single-phase bridge inverter shown in Fig. 7-5a is triggered so as to yield the output waveform shown in Fig. 7-7, a sine-wave resultant relatively low in third and fifth harmonics is produced, in comparison to MPW waveforms shown in Fig. 7-5b. Actually, by proper pulse-width control, output waveforms are possible, having no harmonics below the eleventh, as shown by F. G. Turnbull, with the following advantages:

* *Ibid.*

† F. G. Turnbull, "Selected Harmonic Reduction in Static dc-ac Inverters," IEEE Transaction Paper 63–1011, June 1963.

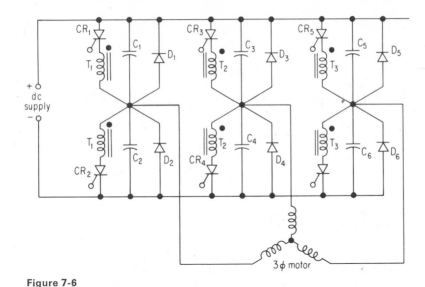

Figure 7-6

MPW control of output waveform of
3-phase bridge inverter. *LC* switching by
load-carrying SCRs, commercial
McMurray-Bedford converter.

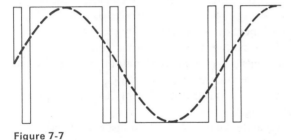

Figure 7-7

Synthesis of ac waveform by
selected harmonic reduction.

1. The triggering is considerably simplified compared to MPW waveform synthesis using the McMurray-Bedford inverter.

2. Complete control of the fundamental from zero to maximum without generation of harmonics (in the McMurray-Bedford inverter, harmonics are produced when the amplitude of the fundamental is varied).

The technique of selected harmonic reduction lends itself nicely to control of the three-phase multiple inverter shown in Fig. 7-3b using only 6 SCRs, although when 12 SCRs are used, all harmonics below the eleventh are suppressed.

7-2.3
Inverter Regulation

It should be noted that a dc supply is specified for the inverter configurations shown in Figs. 7-3, 7-5, and 7-6. This dc supply is usually obtained from a three-phase 60-Hz supply using the three-phase, full-wave SCR circuit shown in Fig. 6-14c, which uses 6 SCRs. By phase-controlling the SCR conduction of the rectifiers shown in Fig. 6-14c, the dc output waveform and dc inverter input may be varied

161

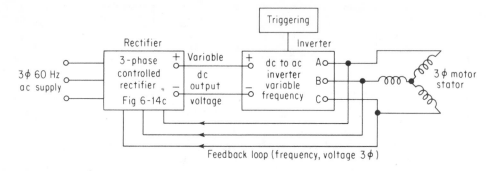

Figure 7-8

Complete inverter circuit, 3-phase, 60-Hz input;
3-phase, variable frequency, variable voltage
output.

from zero to maximum. This method of control (which involves regulation of the dc supply voltage to the inverter), as shown in block-diagram form in Fig. 7-8, is perhaps the most common technique of providing a variable frequency output, at a variable voltage to the stator of a polyphase SCIM, WRIM, or synchronous motor. Since it is necessary that a reduction in frequency is accompanied by a proportional reduction in voltage, the frequency reduction sensed at the load is fed back to the phase-controlled rectifier shown in Fig. 7-8, which in turn reduces the dc input voltage to the inverter in proportion to the frequency change.

Each of the methods of synthesis of the output voltage waveform previously covered also permits some control of output voltage and, in some solid-state three-phase rectifier-inverter packages, the output frequency and voltage are fed back to triggering circuitry to provide proper regulation of the output voltage.

It is also possible to regulate the inverter output just prior to connecting it to the stator of the three-phase motor by a variety of transformer techniques (induction regulators, magnetic amplifiers, or phase-controlled SCRs).

Of these three alternatives, the method shown in Fig. 7-8 is most commonly used and will be assumed below for purposes of comparison.

7-2.4
Comparisons Between
Cycloconverter and
Rectifier-Inverter for
Solid-State ac
Polyphase Motor
Control

The commercial cycloconverters in solid-state ac drive packages employ 36 SCRs (shown in Fig. 7-2), although only 18 SCRs are required as a minimum, as shown in Fig. 7-1a. Similarly, in rectifier-inverter combinations, a minimum of 6 SCRs are required in the rectifier and a minimum of 6 SCRs in the inverter. If we assume minimum requirements for both, the following comparisons indicate the choices currently facing designers of ac solid-state drives for control of polyphase motors, summarized in Table 7-1. Table 7-1 shows that there is no clear choice, in the selection

TABLE 7-1 COMPARISONS BETWEEN RECTIFIER-INVERTER
AND CYCLOCONVERTER PACKAGES FOR
FREQUENCY-VOLTAGE CONTROL OF POLYPHASE ac MOTORS

Rectifier-Inverter Package	Cycloconverter Package
Minimum of 12 SCRs required but only 6 are load-carrying	Minimum of 18 SCRs required, all load-carrying
Very wide frequency range, from 0 to well above 60 Hz	Limited to frequencies below 1/3 of source frequency, using 18 SCRs; doubling SCRs permits both wider frequency range and reversal possibilities
Regeneration possible in both rectifier and inverter stages, if rectifier bridge is full wave (bidirectional)	Regeneration possible only using 36 SCRs
Sensitive to load power factor	Loads of any power factor may be used
Requires additional commutation circuitry and triggering circuitry for PWM or harmonic reduction	Operates using line frequency commutation
Exceeding PRV of any one SCR shuts down system, requiring replacement of SCR	If current surge does not produce SCR damage and is not recurrent, system is not shut down

and design of ac variable frequency solid-state converters for ac polyphase motor control, between the cycloconverter and the rectifier-inverter package, as of this writing.

Approximately 80 per cent of all solid-state static adjustable-frequency drives are currently manufactured in both forms for use with ac polyphase motors below 10 hp. At these relatively low currents, the solid-state cycloconverter competes favorably with the rectifier-inverter package and no clear advantage emerges for either type.

Finally, it should be noted (as of this writing) that in the high-horse-power range from 100 to 10,000 hp, control of ac motors is exclusively restricted to *rotary* methods described in Sections 7-2, and 7-4 through 7-9. Only the advent of higher-current SCRs may displace such rotary drives, in time, with solid-state packages.

7-3.
CHANGE IN
NUMBER OF POLES
OF THE STATOR
WINDING OF A
POLYPHASE MOTOR

It is possible to design the primary stator winding of any (polyphase) induction-type motor so that the number of poles may be changed by means of manual or automatic switching. For a SCIM rotor, the number of poles of the excited secondary windings depends on and is the *same as* the number of poles of the *primary stator* windings, because of the induction principle.*

Thus, changing the number of poles of the primary winding of a polyphase SCIM by means of suitable and simple switching methods, will change the number of squirrel-cage rotor poles *automatically.*

* Kosow, *Electric Machinery and Transformers.*, Sec. 9-4.

163

Relatively satisfactory operation results with pole changing, since the number of poles of *both* the stator and rotor have been changed. Such squirrel-cage polyphase and single-phase motors are called *multispeed induction motors*. These motors have stator windings, specifically designed for pole changing by manual and/or automatic switching methods, in which various primary stator windings are connected in series and parallel combinations. Multispeed induction motors are available in two-speed or four-speed synchronous speed combinations, by pole changing. Complications both in stator design and in switching connections have tended to limit the production of squirrel-cage motors with more than four speeds, although they are technically possible.

In the case of the polyphase *synchronous* motor and the WRIM (wound-rotor induction motor), however, speed control by pole changing is impractical. The number of poles of the stator and of the rotor of these machines are *independently* wound. In the case of the synchronous motor, the numbers of both the ac stator winding poles and the dc field poles would have to be changed simultaneously to produce satisfactory operation. This would involve additional slip rings and complex field windings. In the case of the WRIM, the number of poles of the secondary wound-rotor winding depends on the particular winding design used for the secondary rotor winding. Pole changing of the stator by switching methods would involve commensurate switching of the rotor winding in order to change the number of poles of both simultaneously.*

Pole changing as a method of speed control is, therefore, limited to polyphase (and single-phase) squirrel-cage rotor induction motors, exclusively. As a method of speed control it may only be used to produce relatively fixed speeds (e.g., 600, 900, 1200, or 1800 rpm) for an induction motor whose speed changes only slightly (from 2 to 8 per cent) from no load to full load. Pole changing as a method of speed control has the advantages of (1) high efficiency at any speed setting, (2) good speed regulation at any speed setting, (3) simplicity of control in obtaining any given speed by manual or automatic switching, and (4) relatively inexpensive auxiliary speed controllers in association with the motor. Pole changing is used primarily where it is desired to have the versatility of two or four relatively constant speeds that are widely separated. It is used, for example, in drill-press motors for drilling materials of differing hardness and thickness. Its major disadvantages are: (1) a special motor is required, having the necessary windings and terminals brought out of the stator for interchanging poles; and (2) gradual and continuous speed control cannot be attained, although speed variation is possible within a given speed range by voltage variation and/or frequency variation (Section 7-2).

Two-speed SCIMs are designed with either one or two windings. Dual speeds are obtained with the one winding design by means of the

---

* The foregoing analysis once again points up the importance of the theory of the doubly excited dynamo; see Kosow, *op. cit.*, Secs. 5-3, 8-1, and 9-1.

164

consequent-pole method shown in Fig. 7-9. In this method, two operating speeds are provided by changing the external connections of each phase, using either manual or automatic switching to obtain a higher speed which is twice that of the lower speed.

As shown in Fig. 7-9a, the lower speed is obtained by closing the *L* contacts. This connects each coil of each phase belt in such a manner that the current directions are reversed from coil to coil; i.e., the current direction is clockwise in the first coil, counterclockwise in the second coil, clockwise again in the third coil, and counterclockwise in the fourth coil, thus producing four poles SNSN, as shown in Fig. 7-9a. When the *L*

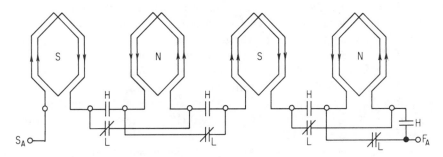

(a) Four-pole, low speed connection.

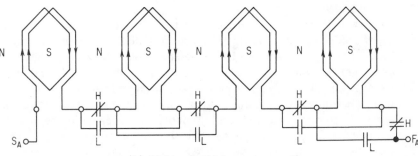

(b) Eight-pole, high speed connection.

Figure 7-9

Phase *A* of a 3-phase winding showing consequent pole connections for multispeed induction motors.

contacts are opened and the *H* contacts are closed, as shown in Fig. 7-9b, the current directions in all the coils are clockwise, producing S poles at the centers of these coils. The magnetic field paths instantly redistribute themselves to create consequent N poles between the centers of the coils. In consequence, *eight* poles are created by this unique switching arrangement, where previously there were four. (For a single-phase machine

165

employing temporary or permanent split-phase technique, it is necessary to switch both the starting and the running windings, although special switching arrangements are available for switching the running winding only, in the case of the former.) For three-phase SCIMs it is necessary to switch the coils in all three-phase belts, connected either in delta or wye. Larger three-phase multispeed induction motors, having many coils, employ more intricate winding schemes in which the coil groups of each phase are connected in various series-parallel arrangements to control the current directions in the various coils and thus obtain the desired coil polarity.

A disadvantage of the consequent-pole method is that the speeds obtained are in the ratio of 2:1, and no intermediate speeds are available by switching techniques. This disadvantage is overcome by using *two separate* windings, each creating a separate field and a separate total number of poles. Figure 7-10a shows a two-winding, three-phase induction motor, in which one winding (winding *A*) is wound for a specific number of poles (say four poles) and another winding (winding *B*) is wound for a different specific number of poles (say six poles). Thus, winding *A* will produce a high speed of 1800 rpm at 60 Hz, whereas winding *B* will produce a low speed of 1200 rpm.

When the principle of the two-winding, multispeed induction motor is combined with the consequent-pole method of switching, described above and shown in Fig. 7-9, a total of *four* synchronous speeds is obtained (1800, 1200, 900, and 600 rpm).

Although the two-speed design shown in Fig. 7-10 provides the advantage of producing speeds having lower ratios than 2:1, the disadvantages of such a motor compared to the consequent-pole type are (1) larger size and weight for the same horsepower output (since only one winding is used at a time), (2) higher cost because of the larger frame size, (3) higher leakage reactance because the deeper slots required for two windings produce poorer power factors, and (4) poorer speed regulation because of the higher reactance of either winding. Furthermore, as shown in Fig. 7-10a, such windings *must* be wye connected to prevent the energized winding from inducing a voltage and a circulating current in the winding which is not being used. The wye or star connection is seldom used in induction motor winding, since it does not produce a symmetrical stator construction and numerous jumpers are required to connect common points. These disadvantages, in addition to the more complex switching, tend to discourage the manufacture and use of multiwinding, multispeed, polyphase SCIMs, in general, and motors of more than four speeds, in particular.

In addition to the availability of two synchronous speeds, reversal of these motors (and this is true for all three-phase stators) involves the reversal of any two line leads to produce a reversed phase sequence and a reversed primary rotating field. Figure 7-10b shows the control circuit

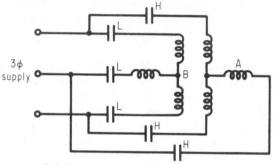

(a) Two winding, multispeed induction motor.

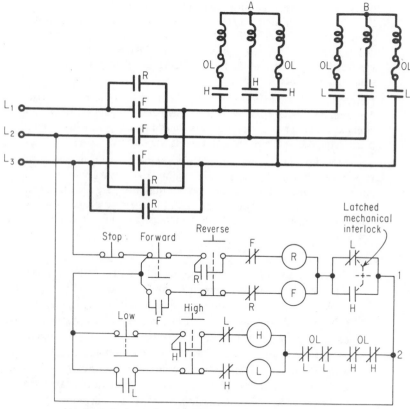

(b) Control circuit for two winding multispeed induction motor with reversing at either speed.

Figure 7-10

Two-speed, two-winding, 3-phase induction motor.

SEC. 7-3 / *Change in Number of Poles of the Stator Winding of a Polyphase Motor*

of a typical automatic across-the-line two-winding, multispeed induction motor with reversing at either high or low speeds. The speed controller of Fig. 7-10b operates in the following manner:

1. The forward and reverse relays in control line 1 are mechanically and electrically interlocked by means of their momentary contact push-buttons and their n.c. contacts R and F, respectively, in series with their control relays. Depressing the momentary-contact FORWARD push-button energizes the F relay through a closed LOW or HIGH contact. Contacts L and H in control line 1 are mechanically interlocked and latched so that either contact is in the closed position while the other is in the open position. The latched contact cannot be switched over until the relay controlling the open contact is energized.

2. For the latched position shown in the figure, the motor starts in the FORWARD direction at low speed. In order to obtain high speed, it is necessary to depress the HIGH pushbutton in control line 2, energizing the H relay and closing the H contact in control line 1 (simultaneously opening the L contact). When the H relay is energized, its pushbutton deenergizes all L contacts. The contacts to winding B, the low-speed winding, are opened, and the line contacts to winding A, the high-speed winding, are closed.

3. Reversal is accomplished by pressing the momentary REVERSE contact, deenergizing the FORWARD relay, and energizing the REVERSE relay. Line contacts R in the power circuit reverse lines 1 and 2, producing reversed phase sequence and reversed rotation of the motor.

4. The FORWARD and REVERSE pushbuttons are mechanically or electrically interlocked, and line L_3 contacts are required to prevent accidental short circuit of the lines in the event that both pushbuttons are simultaneously depressed. Similar protection is provided in the case of the HIGH and LOW pushbuttons, so that both windings are not simultaneously energized.

**7-4.
SECONDARY
RESISTANCE
CONTROL OF
WOUND-ROTOR
INDUCTION
MOTORS (WRIMs)**

The insertion of added rotor resistance in a WRIM produces a proportional increase in rotor slip.* This method is applicable to WRIMs, primarily and (in addition to its high *starting* torque) it has the advantages of (1) variation of speed over a wide range *below* the synchronous speed of the motor; (2) simplicity of operation, both from a manual and an automatic point of view; and (3) low initial as well as low maintenance costs for manual and automatic controllers. It has the disadvantages, however, of (1) low efficiency, because of the increased rotor resistance losses (at large values of slip, these losses are almost the total loss), and (2) poor speed regulation at any given rotor resistance setting.

Speed control by manual or automatic rotor *rheostatic* methods is

---

* Kosow, *Electric Machinery and Transformers*, Sec. 9-11.

hardly economical for *continuous* operation at speeds below synchronous speed. The WRIM is used extensively with secondary resistance control for loads of an *intermittent* nature, requiring high starting torque and relatively rapid acceleration and deceleration, such as foundry or steel mill hoists and where a high starting current causes serious line disturbances.

Because the speed and slip of a WRIM are proportional to the rotor resistance,* the method of speed control by the variation of secondary rotor resistance is sometimes called *slip control*. Two other methods of slip control, involving the change of slip for a given load, are (1) secondary "foreign" voltage control, and (2) variation of the primary line voltage. These methods, as well as devices employing these principles, are discussed in subsequent sections.

7-5.
CONCATENATION

A fundamental and early method of slip control by secondary "foreign" voltage control is called *concatenation*. A WRIM and a SCIM are coupled mechanically to the same shaft and load, and are connected electrically so that the second element, the SCIM, receives its stator excitation input from the open-circuited rotor of the first, a WRIM. Such a *cascaded* arrangement, in which the output of one machine serves as the input for the second, is shown in Fig. 7-11a, where the excited second motor is a squirrel-cage induction motor. If the first motor (or WRIM) has P_1 poles, and the second motor has P_2 poles, the synchronous speed of the two motors at frequency f_1 is

$$S_1 = \frac{120f_1}{P_1} \quad \text{and} \quad S_2 = \frac{120f_1}{P_2}$$

If the slip of each motor is designated as s_1 and s_2, respectively, and if the shaft speed of the two motors is the same, since they are coupled mechanically, then the common shaft speed S is

$$S = \frac{120f_1}{P_1}(1 - s_1) = \frac{120f_2}{P_2}(1 - s_2) = \frac{120f_1 s_1}{P_2}(1 - s_2)$$

Dividing by $120f_1$, we get

$$\frac{1 - s_1}{P_1} = \frac{s_1 - s_1 s_2}{P_2}$$

The product $s_1 s_2$ may be dropped, since it is negligible (even at full load), and so

$$s_1 = \frac{P_2(1 - s_1)}{P_1} = \frac{P_2}{P_1 + P_2}$$

* *Ibid.*, Sec. 9-11.

169

But

$$S = \frac{120f_1}{P_1}(1 - s_1) = \frac{120f_1}{P_1}\left(1 - \frac{P_2}{P_1 + P_2}\right)$$

and, therefore, the synchronous speed of the shaft is

$$S = \frac{120f_1}{P_1 + P_2} \qquad\qquad (7\text{-}1)$$

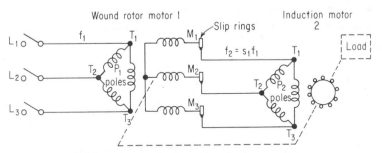

(a) Cascade–wound rotor and squirrel-cage motors.

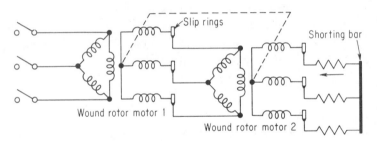

(b) Two cascaded wound rotor motors.

Figure 7-11

Concatenation, secondary foreign-voltage
slip control.

It is possible, therefore, by means of two coupled motors each having different pairs of poles, to obtain three specific *synchronous* speeds in the following manner:

1. For motor 1, connected to the line with its rotor short-circuited,

$$S_1 = \frac{120f_1}{P_1}$$

2. For motor 2, connected to the line,

$$S_2 = \frac{120f_1}{P_2}$$

3. For motors 1 and 2, in *direct* cascade concatenation,

$$S = \frac{120f_1}{P_1 + P_2}$$

It should be noted that the above speeds are synchronous speeds and that the actual concatenated motor speeds would be reduced slightly by the factor $(1 - s)$, as indicated in the above derivation.

In a similar manner it can also be proved that, when operating in cascade, the torque produced by each motor is divided in the ratio*

$$\frac{T_1}{T_2} = \frac{P_1}{P_2} \qquad (7\text{-}2)$$

The following example illustrates the application of concatenation as a speed-control method.

EXAMPLE 7-1

Two 60-Hz polyphase induction motors are connected as shown in Fig. 7-11a to drive a load at a full-load slip of 3 per cent. When connected separately, motor 1, a six-pole motor, has a slip of 5 per cent; and motor 2, a four-pole motor, has a slip of 4 per cent.
a. The speed of motor 1 when driving the load.
b. The speed of motor 2 when driving the load.
c. The speed of the two motors in direct cascade concatenation.
d. The per cent of full-load torque carried by each motor in part c.

Solution

a. $S_1 = \dfrac{120f_1}{P_1}(1 - s_1) = \dfrac{120 \times 60}{6}(1 - 0.05) = 1200 \times 0.95 = \mathbf{1140\ rpm}$

b. $S_2 = \dfrac{120f_1}{P_1}(1 - s_2) = \dfrac{120 \times 60}{4}(1 - 0.04) = 1800 \times 0.96 = \mathbf{1728\ rpm}$

c. $S = \dfrac{120f_1}{P_1 + P_2}(1 - s) = \dfrac{120 \times 60}{6 + 4}(1 - 0.03) = 720 \times 0.97 = \mathbf{698\ rpm}$

d. $\dfrac{T_1}{T_2} = \dfrac{P_1}{P_2} = \dfrac{\mathbf{6}}{\mathbf{4}}$

Motor 1 carries **60** per cent of the load, and motor 2 carries **40** per cent of the load.

The speed regulation of the motors, whether connected individually or in cascade as shown in Fig. 7-11a, is relatively good to excellent, and the two motors are capable of producing three distinct and separate speeds. An extremely *wide* range of speed control may be obtained by using *two* wound-rotor induction motors as shown in Fig. 7-11b. Thus, using the data of Ex. 7-1, motor 2, when used individually with rotor resistance, is capable of speed control from 1728 rpm down to (say) 730 rpm; motor 1, when used individually with rotor resistance, is capable of speed control from 1140 rpm to 700 rpm; and the two machines in concatenation, with added rotor resistance on motor 2, will provide speeds from 698 rpm to

---

* L. V. Bewley, *Alternating Current Machinery* (New York: The Macmillan Company, 1949), pp. 184–186.

zero speed. The direct cascade combination of Fig. 7-11b is thus capable of a smooth speed range from zero up to the synchronous speed of the machine with a smaller number of poles.

If the phase sequence of the "foreign" voltage applied to motor 2 in Fig. 7-11 is reversed, so that the rotation and torque of motor 2 tend to *oppose* those of motor 1, the connection is called a "differential cascade," and the synchronous shaft speed is

$$S = \frac{120f_1}{P_1 - P_2} \tag{7-3}$$

The differential cascade connection yields one additional speed above the synchronous speed of the motor with fewer poles, but that arrangement is hardly practicable because of the low developed torque [the difference in torque produced by each machine in Eq. (7-2)].

The motor combination *cannot* be started in differential cascade (because of the low starting torque) and usually requires that the load be brought up to speed by using the motor with the fewer poles and then switching to the differential connection. If the applied load is too great, the differential cascade set will tend to break down. While theoretically possible, the differential cascade connection is seldom used.

Concatenation, as a method of speed control, possesses the advantage of wider and smoother speed variations of induction motors above and below synchronous speeds, in comparison to pole changing and secondary resistance control. The disadvantages, however, are (1) low efficiency during cascade operation, since the losses of two dynamos are involved and each is not loaded fully; (2) poor speed regulation with added rotor resistance; and (3) low starting and pullout torque in direct cascade (and even lower in differential cascade), because of the reduced excitation of each machine and the increased reactance produced by the combination. It is important, however, because the concatenation principle of speed control by introduction of a foreign voltage at a different frequency $(f_2 = s_1 f_1)$ is used in the methods to be presented subsequently. Finally, it should be noted that despite the wide speed variation possible with Fig. 7-11b, a good deal of energy is dissipated in the added rotor resistance. Since these rotary techniques are used with larger polyphase motors, significant amounts of slip energy are lost. The methods presented below overcome this disadvantage.

7-6.
THE LEBLANC
SYSTEM

The Leblanc system is similar to dual-cascade wound-rotor motor concatenation, with the substitution of an induction frequency converter* for one WRIM and a three-phase adjustable transformer (variac) for the variable secondary resistor control. The connections are shown in Fig. 7-12. The Leblanc frequency

* Kosow, *Electric Machinery and Transformers*, Sec. 9-23.

converter is a special dynamo similar to a synchronous converter,* with a commutator at one end and slip rings at the other. The commutator is provided with three sets of brushes per pair of poles, displaced from each other by 120 electrical degrees. The stator of the converter is a smooth laminated steel cylinder, without any winding, which provides a low-reluctance path for the flux produced by the rotor winding. The slip rings of the converter are connected to the three-phase line through an adjustable transformer, and the brushes are connected to the slip rings of the main wound-rotor induction motor driving the load. The reduced polyphase voltage of line frequency (f_1) applied (through the polyphase auto-transformer) to the converter rotor armature winding produces a rotating magnetic field which rotates at synchronous speed with respect to the space around the armature iron.

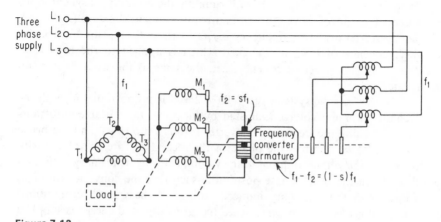

Figure 7-12

Leblanc speed-control system; secondary foreign-voltage slip control.

The phase sequences of the WRIM and the reduced polyphase voltage applied to the converter are reversed with respect to each other. When the WRIM is energized, therefore, it will drive the armature of the frequency converter in a direction *opposite* to the direction of the converter's rotating magnetic field.

When the WRIM is driven at a synchronous speed (slip equal to zero), the frequency-converter rotor is driven at a synchronous speed in a direction of rotation opposite to its rotating magnetic field. The net effect is to produce a stationary magnetic field in space. The armature conductors, rotating at synchronous speed with respect to the stationary flux, will have a voltage induced in them of the same frequency as that of the line (f_1).

* *Ibid.*, Sec. 11-7.

173

Similarly, when the rotor is driven at less than synchronous speed (in the opposite direction, by the WRIM), a magnetic field is produced which rotates (in the same direction as the armature rotating field) at a speed which is the difference between the synchronous speed S and the WRIM speed, $S(1 - s)$. Thus, at any slip s, the frequency of the WRIM, f_2, equals sf_1, and the frequency of the frequency-converter rotating field at this slip s is the difference between the line frequency f_1 and the armature frequency $f_1(1 - s)$, or

$$f_2 = f_1 - f_1(1 - s) = sf_1 \qquad (7\text{-}4)$$

The significance of Eq. (7-4), therefore, is that the frequency of the frequency converter is *always the same as* the frequency of the wound rotor of the WRIM. The validity of this statement may be proved easily by the condition when the WRIM is running (theoretically) at synchronous speed at a slip of zero. The wound-rotor frequency is zero ($f_2 = sf_1$). The converter armature frequency is $f_1(1 - s)$, or f_1 at a slip of zero, and the frequency of the rotating field which is applied to the brushes is $f_1 - f_1(1 - s)$ or zero frequency, i.e., the same as the wound-rotor frequency.

In the Leblanc system, the WRIM speed is proportional to both the secondary foreign voltage (adjusted by means of the variable autotransformer feeding the frequency converter through slip rings) and the brush position which controls the phase of the secondary foreign voltage with respect to the voltage induced in the wound rotor by its primary excitation flux. When the brush phase position is such that the converter (foreign) voltage *aids* the secondary induced voltage of the wound-rotor motor, the speed is *above synchronous speed* (negative slip) and energy is supplied to the rotor inductively from its own primary and conductively from the converter. When the brush phase position is such that the converter (foreign) voltage *opposes* the secondary induced voltage of the wound-rotor motor, the speed is *below synchronous speed* (less energy in the rotor, less torque) and the wound-rotor primary supplies energy to both its own rotor and to the converter. The latter regenerates the majority of this energy (less losses) to the ac supply by transformer action. At any given brush position, the magnitude of aiding or opposing foreign voltage is governed by the variable autotransformer.

The Lebranc system is, thus, a much more efficient method of foreign voltage control for motors of high horsepower rating.

7-7.
KRAMER CONTROL SYSTEM

A somewhat more flexible and sophisticated system of speed control, based on the principle of the Leblanc system, is the Kramer speed-control system, shown in Fig. 7-13 with two of its modifications. The basic system is shown in Fig. 7-13a. The brush output of the (driven) frequency converter (also displaced 120°)

is fed to the slip rings of the wound rotor as in the Leblanc system. But the variable transformer is connected to the wound-rotor slip rings instead of to the line. In effect, the frequency-converter rotor winding output at the brushes is connected back on itself (closed-loop feedback) through a variable transformer to its input, and the converter is driven by the WRIM.

As in the case of the Leblanc system, the wound-rotor frequency, f_2, is also the frequency of the voltage from (and at) the commutator, and the relations of Eq. (7-4) apply to the Kramer system as well. Shifting the phase position of the brushes will alter the power factor of the wound-rotor motor, and increasing the magnitude of the voltage applied to the converter rotor will increase the speed, i.e., when the phase of the converter brush output voltage aids the wound-rotor voltage. Speeds above and below synchronous speed are possible. In addition, the frequency converter also acts as a motor in converting the power supplied to it through the autotransformer to mechanical power available at the shaft.

The basic Kramer drive system shown in Fig. 7-13a is not used extensively because (1) it requires a larger and more expensive frequency converter at large values of slip, (2) like the Leblanc converter, the slip is affected both by the power factor control and the voltage (or speed) control, and (3) a singly excited converter is not an efficient torque-producing machine. The basic drive is called a constant-horsepower drive, because the converter electric and mechanical output is being utilized on the main drive shaft to supply useful horsepower constantly at all values of slip.

One variation of the Kramer system is the constant-horsepower drive using a *separate* synchronous converter in lieu of a frequency converter. Coupled to the shaft of the WRIM is a dc dynamo, supplied from a synchronous converter which receives its ac excitation from the wound-rotor slip rings. Both the dc machine and the synchronous converter are separately excited from a dc source or an exciter on the same shaft as the dc dynamo. For a given wound-rotor motor speed, the commutator output voltage of the dc dynamo is a function of its field excitation.

The synchronous converter (operating as an inverted converter) converts the direct current at its brushes to alternating current at its slip rings; this alternating current in turn is impressed on the wound rotor. The speed of the synchronous converter is determined by the frequency at its slip rings, i.e., the frequency of the wound-rotor motor ($f_2 - sf_1$). The speed of the main wound-rotor motor is determined by the polarity of the dc field of the dynamo and its excitation.

If the field excitation of the dc dynamo is *increased*, the synchronous converter dc input and ac output voltages are *increased*, causing an increased voltage opposition to the secondary induced emf in the rotor winding and a decrease in rotor current, flux, and torque, as well as a

175

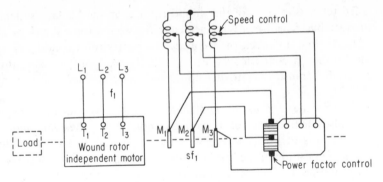

(a) Constant hp drive.

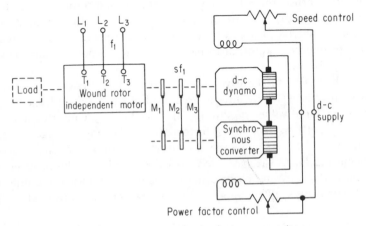

(b) Modified drive using seperate synchronous converter.

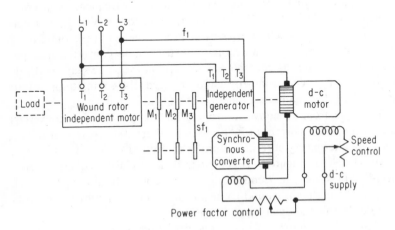

(c) Constant torque drive.

Figure 7-13

Kramer speed-control system; secondary foreign-voltage slip control.

decrease in speed. If the dynamo excitation is increased sufficiently, the speed decreases to a point where the synchronous converter voltage exceeds the dc dynamo voltage, and the dynamo runs as a dc motor. When the dc dynamo runs as a motor, as shown in Fig. 7-13b, its output is used to drive the main shaft, and a constant-horsepower drive is produced.

If, however, an induction generator* is coupled to the motor and to the main shaft as shown in Fig. 7-13c, and if a portion or all of the energy is returned to the power system so that the torque is a function only of the torque produced by the main wound-rotor motor, then a constant-torque drive system is produced.

Higher speeds are obtained by decreasing the dc dynamo excitation. Decreased dc dynamo excitation reduces the converter output and its opposition to the induced wound-rotor voltage, producing higher induction motor speeds. When the dc dynamo excitation is reduced to zero, the synchronous converter and the wound-rotor motor run practically at synchronous speed, since the only voltage generated is due to induced secondary voltage and the impedance drop across the converter armature is relatively small. Reversing the polarity of the dc field excitation will produce a converter voltage which aids the induced secondary rotor voltage, increasing speed above the synchronous speed. A field rheostat to control the excitation of the synchronous converter varies its power factor† (but not its speed), changing the phase of the voltage applied to the rotor of the wound-rotor machine. The Kramer system permits independent control of the speed as well as the power factor of the induction motor, for speeds well below to well above synchronous speed.

The major applications of such drive systems as the Leblanc and Kramer speed-control systems is for *extremely large* wound-rotor motors from 500 hp to about 3000 hp. Rheostatic wound-rotor resistance control would be extremely inefficient and expensive to operate because of the power losses. The advantages of returning the energy to the system, plus the low power losses in the converter (a synchronous converter has an extremely high efficiency) and the autotransformer (the most efficient piece of electric apparatus yet developed), combined with the advantage of power-factor correction, make the Kramer system particularly useful for heavy-power speed-control applications. Its major disadvantage is its high initial cost.

7-8.
SCHERBIUS SYSTEM

An outgrowth of the Leblanc system, and similar to the Kramer system, is the *Scherbius* system, shown in Fig. 7-14. This system is a completely ac system employing a variable frequency converter (a so-called *ohmic-drop exciter*) directly coupled to the WRIM.

* Kosow, *Electric Machinery and Transformers*, Sec. 9-22.
† *Ibid.*, Secs. 11-6 and 11-7.

177

The output of the three brushes of the variable frequency converter is fed to a regulating frequency converter coupled to a squirrel-cage induction dynamo whose stator is connected to the ac polyphase supply. The brushes of the regulating frequency converter (sf_1) are directly connected to the slip rings of the wound-rotor machine $(f_2 = sf_1)$ and also to the field through the adjustment taps of a speed-control autotransformer.

Since both the rotor and the stator of the regulating converter are supplied with the slip frequency (sf_1), the Scherbius system cannot operate or pass through the synchronous speed, because the slip-ring voltage would be zero at zero slip. An ohmic-drop exciter or variable frequency converter driven by the induction motor is used to carry the system through its unstable speed range slightly below and above synchronism.

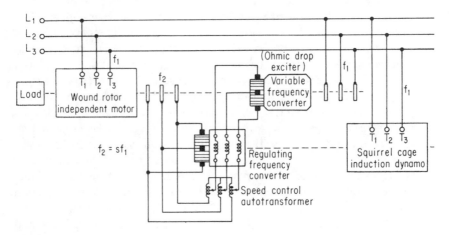

Figure 7-14

Scherbius speed control system; secondary
foreign-voltage slip control.

The ohmic-drop exciter is essentially a synchronous converter in which the slip ring and commutator voltages have a definite ratio regardless of speed.* The exciter is phase-connected in such a manner that its rotating magnetic field rotates in the opposite direction to the main motor shaft rotation. When the WRIM is rotating at synchronous speed, therefore, the magnetic field is stationary in space (see Section 7-6) and the voltage at the commutator has zero frequency; i.e., a dc voltage is produced. At the synchronous speed, therefore, the ohmic-drop exciter furnishes a dc voltage to and excites the field of the regulating frequency converter. The dc field excitation is sufficient to generate a rotor voltage to supply sufficient current to the wound-rotor motor to drive it through the unstable region at or near synchronism. At synchronism, the only

* *Ibid.,* Secs. 11-6 and 11-7.

voltage drop opposing the dc voltage, generated by the variable frequency converter to the slip rings of the main induction motor, is the "ohmic drop" of the regulating frequency converter; hence, the name ohmic-drop exciter.

The regulating frequency converter operates at an approximately constant speed (since it is driven by a squirrel-cage induction motor), and therefore its brush voltage is practically a function of its excitation, as controlled by the autotransformer exciting its field.

The Scherbius system, in comparison to the Kramer system, has the advantage of requiring no dc power whatsoever. It has disadvantages, however: (1) It has no power-factor adjustment, and (2) it requires special machines (rather than general-purpose ones) which are *not* readily available commercially, and is, hence, more expensive, particularly when specified for heavy drive systems. Like other foreign voltage slip-control systems, energy in the Scherbius machine is returned to the line through the induction dynamo (regeneration) when the regulating frequency converter is operating at speeds well above synchronous speed.

7-9.
SCHRAGE BRUSH-SHIFTING (BTA) MOTOR

The three essentials of the Leblanc system (a WRIM, a frequency converter, and a three-phase variable transformer, as shown in Fig. 7-12) were first combined in a single dynamo by Karl H. Schrage of Sweden. The same speed-controlled motor is also produced in the United States as the *BTA brush-shifting motor* (manufactured and distributed by General Electric Co.). The longitudinal and transverse cross sections of the motor are shown in Figs. 7-15a and b. As shown, the primary excitation winding of the motor is located on the rotor (the reverse of a wound-rotor motor) and is excited through slip rings. The wound-rotor secondary is on the stator, with brushes located on the commutator 120 electrical degrees apart exciting the stator winding (which corresponds to the rotor of the WRIM in a Leblanc system).

The auxiliary winding, connected to the commutator and excited through the primary rotor winding, corresponds to the converter. When the two brushes of each phase winding (S_1, S_2, and S_3) of the stator secondary are in contact with the same commutator bar, i.e., shorting the phase winding or secondary, there is no foreign voltage induced in the secondary and the motor operates as a shorted wound-rotor induction motor. If the brushes are separated, by means of a mechanical handwheel, in such a way that a foreign voltage is *conductively* introduced into the rotor from the converter winding which opposes the secondary emf induced from the rotor primary, the rotor current, flux, and torque decrease, and the speed drops. If the brushes are separated by means of the handwheel in the *opposite* direction, so that the foreign voltage *aids* the secondary induced emf, the motor current, flux, and torque increase *above synchronous speed.*

179

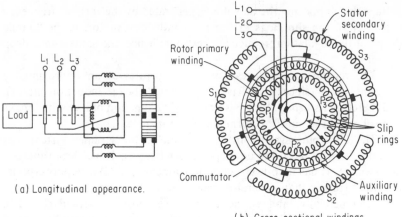

(a) Longitudinal appearance.

(b) Cross-sectional windings.

Figure 7-15

Schrage brush-shifting (BTA) motor.

It is also possible to rotate the *entire* brush rigging structure to provide power-factor adjustment as well. Unfortunately, the machine does *not* operate satisfactorily at synchronous speeds (no voltage is induced in the stator) and for this reason cannot be readily constructed in the larger horsepower ratings. The torque instability at synchronism occurs because of the inertia of heavy loads. This is its major disadvantage. It is available in sizes from 10 to 50 hp in voltage ratings of 220, 440, and 550 V, where its operation is quite satisfactory. The speed is entirely variable over a range of from 3:1 to 4:1, depending on the size and the number of poles (usually four- or six-pole machines are commercially available). It has the advantages of (1) relatively good starting and maximum torque (1.5 and 2.5 times rated torque, respectively) for a variable speed motor; (2) exceedingly smooth speed control above and below synchronous speed; (3) extremely high efficiency; (4) excellent speed regulation; and (5) it combines all three devices of the Leblanc system in a single machine with the same advantages but for machines below 50 hp.

7-10.
ELECTRONIC SLIP-POWER CONTROL

As indicated earlier (Section 7-2), the emergence of high-current SCRs in variable-frequency static converters has encouraged development of solid-state packaged drives for ac polyphase motors below 100 hp. In competition with the Schrage brush-shifting (BTA) motor described in the previous section is the automatically torque-regulated ac adjustable-speed drive shown in Fig. 7-16. The solid-state electronic package shown replaces a three-phase variable rheostat in the secondary circuit of a WRIM. There is no loss of slip power, however, in the electronic package compared to the rheostat, because the electronic package is *regenerative*.

CHAP. SEVEN / *Manual and Automatic Speed Control of ac Polyphase Motors*

The ac voltage induced in the WRIM rotor is rectified in the full-wave (6-diode) bridge section to dc which is fed via a filter directly to a three-phase, bridge-type inverter (Section 7-2) consisting of 6 SCRs. The inverter output is paralleled to the three-phase 60-Hz line.

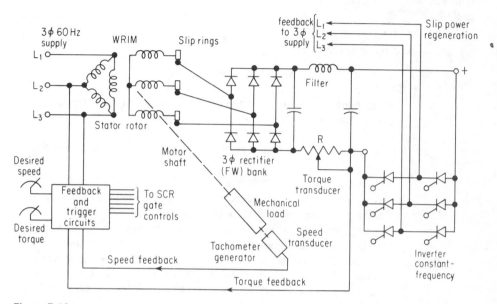

Figure 7-16

Electronic slip power control with speed and torque feedback loops.

The SCR inverters are triggered at an inversion rate established by the frequency of the three-phase line so that the inverter output is maintained at a constant frequency. Two types of feedback sensors furnish signals to the trigger circuitry controlling the inverter regenerative output voltage and power. An ac single-phase tachometer alternator provides an indication of output motor speed. A tap on a resistor which senses (rectified) current in the rotor secondary circuit provides a measure of the torque developed by the WRIM.

Potentiometers on the control panel of electronic package permit adjustment of desired speed and torque. Since horsepower is proportional to the torque-speed product (hp = kTS), the feedback circuitry acts in such a manner to maintain a constant speed-torque relationship. Should the applied torque (of the load) momentarily decrease (tending to increase speed), the gating of the SCRs acts to regenerate more ac power to the three-phase line, thereby restoring the applied torque to the WRIM rotor and maintain the same speed and torque on the motor.

It is anticipated that state-of-the-art device developments in increasing the current ratings of SCRs will encourage extended use of static packages shown in Fig. 7-16 upward beyond the 100-hp range.

7-11.
LINE VOLTAGE
CONTROL

An examination of Eq. (7-5) for any induction motor reveals that slip s is

$$s = \frac{S - S_r}{S} \tag{7-5}$$

and rotor speed S_r is

$$S_r = S(1 - s) = \frac{120 f_1}{P}(1 - s) \tag{7-6}$$

The methods of (rotor) speed control (covered thus far) emerging from Eq. (7-6) are:

1. A change in f_1, the frequency of the voltage applied to the stator (Section 7-2).
2. A change in P, the number of poles of the stator winding (Section 7-3).
3. A change in slip, s, produced by a change in ohmic rotor resistance using secondary resistance control (Section 7-4).
4. A change in slip, s, by the insertion of a foreign voltage in the secondary circuit by concatenation (Section 7-5), by the Leblanc system (Section 7-6), by the Kramer system (Section 7-7), by the Scherbius system (Section 7-8), by the Schrage motor (a modification of the Leblanc system, Section 7-9), or by electronic slip power control (Sec. 7-10).

The SCIM torque under starting and running conditions varies as the square of the voltage impressed on the stator primary $(T = kV_p^2)$. For any given load, reducing the line voltage will reduce the torque by the square of the reduction in line voltage, and the reduction in torque should produce an increase in slip, s. Although reducing the line voltage and the torque as a method of increasing slip will serve to control the speed to a degree in split-phase single-phase motors, particularly, and in small induction-type motors, generally, it is a *most unsatisfactory* method of speed control for polyphase motors. Figure 7-17 shows the typical torque-slip curve for an induction motor at the rated voltage and at one-half the rated voltage. The maximum torque at one-half the rated voltage is one-fourth of the maximum torque at the rated voltage. It is no longer possible to obtain the rated torque, or even one-half of the rated torque, because the motor speed drops rapidly and the motor stalls before

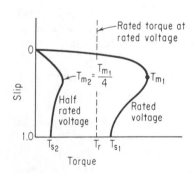

Figure 7-17

Variations of torque with impressed stator voltage.

it will develop the rated torque. Thus, if reduced voltage is used as a method of speed control, it is necessary that the applied load torque be reduced considerably as the stator voltage and speed are decreased. This method of control, therefore, will work reasonably well for an induction motor which is lightly loaded. The instant any appreciable load is placed on the motor, the speed regulation is very poor (as may be seen from Fig. 7-17) and the motor stalls.

This method of speed control, in which a change of slip is produced by a change of primary voltage, while feasible for small single-phase fans or blowers, where the applied torque (windage) is low at low speeds, is not applicable to polyphase motors.

7-12.
THE ROSSMAN
DRIVE

This method of speed control, as mentioned in Section 7-1, might be considered an electromechanical method, as distinct from the electric methods discussed previously and the subsequent mechanical methods below. The Rossman drive motor has a stator and rotor, *both* of which are capable of rotation as shown in Fig. 7-18 (in this respect it is similar to a dynamometer and also to the supersynchronous motor).* The stator is (usually) driven by a speed-controlled dc motor (capable of speed variation from zero to induction-motor synchronous speed) by any of the suitable methods covered in Chapter 6. From an ac supply, it is usually convenient to use the Ward-Leonard method for the larger horsepower motors, or electronic methods (Section 6-12) for the stator drivers of smaller motors used to drive the special Rossman drive motor. The principle of the Rossman drive stems from the fact that the rotor speed of any induction motor, S_r, is the product of the synchronous speed S times a factor $(1 - s)$. The synchronous speed, S, is the speed of the rotating field with respect to the stationary stator. Assuming, however, that a stator is capable of rotation in either

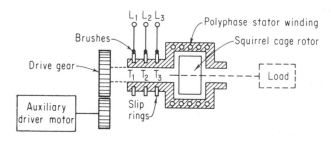

Figure 7-18

The Rossman drive.

* Kosow, *Electric Machinery and Transformers*, Sec. 8-24.

the same or the opposite direction as the rotating magnetic field at a speed, $\pm S_2$, the rotor speed is

$$S_r = S_1(1 - s) \pm S_2 \tag{7-7}$$

where S_1 is the standstill synchronous speed and $\pm S_2$ is the stator speed in the same direction as the rotating field.

The Rossman drive is a constant-torque drive (torque is a function only of the torque produced by the induction motor) and therefore the countertorque or reaction torque produced by the rotor against the stator must be absorbed by the stator drive motor. In addition to the torque required to drive the (unexcited) stator, the horsepower required of the motor is $kT(S_1 \pm S_2)$. In terms of the main Rossman drive motor, therefore, the horsepower of the drive motor or auxiliary stator driver is

$$\text{hp} = \text{HP}_M \frac{S_1 - S_2}{S_1} \tag{7-8}$$

where HP_M is the horsepower of the main Rossman drive motor and $S_1 - S_2$ is the difference in speed of the stator from its synchronous speed.

Rossman drive units are manufactured in sizes up to 3000 hp and are used primarily in boiler feed pumps in power plants. The drive unit is started by using either manual drum or automatic starters, as described in Chapter 5. The disadvantage of this method of speed control is that a special mechanically constructed squirrel-cage induction motor is required. The advantages of the method are (1) excellent speed regulation over (2) a fairly wide range of speed above and below synchronous speed within the limits of the horsepower capacity of the auxiliary driver [Eq. (7-6)]. The complete drive unit is competitive in price with other large-capacity speed-control systems such as the Leblanc, Kramer, and Scherbius systems. As a constant-torque drive, means have been found to return the excess power to the supply mains through the auxiliary drive, reducing the operating cost by a saving in consumed power.

7-13.
SPEED CONTROL
BY MECHANICAL
MEANS

Polyphase induction and synchronous motors, having essentially constant speed characteristics at the rated voltage, are assembled in packaged drive units to drive mechanical arrangements of gears, cylindrical and conical pulleys, and even hydraulic pumps to produce a variable-speed output. Some of these units employ magnetic slip clutches and solenoids to control the various mechanical or hydraulic arrangements, through which a relatively smooth control of speed may be achieved, in addition to reversal of rotation. A discussion of the various mechanical and hydraulic techniques employed is beyond the scope of this volume.

Reversing the direction of rotation of a polyphase induction motor was mentioned in connection with speed control by pole changing (Section 7-3), and was shown in Fig. 7-10b. The phase sequence of *any* polyphase stator (either a synchronous or induction dynamo) for whatever purpose may be reversed by closing either the *F* or *R* contacts shown in the primary power circuit of Fig. 7-19a. This may be accomplished manually by using drum or can switches. It also may be accomplished automatically, as shown in Fig. 7-19b, in which the pushbuttons are electrically and mechanically interlocked (the former through n.c. *R* and *F* contacts and the latter through the dual-contact pushbuttons). The line (forward and reverse relay) contactors are also mechanically interlocked.

If other independent operations in addition to reversing (such as braking or resistance starting in primary and secondary circuits) are to be automatically controlled, it is customary to have a separate line contactor, *M*, to initiate and control the sequence, as shown in Fig. 7-19c.

7-15.

PLUGGING

Dynamic braking of ac motors, both polyphase and single phase, SCIM, and synchronous motors is accomplished in much the same manner as with dc motors. Thus, any control system which is capable of reversing the motor may also be used for braking by plugging (Section 6-7). As in the case of the dc motor, it is necessary to disconnect the motor from the line before the direction of rotation of the motor is reversed (see Section 6-7).

Unlike the dc motor and synchronous motor, however, which have a constant field or excitation flux during the braking period when the armature is decelerating, the excitation flux of a polyphase SCIM stator varies from maximum (at the instant before phase reversal is initiated) to zero (at the instant when the stator currents are reduced to zero by line reversal and are about to increase in the opposite direction). The magnitude of the resultant rotating ($120f/P$) magnetic field, therefore, is diminishing rapidly as the line connections are reversed. Since the torque developed by the SCIM rotor is $k_t\phi I_r \cos\theta_r$, and since the rotor current is dependent on the excitation flux, the developed torque drops rapidly as the flux, ϕ, and rotor current, I_r, diminish rapidly. As the applied load torque exceeds the developed torque, the load itself assists in decelerating the motor rapidly to a stop. Thus, the reversal by plugging of a SCIM or WRIM at full voltage will *not* exceed the normal starting current (as it does in a dc motor), and no additional reversing precautions or protective devices are required.

It may be stated simply that, where an induction motor is started across the line, it may be reversed by plugging (reversing the line connections) in the same manner; and, where it is started using auxiliary, manual, or automatic starting methods (see Chapters 3 and 5), these must also be

185

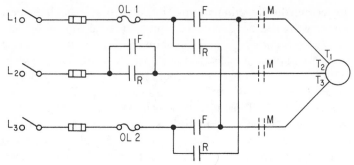

(a) Power circuit for manual or automatic reversal and plugging.

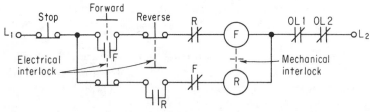

(b) Control circuit with electrical and mechanical interlocks.

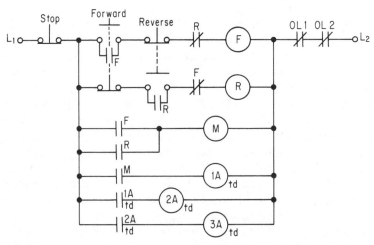

(c) Control circuit with main contactor for initiating timed relay sequence.

Figure 7-19

Reversing rotation and phase sequence
of polyphase stator for reversing and/or
plugging braking.

used for reversing. Since braking by plugging to unity slip (standstill) of a SCIM or WRIM is accompanied by a reduction of line current, no additional protective devices are required except a reverse-current centrifugal switch or a plugging relay to disconnect the motor from the reversed lines.

Figure 7-20a shows the basic plugging and braking to a stop (plug-stop braking) control circuit for a polyphase or single-phase (split-field) induction motor having a squirrel-cage rotor. Relay *PR* (not shown) is a reverse-current relay (Sections 1-12 and 1-13) connected in the primary power circuit. The control circuit of Fig. 7-20a may be used with the power circuit of Fig. 7-19a as a plug-stop controller which operates in the following manner:

1. The motor is started in a specific direction (determined by the phase sequence of the line connections) by pressing the RUN momentary contact. Relay *F* is energized in control line 1, and relay *M* is energized in control line 3, through n.c. reverse-current relay (*PR*) contacts.

2. When the STOP button is pressed, relay *F* is deenergized in control line

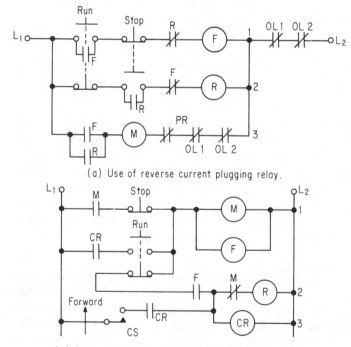

(a) Use of reverse current plugging relay.

(b) Alternate circuit using centrifugal friction switch.

Figure 7-20

Plugging braking of an induction
motor.

187

1, and relay R is energized through n.c. F in control line 2. The closing of R contacts in the power circuit (Fig. 7-19a) reverses the two line connections to plug the motor rapidly to a stop. When the line current attempts to reverse, PR opens the line connection by deenergizing relay M in control line 3.

The n.c. contacts of the reverse-current plugging relay, PR, could also be a centrifugal switch which opens whenever the motor is stopped or approaches standstill. Figure 7-20b shows an alternative control circuit using such a device which operates as follows:

1. When the RUN contact is depressed, the motor is started by energizing line contactors M and F in control line 1 and in Fig. 7-19a. At the same time, the centrifugal switch, CS, in control line 3 closes when the motor accelerates in the forward direction. Releasing the RUN button also energizes control relay CR in control line 3 through n.o. F contacts.
2. When the STOP button is depressed, relays M and F are deenergized, but relay CR in control line 3 still remains energized through the centrifugal switch with rotation in the FORWARD direction. The centrifugal switch CS also energizes the reverse relay R in control line 2 through n.o. CR and n.c. M contacts. (The main reverse contacts R in the power-line circuit are connected in such a manner that they bypass the M contacts).
3. Plugging occurs as the line connections are reversed. As the motor approaches standstill, the centrifugal switch opens, deenergizing the reverse relay R and control relay CR, and disconnecting the motor from the line.

In the case of split-phase *single-phase* induction-principle motors, the R contacts are connected in such a manner as to reverse the instantaneous polarity of one of the windings so that a rotating magnetic field is produced in the opposite direction and the motor is brought to a standstill by plug-stop braking. The same control circuits as those shown in Fig. 7-20 may be used for plug-stop braking of single-phase motors as well.

Magnetic brakes (Section 6-10) are also used to set the rotor when power is removed from polyphase or single-phase SCIMs.

7-16.
DYNAMIC BRAKING
Unlike the dc motor or synchronous motor, in which the armature is disconnected from the line and connected across a resistor as a generator (see Section 6-8), there is no way to disconnect the primary polyphase armature of SCIM and still maintain excitation of the rotor secondary in the same manner. Dynamic braking nevertheless *is* possible, however, if the primary ac polyphase excitation is removed and if the *stator is excited with direct current* instead. The constant, unidirectional direct current will produce fixed electromagnetic poles on the stator. The squirrel-cage rotor conductors will have an alternating

emf induced in them as they pass fixed N and S stator poles. The alternating rotor emf is short-circuited, producing high rotor currents and fluxes which react (in accordance with Lenz's law) against the strong fixed dc stator field to bring the motor rapidly to a stop. A high I^2R loss, produced internally in the rotor, dissipates the rotational energy in the form of heat. In effect, the SCIM becomes a generator having a short-circuited armature which results in a heavy rotor current and rapid braking.

Unlike dc motors (where dynamic braking produces maximum braking at high speeds and minimum braking as the motor approaches standstill), induction motor dynamic braking action increases from a low value at maximum speeds to a maximum as the speed drops, but it decreases to a minimum near standstill. When braking is first initiated by applying direct current to the stator, the slip of the rotor is a maximum, producing high rotor reactance and low rotor power factor. The high rotor impedance causes the rotor currents to be limited. Furthermore, the lagging power factor of the rotor produces a highly *demagnetizing* armature reaction flux* which reduces the net dc field flux considerably. As the speed drops, however, the slip and rotor reactance decrease, increasing the rotor current (improving the power factor) and the net dc field flux as well, so that the braking action is increased. At very low speeds, however, *despite* the small slip and low reactance, the emf induced in the rotor is small and the braking action tends to decrease once more. The use of direct current on stators of polyphase (or single-phase) induction motors is, nevertheless, effective as a means of dynamic braking in bringing these motors to a stop.

The switching for manual or automatic dc dynamic braking of a three-phase motor is shown in Fig. 7-21a. When the M contacts are closed, the motor is started, and it runs as a three-phase induction motor. When the M contacts are opened and the braking B contacts are closed, the dc circuit, energized by means of a transformer and a full-wave rectifier, places direct current across the primary stator armature terminals T_1, T_2, and T_3 in a series-parallel arrangement. The dc excitation is controlled by means of a variable resistor which serves to limit the excitation and to protect the rectifiers as well. When automatic control is used (since direct current for braking purposes is already available, and dc relay operation is superior to ac operation), the direct current produced is fed directly to the control circuit and taken from points $X–X'$ which are continuously energized.

A similar circuit for a single-phase induction-type motor, shown in Fig. 7-21b, uses a full-wave bridge-type rectifier for dc excitation of a single-phase stator.

The control circuit, taken from points $X–X'$, works equally well for

---

* Kosow, *Electric Machinery and Transformers*, Sec. 5-9.

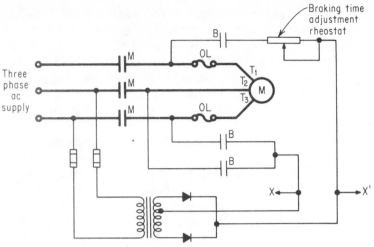

(a) Three phase braking power circuit

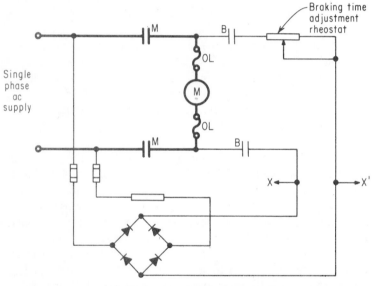

(b) Single phase braking power circuit

Figure 7-21

Dynamic braking for manual or
automatic switching of polyphase and
single-phase induction motors.

three-phase or single-phase motors, as shown in Fig. 7-22. It operates as
follows:

    1. Depressing the START button energizes line contactor *M* and adjustable
        time-delay control relay *CR* through n.c. braking contact *B*. The closure

of contactor M may also be used to energize various starting and accelerating relays (not shown), if required for reduced voltage ac starting, speed control, part windings ,etc.

2. The motor starts and runs as an ac motor with M energized through auxiliary M bypassing the START button. In time, control relay CR, simultaneously energized with M, closes its n.o. CR contact in series with the braking relay B, which is not energized because of open n.c. contact M.

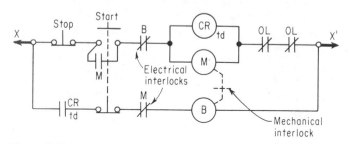

Figure 7-22

Control circuit for automatic
dynamic braking of motors in Figs. 7-21 and 7-24.

3. In the event of either an overload or the depressing of the STOP button, M and CR are deenergized, energizing braking relay B. When B closes, direct current is applied to the field, producing dynamic braking. Time-delay contact CR remains closed during the entire braking period. Adjustable relay CR is adjusted so that contact CR opens when the motor stops.

The control circuit of Fig. 7-22 has several mechanical and electric interlocks to prevent alternating and direct current from being applied simultaneously to the stator windings. Pressing the START button opens the control line to braking relay B. The electric n.c. interlocks, B and M, respectively, in series with the opposite line contactors (M and B, respectively), ensure that when one relay is energized, the other is not. A final precaution is the mechanical interlocking of line contactors M and B so that when one is operative, the other is not.

There is little danger that a motor directly coupled (nongeared) to a load may be harmed, as long as the rotor and stator powers dissipated during the braking does not damage the conductors by overheating. If the braking is very sudden, however, some harm may result to the mechanism of the connected load. In the case of geared motors driving loads of considerable inertia, the gearing may be damaged by sudden braking. A good rule to use in applying dc to the stator during braking is to *use the minimum power that obtains the desired results.* This rule results in cooler stator and rotor operation and places less stress on the coupling, gearing, and mechanical components within the load driven by the motor.

191

7-17.
REGENERATIVE
BRAKING

As in the case of dc motors (see Section 6-9), regenerative braking is possible and occurs automatically whenever the motor speed is excessive, for example, in the case of overhauling loads (i.e., where the load tends to drive the prime mover in the same direction at an excessive speed). The polyphase (and single-phase) SCIM, when driven at speeds that exceed synchronous speed, automatically operate as induction generators, producing braking action and tending to restore the motors to normal speed. At the same time, the braking energy is regeneratively returned to the ac line so that there is no waste of power. Like its dc counterpart, regenerative braking cannot bring the motor to a stop; but it serves to limit excessive motor speed without the necessity for mechanical braking and with little energy loss. It is particularly applicable to such loads as hoists, cranes, lifts, elevators, and traction applications, where the inertia of the load or the force of gravity tends to increase the motor speed excessively.

7-18.
JOGGING AND
MECHANICAL
BRAKING

Jogging, used for the reasons indicated in Section 6-11, in association with electromagnetic braking, described in Section 6-10, may also be used in conjunction with ac polyphase (and single-phase) motors in a similar manner. A typical control circuit, adaptable for either polyphase (or single-phase) motors, shown for a nonreversing motor in Fig. 7-23a, operates in the following manner:

1. Depressing the START button energizes jog relay J in control line 1. Relay J energizes line contactor M in control line 2 through n.o. J contacts. Simultaneously, relay B is energized through n.c. M contacts in control line 3, to short out the series resistance for the electromagnetic brake.

2. A dc full-wave rectifier circuit, in control line 4, continuously energized through the control circuit transformer, immediately releases the fast-acting magnetic brake when energized through n.o. M contacts at reduced voltage (see Section 6-10 for a detailed explanation of instantaneous brake action at conditions of overvoltage).

3. The motor starts with the brake released when the slower acting M contacts in the power line are closed. *Note:* If reduced voltage starting is required, the M contactor may also be used to initiate the control circuitry for accelerating the motor (described in Chapter 5).

4. Time-delay n.c. contact M in control line 3 opens, deenergizing relay B, and inserting a protective resistor R in series with the brake coil in control line 4 to prevent the coil from overheating and to decrease the coil time-constant when the brake is deenergized.

5. Pressing the STOP button deenergizes relay M in control line 2, instantly deenergizing and setting the brake coil, which serves to stop the motor rapidly.

6. When it is desired to jog the motor, pressing the JOG control energizes relay M in control line 2 but does not energize relay J in control line 1.

192

CHAP. SEVEN / *Manual and Automatic Speed Control of ac Polyphase Motors*

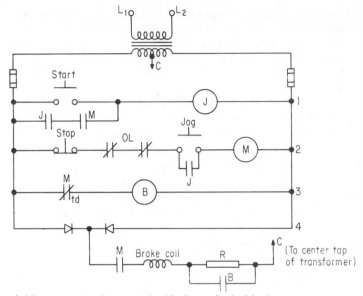

(a) Non-reversing jog control with dc mechanical brake.

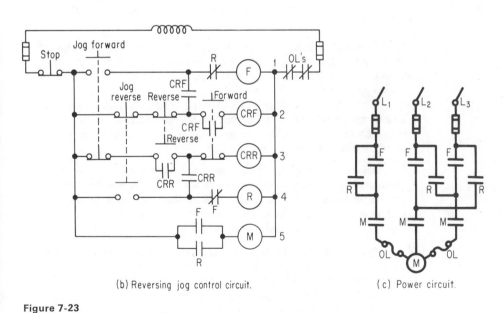

(b) Reversing jog control circuit.

(c) Power circuit.

Figure 7-23

Jog control circuits for polyphase SCIMs and
WRIMs.

193

Coil M is energized only as long as the button is depressed. Similarly, the brake is released only as long as the JOG button is depressed. The motor is thus jogged or inched with the full protective starting series line reactance or resistance (not shown) and is immediately stopped by braking whenever the JOG control is released.

The control circuit shown in Fig. 7-23a is adapted for a motor which runs in one direction only. For use with reversing control circuitry, the control circuit shown in Fig. 7-23b permits jogging in either the forward or the reverse direction. The F and R relays (control lines 1 and 4, respectively), as well as the jog relays, are interlocked in the usual manner, to prevent accidental short circuits across the line when both buttons are depressed. When it is desired to inch or jog the motor, the separate JOG controls (JOG FORWARD or JOG REVERSE) energize the M contactor independently of their pushbutton interlocks in the same manner as in Fig. 7-23a, only for that time during which the pushbutton is depressed. The customary reversing power circuit for the polyphase motor is shown in Fig. 7-23c.

7-19.
POLYPHASE
SYNCHRONOUS
MOTOR
REVERSING AND
BRAKING

Reversing. Polyphase synchronous motors are reversed using the same techniques described for SCIMs and WRIMs in Section 7-13 and shown in Fig. 7-19. Although it is possible to accomplish such reversals, the application situation requiring a *constant* speed motor having the capability of rapid and frequent reversal is extremely rare. If it does occur, mechanical means are used (Sections 7-12, 13).

Plugging. A polyphase synchronous motor may be brought quickly to a stop using the method described in Section 7-14. The dc field is simultaneously disconnected from the dc supply and short-circuited to assist the braking action.

As in the case of plugging SCIMs and WRIMs, a mechanical brake is set by means of a centrifugally operated plugging relay which deenergizes the reversing contactors, disconnecting the stator from the reversed lines, as the motor approaches zero speed.

Regenerative braking. Regenerative braking may be used with synchronous motors, particularly with overhauling loads that tend to drive the synchronous motor at speeds greater than synchronous speed. In such cases, the synchronous motor acts as an alternator in parallel with the three-phase supply and no special equipment is necessary.

Dynamic braking. In *dynamic* braking (Section 6-8), the synchronous motor is brought to a standstill quickly by dissipating the inertia of its rotor in braking resistors. Unlike dynamic braking of SCIMs and WRIMs (Section 7-15), the polyphase synchronous motor is used in its alternator mode for rapid dynamic braking. The dc field circuit is continuously energized but the polyphase stator is disconnected from its lines (M

194

contacts opened) and connected to a three-phase resistive load, as shown in Fig. 7-24, via *B* contacts closed.

The control circuit for Fig. 7-24 is the same as that shown in Fig. 7-22. It is usually customary to also include a timing relay (not shown) which disconnects the dc field circuit after the motor has stopped.

The three-phase braking resistor shown in Fig. 7-24 may be intermittently rated, depending on the frequency of stopping and starting, the inertia (size) of the rotor, and the time required to bring the motor to

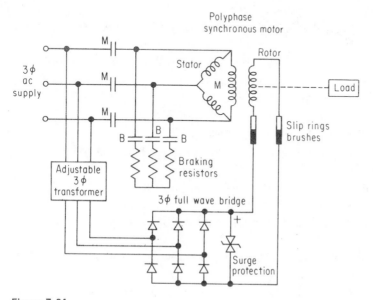

Figure 7-24

Dynamic braking of synchronous motor
with static field excitation. (see Fig.
7-22 for control circuit.)

a standstill. A short circuit across the stator (of course) brings the rotor to a standstill most rapidly, but this produces excessively high stator currents. The braking resistance values are selected to produce no more than from 200 to 300 per cent of the rated stator current.

Note that Fig. 7-24 also shows the dc field excitation supplied from an adjustable three-phase transformer and full-wave, three-phase bridge rectifier, with voltage-surge protection. Such static solid-state packages have replaced the older shunt-generator exciters on the rotor shafts and represent an intermediate step in the development of the *brushless synchronous motor.**

* Kosow, *Electric Machinery and Transformers*, Secs. 8-30 and 8-31.

BIBLIOGRAPHY

Alger, P. L. *The Nature of Polyphase Induction Machines* (New York: John Wiley & Sons, Inc., 1951).

―――― and Y. H. Ku. "Speed Control of Induction Motors Using Saturable Reactors," *Electrical Engineering* (February 1957).

Anderson, E. "The Schrage Motor for Wide-Range Adjustable Speed from 'Straight a-c,'" *Electrical Manufacturing* (January 1957).

Baude, J. "Multi-Function Magnetic Amplifier Control for Large Motors," *Electrical Manufacturing* (October 1957).

Bewley, L. V. *Alternating Current Machinery* (New York: The Macmillan Company, 1949).

Dunigan, F. P. "Load Control System," *Electrical Manufacturing* (April 1959).

Electric Clutch and Brake Engineering Handbook (Warner Electric Brake and Clutch Co., Beloit, Wis., 1965).

"Electronic Standards for Industrial Equipment" (General Motors Technical Center) ,*Electrical Manufacturing* (August 1958).

Evert, C. F. "Dynamic Braking of Squirrel-Cage Induction Motors," AIEE Paper 54–3.

Graphical Symbols for Electrical Diagrams (ASA Y32.2) (New York: American Standards Association), IEEE Std. No. 315-1971.

Gutzwiller, F. W. "The Silicon Controlled Rectifier," *Electrical Manufacturing* (December 1958).

Harwood, P. B. *Control of Electric Motors*, 4th ed. (New York: John Wiley & Sons, Inc., 1970).

Heumann, G. W. *Magnetic Control of Industrial Motors*, 3 vols. (New York: John Wiley & Sons, Inc., 1961).

Industrial Control Equipment (Group 25) (ASA C42.25) (New York: American Standards Association).

Industrial Control Equipment (UL508) (Chicago: Underwriters Laboratories, Inc.).

James, H. D., and L. E. Markle. *Controllers for Electric Motors* (New York: McGraw-Hill Book Company, 1952).

Jones, R. W. *Electric Control Systems* (New York: John Wiley & Sons, Inc., 1953).

Kosow, I. L. *Electric Machinery and Transformers* (Englewood Cliffs, N.J.: Prentice-Hall, Inc., 1972).

Laithwaite, E. R. "Two New Ways to Vary Induction Motor Speed," *Control Engineering* (July 1960).

Liwschitz, M. M., and L. A. Kilgore. "A Study of the Modified Kramer or Asynchronous-Synchronous Cascade Variable-Speed Drive," *Trans. AIEE*, Vol. 61 (May 1942).

Marx, C., and Dessmer, R. "Electronic ac Adjustable Speed Drive," *Electrical Manufacturing* (May 1958).

Motor and General Standards (MG1) (New York: National Electrical Manufacturers Association).

Ogle, H. M. "The Amplistat and Its Application," *General Electric Review* (February, August, and October 1950).

Onjanow, N. "A-C Drive Offers System Design Flexibility," *Electro-Technology* (December 1960).

Press, V. W., and W. R. Jones. "Rugged Adjustable Speed Drives Use Magnetic Amplifiers," *Electrical Manufacturing* (November 1958).

Schohan, G. "Static Frequency Multipliers for Induction Motors," *Electrical Manufacturing* (April 1956).

Schwarz-Kast, E. L. "Synchronized Drives with Standard Electric Motors," *Machine Design* (April 1950).

Siskind, C. S. "A-C Motors for Adjustable-Speed Systems," *Control Engineering* (June 1959).

Storm, H. F. *Magnetic Amplifiers* (New York: John Wiley & Sons, Inc., 1955).

Taylor, E., and J. Burnett. "Use Thyratrons to Control Higher Power A-C Servomotors," *Control Engineering* (April 1959).

Trofimov, L. A. "Wide Speed Range Electrical Drives," *Electrical Manufacturing* (March 1954).

Uri, J. Ben. "Variable Speed Control Systems," *Electro-Technology* (March 1961).

Vedder, E. H., and J. M. Cochran. "Silicon Rectifier Power Drives," *Electrical Manufacturing* (March 1959).

Wang, A., and G. Y. Chu. "Pulse Generator Replaces Gearbox of Numerically Controlled Lathe," *Electrical Manufacturing* (December 1957).

Winsor, L. P., and E. E. Moyer. "Adjustable Speed Drives," *Electrical Manufacturing* (November 1952).

Woll, R. F. "Applying the Wound-Rotor Motor," *Westinghouse Engineer* (March 1953).

Wulfken, H. J. "Magnetic-Amplifier Regulators in Adjustable-Speed Drives," *Electrical Manufacturing* (May 1955).

Zollinger, H. A. "Reactor Control of Induction Motors," *Electrical Manufacturing* (January 1960).

PROBLEMS AND QUESTIONS

7-1.   a. List two methods for changing speed of a polyphase synchronous motor
      b. Repeat part a, giving five methods of changing speed of a polyphase induction motor.

197

7-2. Changing the frequency of the voltage applied to the stator is a method of speed control applicable both to synchronous and induction motors. Discuss a number of limitations of this "universal" method of speed control.

7-3. Changing the number of poles is a second method of speed control applicable, theoretically, to all motors. Discuss advantages as well as disadvantages of this "universal" method of speed control.

7-4. Based on your answers to Problems 7-2 and 7-3, discuss the possibilities of a *single* universal method of ac motor speed control.

7-5. Design a control circuit, similar to the one shown in Fig. 7-10b, for providing automatic two-speed forward-reverse switching of a consequent pole (four/eight pole) induction motor. Include both mechanical and electrical interlocks in the forward-reverse and low-high relay circuits.

7-6. Using as models the wound-rotor motor starter circuits shown in Figs. 5-5 through 5-8, design a three-speed (low-medium-high) automatic controller for the concatenated wound rotor and induction motor combination developed in Ex. 7-1. Show all power and control switching required to develop the three speeds.

7-7. Give one advantage and three disadvantages of concatenation.

7-8. a. Explain why the frequency of the frequency converter is always the same as the frequency of the wound-rotor induction motor in the Leblanc method of secondary foreign voltage speed control
b. What governs the speed of the induction motor and how is its direction reversed?

7-9. a. What are the differences between the Leblanc and Kramer methods and what disadvantages exist for the basic Kramer drive system (constant horsepower)?
b. Explain how the modified Kramer drives overcome these disadvantages
c. Give some of the major uses and applications of such systems as the Leblanc and Kramer drives.

7-10. Describe the Scherbius system, showing the differences between it and other systems, and discuss its relative advantages and disadvantages compared to the Kramer system.

7-11. a. Show why the BTA (Schrage) motor uses the Leblanc method of speed control, essentially
b. How are speeds above and below synchronous speed obtained with this motor?
c. How is power-factor adjustment obtained and what effect does this have on speed?
d. Up to what horsepower rating and in what voltage ratings is this motor commercially available?
e. Why is there an upper and lower limitation on the horsepower sizes available? Explain fully
f. What are the advantages and disadvantages of this motor over a wound rotor motor of similar horsepower?

7-12. Using the basic speed equation $S = (120f/P)(1 - s)$, make a table listing each of the factors in the equation, and under each factor list the various methods of ac speed control derived from a variation of that particular factor.

7-13. a. Based on the table constructed in Problem 7-12, classify the Rossman drive (Fig. 7-9)
     b. What are the advantages of the Rossman drive system over the Leblanc, Kramer and Scherbius systems?
     c. What are the disadvantages?

7-14. Design a controller capable of starting a polyphase induction motor and able to reverse and brake this motor from either direction to a quick stop by means of plugging. Provide both electrical and mechanical interlocks between forward and reverse relays and include a reverse current plugging relay to prevent accidental reversal.

7-15. Repeat Problem 7-14 using dynamic braking instead of plugging braking. Include a centrifugal friction switch to ensure rapid stopping at low speeds. Use dc relays throughout and include a braking adjustment rheostat.

7-16. Modify the design of Problem 7-14 to provide both jog forward and jog reverse controls in addition to those features already specified.

7-17. Define the following:
     a. Rectifier
     b. Converter (ac)
     c. Inverter
     d. Converter (dc)
     e. Cycloconverter.

7-18. Explain for the cycloconverter why the same triggering cannot be used for SCRs in each phase of Fig. 7-1a.

7-19. Describe three methods of controlling the three-phase, full-wave, bridge (multiple) inverter to vary voltage and frequency across a three-phase stator, including
     a. Method
     b. Schematic
     c. Frequency range
     d. Possibility of regeneration
     e. Disadvantages of each method to advantages of each method.

7-20. Compare ratifier-inverter vs. cycloconverter packages and indicate your preferences for either type as a means of controlling speed of polyphase synchronous and induction machines.

7-21. Compare electronic slip-power control of WRIMs with the conventional LeBlanc and BTA motor systems with respect to efficiency, speed regulation, and horsepower rating. (See Problem 7-11.)

7-22. a. Show how braking of a synchronous motor is accomplished (draw diagrams) using (1) plugging, (2) regenerative braking, (3) dynamic braking

b. Draw complete motor and control circuits for starting and braking a synchronous motor by each of the methods above.

7-23. Investigate the brushless polyphase synchronous motor and describe its construction and advantages over the conventional polyphase synchronous motor.

speed control
of single-phase motors

8-1.
DEFINITIONS

The following definitions apply to the various motors, poly-phase, single-phase, and dc, covered throughout this work. They are introduced here because of their frequent application to single-phase motors.

Single-phase motor. Any motor capable of starting and running from a single-phase ac supply, regardless of principle employed. Single-phase motors are grouped under three major headings: commutator type, induction type, and synchronous-type single-phase motors.*

Fractional-horsepower motor. Any motor built in an open frame developing less than 1 hp at a speed of 1700 to 1800 rpm. All fractional horsepower motors are termed "small motors" by the ASA.†

---

* For a complete list of the various individual types of single-phase motors classified under each of the three major headings, see I. L. Kosow, *Electric Machinery and Transformers* (Englewood Cliffs, N.J.: Prentice-Hall, Inc., 1972), Sec. 10-18.

† A 1.5-hp, 3600-rpm motor is considered a fractional-horsepower motor. At a speed of 1800 rpm this motor develops 0.75 hp; see Kosow, *op. cit.*, Sec. 10-1.

Integral-horsepower motor. Any motor built in an open frame developing more than 1 hp continuously at a speed of 1700 to 1800 rpm.

Split-phase motor. A single-phase motor of the induction type having two separate distributed stator windings connected in parallel to the single-phase supply. The rotor of a split-phase motor is exactly the same as the polyphase squirrel-cage induction motor (SCIM). This classification includes resistance-start motors, capacitor-start motors, capacitor motors, and shaded-pole motors.*

Constant-speed motor. One whose speed varies a relatively small amount from no load to full load. While no definite limit has been set, it is usually considered that a speed regulation of approximately 20 per cent or better (less) is acceptable. This class includes shunt motors, squirrel-cage induction motors, synchronous motors, and various single-phase motors of the induction and synchronous types.

Varying-speed motor. One whose speed varies considerably from no load to full load, i.e., those whose speed regulation exceeds and is poorer than 20 per cent. Series motors, some compound motors, repulsion motors, and repulsion-induction motors fall into this category.

Adjustable-speed motor. One whose speed can be adjusted gradually over a considerable range, i.e., higher or lower than rated, but whose speed for any adjustment (speed regulation) will vary only a relatively small amount from no load to full load. The dc shunt motor is an excellent example of this type of motor.

Adjustable varying-speed motor. One whose speed can be adjusted gradually over a considerable range, but whose speed for any adjustment will vary considerably from no load to full load, i.e., speed regulation poorer than 20 per cent. Series motors, some compound motors, repulsion-induction motors, and wound-rotor induction motors fall into this category.

Multispeed motor. One whose speed can be adjusted for two or more definite values, but whose speed cannot be adjusted gradually and whose speed for any definite adjustment will vary only a relatively small amount from no load to full load. The induction motor, both polyphase and single-phase, having *consequent poles* is an excellent example of this category.

Nonreversible motor. A motor whose direction of rotation cannot be reversed, either while running or almost at standstill. A reluctance-start induction motor is a nonreversible motor.

Reversible motors. A motor which may be reversed by changing certain *external* motor connections, even when the motor is running in one direction, without requiring that the motor stop. A capacitor-start motor is an example.

Reversing motor. A motor which may be reversed at any time under *any* load condition, even when running at rated load rated speed, by changing certain external motor connections.

* *Ibid.,* Chap. 10.

All dc motors are reversing motors, using armature reversal plugging. Of the ac single-phase induction types, only the capacitor motor is a reversing motor. (All polyphase induction motors are reversing motors, by plugging.)

There are numerous occupancies, industrial as well as residential, to which the electric utility has only brought a *single-phase* ac service. In all occupancies, furthermore, there is usually a need for small motors which will operate from a single-phase supply to drive various electric appliances such as sewing machines, drills, vacuum cleaners, air conditioners, etc. Generally, the term "small motor" means a motor of less than 1 hp, i.e., a *fractional-horsepower* motor,* and *most* single-phase motors are fractional-horsepower motors.

8-2.

REVERSAL OF ac
SINGLE-PHASE
MOTORS

In general, permanent-split induction-type motors are either externally reversible or electrically reversible motors by interchanging the line connections of one set of windings (either main or auxiliary starting winding) with respect to the other. Such motors as the single or dual-value capacitor permanent split-phase motor are electrically reversible and are *reversing* motors.† On the other hand, split-phase motors which employ an auxiliary winding for starting purposes only‡, such as the capacitor start motor, are *externally* reversible. Such a motor must be permitted to slow down (after it has been disconnected from the line) until its centrifugal mechanism closes, before the starting winding (which has been reversed with respect to its main running winding) may be reconnected to the single-phase supply. When reconnected at this lower speed in the manner described, plugging occurs and the motor direction is reversed in the normal manner. In effect, this is the same, electrically, as stopping the motor to reconnect its windings and permit reversal to occur when the motor is restarted.

Shaded-pole motors equipped with shading coils are reversed by a

---

* A small motor as defined by the American Standards Association (ASA) and the National Electrical Manufacturers Association (NEMA) is "a motor built in a frame smaller than that having a continuous rating of 1 hp, open type, at 1700 to 1800 rpm." Small motors are generally considered fractional-horsepower motors, but since the determination is based on frame size, the following comparisons are of interest:

1. A $\frac{3}{4}$-hp, 900-rpm motor *is not* considered a fractional-horsepower motor because its frame size, if used for an 1800-rpm motor, would yield a rating of more than 1 hp. Therefore, it is considered an *integral*-horsepower motor of 0.75 hp $\times \frac{1800}{900} = 1.5$ hp.
2. A 1.5-hp, 3600-rpm motor *is* a fractional-hp motor because its frame size, if used for an 1800-rpm motor, would yield a rating of less than 1 hp of 1.5 hp $\times \frac{1800}{3600} = 0.75$ hp.

† See definitions, Sec. 8-1 and Kosow, *Electric Machinery and Transformers*, Secs. 10-7 and 10-8.

‡ *Ibid.*, Secs. 10-5 and 10-6.

203

variety of techniques* which in effect reverse the position of the shading coil with respect to the pole.

Repulsion-type motors actually are capable of reversal by brush shifting, but this is not easily done except on the (rarely manufactured) brush-shifting repulsion motor which has been designed for such reversal.

Universal and ac series motors are reversed in much the same manner as dc motors. Figure 8-1 illustrates the three most common methods of reversing ac series or universal motors. Armature reversal, Fig. 8-1a, is

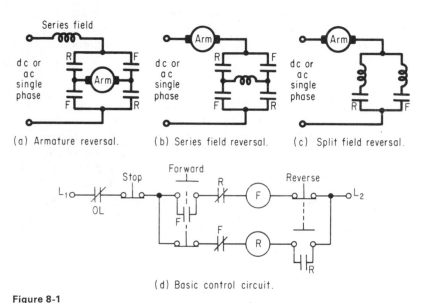

(a) Armature reversal.   (b) Series field reversal.   (c) Split field reversal.

(d) Basic control circuit.

Figure 8-1

Reversing universal or ac series motors.

generally preferable to series-field reversal, Fig. 8-1b, because the reversal is accompanied and accelerated by plugging and there is no danger of runaway due to an open field. Split-field reversal, Fig. 8-1c, is frequently employed, particularly with governors and external microswitches, in small ac servo drives for automatic control purposes. Universal and ac series motors may be reversed manually by using cam or drum operating switches, or may be reversed automatically by using the basic control circuit shown in Fig. 8-1d.

8-3.
SPEED CONTROL
OF ac SINGLE-
PHASE MOTORS

The principal method of speed control used for fractional-horsepower single-phase, induction-type, shaded-pole, reluctance, and even series and universal motors (armature voltage control) is the method of primary line voltage (slip) control discussed in Section 7-11. Briefly, it involves a reduction in the

* *Ibid.*, Sec. 10-9.

voltage applied to the stator winding (of induction-type motors) or to the armature of series and universal motors. In the former case, it produces a reduction of torque and an increase in slip. In the latter case, it is simply a means of controlling speed by armature voltage control or flux control.

The reader may ask why line voltage control lends itself to single-phase induction-type motors and *not* to polyphase SCIMs and WRIMs. The difference lies in the manner in which the rotor of a single-phase induction-type motor develops torque* in comparison to the polyphase asynchronous induction motor. Briefly, the latter, in accordance with doubly excited dynamo theory, tends to maintain excitation of the rotor as the stator voltage is decreased. The result is that the impressed stator voltage across a polyphase induction motor stator must be reduced *considerably* before appreciable changes in slip occur. This, in turn, reduces the torque even more considerably (since torque is proportional to the square of the impressed stator voltage) and greatly reduces the effective horsepower rating of the polyphase SCIM or WRIM.

Since the torque developed by a single-phase induction-type motor is that developed by two oppositely rotating fields, the rotor slip of a single-phase motor is more sensitive to a change in excitation than a polyphase motor.

Figure 8-2a shows the effect of stator field excitation on torque and slip of a split-phase induction motor, at three different values of voltage, applied to the stator. The curves show that for any given value of load, a reduction in stator excitation voltage produces a corresponding reduction in speed (an increase in slip). Since the torque of any induction motor varies as the square of the impressed stator voltage, the torque-slip curve at 50 per cent rated voltage is 25 per cent of that at rated voltage, roughly at all values from starting to breakdown to running. Note that as the load is increased from light load to rated load, the effect of a voltage reduction from rated voltage to 25 per cent of rated voltage is to produce a greater speed drop with voltage change, up to approximately rated load. At loads which are close to breakdown, changes in stator voltage do not produce as marked changes in slip and speed. Operation close to breakdown, however, is impractical because slight increases in load will stall the motor.

Figure 8-2a shows that any single-phase induction-type motor theoretically may be speed-controlled by varying the applied voltage to its stator. However, not *all* rotors are equally adaptable for such speed-control methods. Table 5-1 shows the various SC rotors available and it should be noted that Class C and D rotors are *high-resistance* types. Such rotors, as shown in Fig. 8-2b, tend to produce a *larger* speed variation for a given reduction in stator voltage, for a given load, than a low-resistance rotor, as shown in Fig. 8-2c.

For single-phase induction-type motors, including shaded-pole

---

* Kosow, *Electric Machinery and Transformers*, Secs. 10-3 and 10-4.

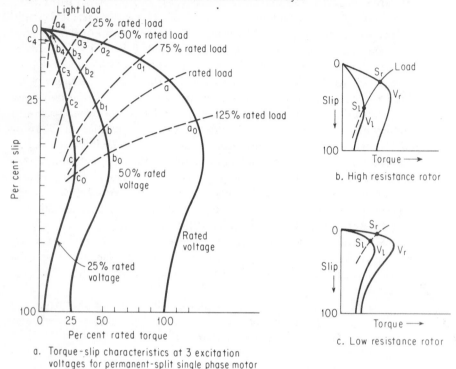

Adjustable speed characteristic of permanent-split
single phase capacitor motor at three different excitation voltages

a. Torque-slip characteristics at 3 excitation
voltages for permanent-split single phase motor

b. High resistance rotor

c. Low resistance rotor

Figure 8-2

Torque-slip curves for single-phase
induction-type motors with primary voltage
variation.

motors, therefore, some recognition of the nature of the load as well as
the class of rotor (Table 5-1) employed is required for proper speed control
using primary voltage variation.

    Some methods of obtaining the required speed control by primary
voltage reduction are shown in Fig. 8-3 for either manual or automatic
switching methods. The tapped winding method, shown in Fig. 8-3a,
indicates that the highest speed is obtained when the full supply voltage
is applied to the smallest section of the main winding. The lowest speed is
obtained when the supply voltage is applied to the entire winding. The
explanation for this relationship stems from the equation ($E_{\text{eff}} = k\phi Nf$),
which states that the excitation flux, ϕ, at a given frequency, depends on
E_{eff}/N, or the effective stator volts per turn. As the number of stator turns
to which the (same) voltage is applied decreases, the excitation flux
increases. In any induction-principle motor, the rotor current (I_r) and
the torque ($T = k\phi I_r \cos \theta$) are a function of the excitation flux. For a

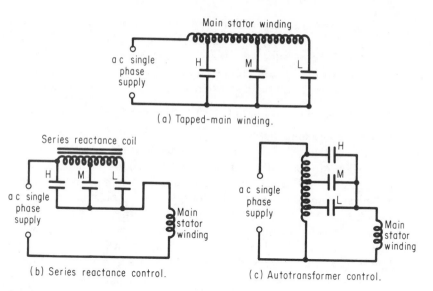

(a) Tapped-main winding.

(b) Series reactance control.

(c) Autotransformer control.

Figure 8-3

Various methods of obtaining primary voltage
speed (slip) control for single-phase motors
(shaded-pole, reluctance, and split-phase types)
from a fixed ac supply.

given load, therefore, more torque and less slip are produced with a higher
ratio of volts per turn. The speed, therefore, is *highest* when the *fewest*
turns are used for a given applied ac stator voltage.

When the n.o. H contact is opened, and either the M (medium speed)
or L (low speed) contacts are closed, as shown in Fig. 8-3a, the motor
speed drops proportionately with the decrease in the developed torque.
The tapped main winding is wound in such a manner that even the small
(high-speed) section is uniformly distributed around the stator, and the
winding is designed to carry the rated voltage without overheating.

The method of using a tapped external *series* reactance coil, shown
in Fig. 8-3b, provides a series voltage drop such that, for any given load,
the entire coil will produce the greatest voltage drop. The highest speed,
obtained by closing the H contacts, is obtained *without* the use of the coil,
and the lowest speed is obtained with the *full use* of the coil. The series
reactance method (Fig. 8-3b) has the advantage of being adaptable to any
single-phase, shaded-pole, induction-type, or universal motor, and no
special winding taps on the motor are required. It has the disadvantage,
however, of poor speed regulation. If the load is increased at any given
low- or medium-speed setting, the load current and voltage drop across
the coil is increased and the speed drops. The speed also decreases as a
result of increased slip at reduced voltage.

A tapped autotransformer, shown in Fig. 8-3c, partly overcomes the

207

speed-regulation disadvantages of the reactance coil method, providing better regulation at low or medium speeds.

It should be pointed out that the three principal speed-control methods shown in Fig. 8-3 are also adaptable to ac series or universal motors (operated on alternating current) as well. The fundamental speed equation* governing these motors is $S = k\dfrac{[V_a - I_a(R_a + R_s)]}{\phi}$. Both the armature voltage and the current are functions of the magnitude of the supply voltage. The drop across the series field, I_aR_s, is usually a small portion of the total armature circuit voltage drop. The method of employing a tapped series field winding is shown in Fig. 8-4a. Usually small universal and series motors of this type are two-pole motors, so that it is unnecessary to have the same number of turns on each field pole. The lowest speed is obtained when using the maximum series field mmf, since speed varies inversely with field flux. As shown in Fig. 8-4a, some voltage drop will occur across the series fields, as well, when the L contacts are closed, tending to reduce the speed. The major speed reduction, however, is produced by increased flux. Opening the L contacts, and closing either M or H, will increase the speed because of the reduction in flux,

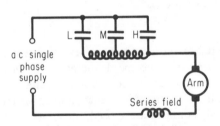

(a) Tapped series field winding.

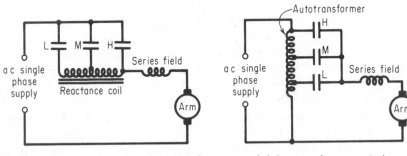

(b) Tapped series reactance coil (or resistor).   (c) Autotransformer control.

Figure 8-4

Various methods of obtaining speed control of ac series or universal motors from a fixed ac supply.

* Kosow, *Electric Machinery and Transformers*, Secs. 10-16 and 10-17.

primarily, and the increased voltage across the armature. The method of tapped field windings has the disadvantage of requiring a special motor in which such taps have been brought out.

A tapped series reactance coil may be used to produce a variable voltage in the armature and to serve as a method of speed control, as shown in Fig. 8-4b. This method has the advantage of being used with any small series or universal motor. In the case of the latter, tapped resistors are also sometimes used. In this method, the voltage across the armature, V_a, is the principal factor being controlled in the basic series motor speed equation. The highest voltage and speed are produced when the reactance coil is not in the circuit, and the lowest voltage and speed are produced with the full reactance coil. This method also has the disadvantage of poor speed regulation, but it is not as serious in the case of a series motor which already has a drooping speed-load characteristic and poor speed regulation, inherently.

The tapped autotransformer is also used, as shown in Fig. 8-4c, and has the advantage of improving the speed regulation somewhat over the reactance coil method.

The reader should compare the methods of Figs. 8-3 and 8-4 and note that the speed results produced by these methods is essentially the same. Reducing the field turns will produce an increase in speed in shaded-pole, reluctance, induction-type, and series-type motors, universally. Similarly, increasing the applied voltage across the motor line terminals will produce an increase in speed universally in *all* these motors, regardless of type. The only exception is the single-phase synchronous motor.

It goes without saying that the only method which may be employed to control the speed of single-phase synchronous or hysteresis motors is generally a change in frequency of the applied voltage (see Section 7-2) using single-phase frequency control.

**8-4.
SATURABLE
REACTOR AND
MAGNETIC
AMPLIFIER
CONTROL**

Because an increase in the ac applied voltage across the line terminals of series, universal, induction, shaded-pole, and reluctance-type single-phase motors results in an increase in speed, the ac output of saturable reactors and magnetic amplifiers may be effectively used as speed-control devices.

The core and windings of a saturable reactor are shown in Fig. 8-5a. The reactor windings, wound on the outside legs of the core, may be connected with respect to each other either in series or in parallel, with the saturating dc control winding separately excited. The series-connected reactor windings, shown in Fig. 8-5b, produce mmf's which are instantaneously in the same direction, and are connected in series with the ac motor. In effect, the saturable reactor is a variable-impedance series reactance coil whose impedance voltage drop controls the applied voltage across the motor terminals, T_1 and T_2. The dc control winding *independently* tends to saturate the core (and the instantaneous

209

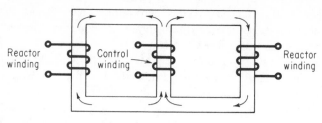

(a) Instantaneous flux relations between
control and reactor windings.

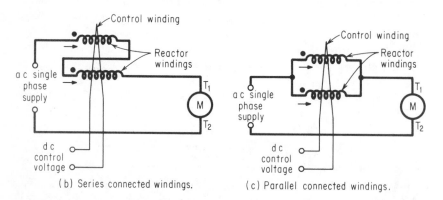

(b) Series connected windings.   (c) Parallel connected windings.

Figure 8-5

Use of desaturating saturable reactor for
voltage control of single-phase motors.

direction of the ac reactor winding mmf with respect to the unidirectional control flux is thus·of little consequence). When the dc control voltage is zero, the iron is unsaturated, and the reactor behaves as an ordinary choke coil having a high self-inductance and a large-impedance voltage drop. The ac voltage across the motor, therefore, is small with little or no dc control voltage. When the control voltage across the highly inductive control coil is increased, saturating the core, the inductive reactance and the impedance of the reactor decrease, producing a large ac output voltage across the motor. Increasing the dc excitation of the control winding correspondingly increases the ac output of the reactor. An extremely large value of ac output current and power may be controlled by a relatively small dc control power when the ratio of control turns to reactor turns is large. In this respect, the saturable reactor may be considered a power amplifier. The reactor coils, wound with a few turns of heavy wire, carry load current with little power loss, despite the large impedance drop produced across them.

Figure 8-5c shows the parallel combination which serves to handle a larger motor load current but produces a smaller impedance voltage drop than the series combination. Suitable switching circuits or potentiometer controls may be used to obtain various dc input voltages required to control the output ac voltage and the motor speed. The various gains of the saturable reactor are

$$\text{Current gain} = \frac{I_L}{I_c} = \frac{N_c}{N_L}$$

where I_L is the ac load (motor) current, N_L is the number of reactor turns, I_c is the dc control current, and N_c is the number of control turns,

$$\text{Voltage gain} = \frac{I_L R_L}{I_c R_c} = \frac{N_c R_L}{N_L R_c}$$

where R_L is the effective resistance of the load (the motor) and R_c is the resistance of the control winding

The power gain, which is the product of the voltage and the current gains in the above expressions, is therefore

$$\text{Power gain} = \frac{I_L^2 R_L}{I_c^2 R_c} = \left(\frac{N_c}{N_L}\right)^2 \frac{R_L}{R_c} \tag{8-1}$$

Equation (8-1) shows that the power gain of the saturable reactor does indeed depend primarily on the (square of the) ratio of the control turns to the reactor winding turns.

A disadvantage of the saturable reactor, however, is that, during the half-cycle when the reactor coil mmf opposes the control coil mmf, the iron is substantially desaturated and the power gain is effectively reduced. This disadvantage of the desaturating saturable reactor is overcome by the addition of a rectifier circuit. When a rectifier is used in conjunction with a saturable reactor to block the desaturating effect of the reactor coils, the device is called a self-saturating magnetic amplifier.

Figure 8-6a shows a full-wave rectifier connected in series with the ac motor load to obtain full-wave direct current to excite a feedback winding, in addition to the dc control winding. If the feedback winding produces mmf which aids the control winding mmf, the feedback is positive; and, conversely, when the feedback mmf opposes the control winding mmf, the feedback is negative. The effects of feedback on magnetic amplifiers are quite similar to corresponding effects in electronic amplifiers.

Another form of feedback, called self-saturation or internal feedback, is shown in Fig. 8-6b. This circuit differs from the external feedback circuit in that the reactor coils in the former carry alternating current (conduct for a full cycle), whereas, in the latter, the reactor coils carry unidirectional current and conduct for a half-cycle. The advantage of the internal feedback arrangement is that by causing the reactor windings to conduct on alternate respective half-cycles, as shown in Fig. 8-6c, the control winding

211

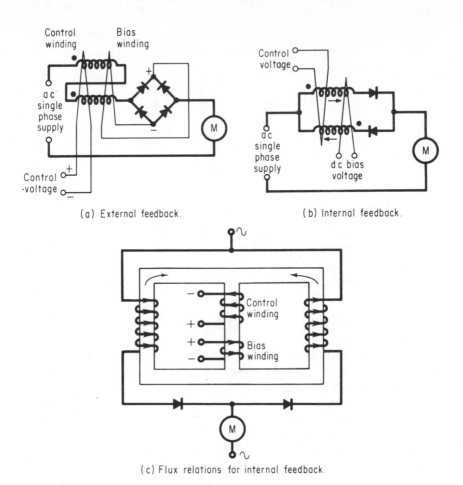

(a) External feedback.

(b) Internal feedback.

(c) Flux relations for internal feedback.

Figure 8-6

Use of self-saturating magnetic amplifier for
voltage control of single-phase motors.

flux is always aiding the reactor core flux, reducing the control current
required for saturation and increasing the power gain [see Eq. (8-1)].

Negative feedback, Fig. 8-7a, reduces the response time and improves
the linearity of the control characteristics as well as the stability of the
amplifier. Positive feedback, Fig. 8-7a, has the opposite effect of lengthen-
ing the response time and increasing the power gain, and it may produce
a useful kind of instability called bistable operation.*

* The amplifier may be made to operate in the same manner as a relay, a flip-flop
circuit, or a switch by proper design of its positive feedback circuit. Thus, it will be
stable only at the upper and lower extremes of its output, representing either an *on*
or *off* condition (i.e., no output or full output) without any intermediate output stability;
see Fig. 8-7a.

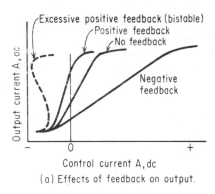 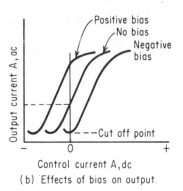

(a) Effects of feedback on output.   (b) Effects of bias on output.

Figure 8-7

Feedback and bias characteristics of
magnetic amplifiers.

The effects of feedback on the output characteristics of a magnetic amplifier are shown in Fig. 8-7a. Without feedback, the control circuit can increase the ac output voltage by increasing the dc control current and mmf in a positive direction; and reversing its polarity will decrease the output, as shown. Negative feedback requires a greater amount of control current to produce saturation, and, for the same control current, the output is reduced proportionately. Positive feedback will improve the amplifier gain so that maximum output is obtained with relatively little control current. The unstable positive feedback bistable condition is shown as a dashed line. Figure 8-7a also shows that the output current and voltage are appreciable (when no feedback or positive feedback is used), even when the control mmf is zero. Since positive feedback is advantageous for increased power gain, it is necessary to reduce the output at zero control mmf. This is accomplished by means of an additional bias winding. The effect of the bias winding is shown in Fig. 8-7b. Note that a negative bias will produce minimum output (cutoff) by shifting the characteristic to the right. Thus, the combination of positive feedback and negative bias will produce (1) maximum gain, and (2) maximum output voltage swing. Its disadvantages, of course, are reduced linearity and the possibility of instability. The core shown in Fig. 8-6c shows the flux relations for negative bias and positive feedback control of speed of a single-phase ac motor.

Although the application stressed above for the magnetic amplifier is its use in ac voltage control for single-phase ac motors, it is obviously possible to convert the ac output to direct current by means of suitable rectifiers in the output circuit, and to use that direct current as a means of "electronic" dc armature voltage (Section 6-12) and field control as well. The magnetic amplifier is also used in the closed-loop dc starter discussed in Section 4-10.

213

As either a dc or ac amplifier, saturable reactors and magnetic amplifiers have been used for lighting dimming, as servoamplifiers, in voltage and speed regulators, and in welding and power supply controls. Applications of magnetic amplifiers in dc and ac feedback control systems will be discussed in Chapter 10.

8-5.
ELECTRONIC
CONTROL OF
SINGLE-PHASE
MOTORS

The speed of a single-phase induction-type motor may be controlled by varying the amplitude of the supply voltage (Fig. 8-2) or the frequency f of the supply (since speed of any ac induction motor is $S = 120f/P$) or *both* (as in the case of the polyphase ac induction motor, Section 7-2). The speed-control methods previously described (Section 8-2) *all* use variable-voltage, *fixed-frequency* techniques. There are a number of electronic techniques employing either transistors or SCRs (or both) which may be used to vary the output ac amplitude to the stator of an ac motor. Some of these are described below, followed by a presentation of methods using variable-voltage, variable-frequency techniques.

8-5.1
Variable-Voltage,
Fixed-Frequency
Control

A number of methods using electronic control of ac output voltage are shown in Fig. 8-8. These methods may be used equally for speed control with universal motors, series ac motors, repulsion motors, shaded-pole motors, or induction-type motors of the permanent split-field type.* Like the methods presented in Section 8-2, the techniques shown are not applicable to synchronous or hysteresis motors whose speeds are a function of frequency.

Figure 8-8a shows an SCR in series with the either a single-phase stator or the ac motor depending on type. The output waveform across the motor, when the SCR is either phase-controlled or triggered, shows conduction only during the positive half-wave portions of the input waveform. Such a waveform contains both a dc component and a high ac component at the fundamental frequency (in addition to higher-order harmonics). The amplitude of the both the ac and dc components is a function of the conduction angle of the SCR. This method of control lends itself to universal, ac series and repulsion motors, primarily, since the dc component produces no useful torque in induction-type or shaded-pole motors.

The use of 2 SCRs, as in Fig. 8-8b, produces a waveform containing no dc component (average value zero). This method of full-wave control may be used for *all* the ac single-phase motors cited above (except the hysteresis and synchronous).

A rather unique circuit is obtained in Fig. 8-8c which uses the circuit

---

* In the case of single-phase resistance-start or capacitor-start induction-type motors, speed control *is* possible *above* the speed at which the centrifugal switch closes. Some motors of these types may be obtained having switching speeds well below the range over which speed control is desired.

of Fig. 8-8a in which the SCR is shunted by a diode-switch combination. With switch S closed, the rms value of the ac output waveform may be varied from $0.707E_m$ to $(0.707/2)E_m$ by shifting the conduction angle from 0 to 180°. With switch S open, the circuit reverts to Fig. 8-8a, in which the **rms** value of the ac output waveform may be varied from $(0.707/2)E_m$ to zero. Thus, the circuit of Fig. 8-8c, using only 1 load-carrying SCR and 1 diode, performs the same control of ac output voltage obtainable with Fig. 8-8b requiring 2 SCRs. It does have the disadvantage, however, of containing a dc component, noted in discussing Fig. 8-8a.

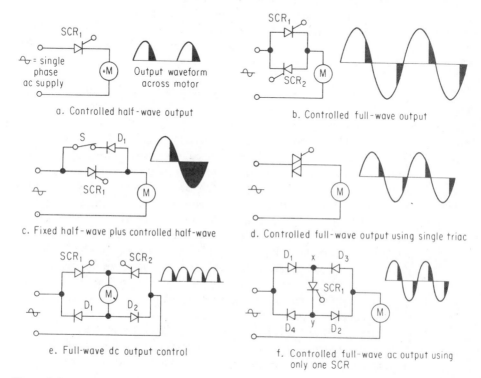

a. Controlled half-wave output

b. Controlled full-wave output

c. Fixed half-wave plus controlled half-wave

d. Controlled full-wave output using single triac

e. Full-wave dc output control

f. Controlled full-wave ac output using only one SCR

Figure 8-8

Basic forms of ac single-phase controlled output voltage, fixed frequency.

A single semiconductor device, called a *triac*, performs the same function as the 2 (back-to-back) SCRs in Fig. 8-8b. The triac is a single-gate bidirectional triode thyristor, capable of conduction in both the first and third quadrants. In operation and function, the triac, shown in Fig. 8-8d, is similar to two SCRs connected in an inverse parallel (back-to-back) arrangement shown in Fig. 8-8b, with the only exception that one gate is required. (The triac conducts with either a positive or negative gate signal in the first and third quadrants, respectively.)

215

The output waveform of Fig. 8-8d is identical to Fig. 8-8b. Since it contains no dc component whatever, the circuit employing the triac is, perhaps, the simplest and most convenient method for control of output ac voltage for motors, light dimming, electric heaters, or a variety of applications requiring full voltage control of an ac supply.

A somewhat more sophisticated configuration employing 2 diodes and 2 SCRs is shown in Fig. 8-8e with its output waveform, primarily for reference. Such an output waveform producing a full-wave rectified output, is suitable for universal motors and ac series motors, only.

The circuit shown in Fig. 8-8f produces the same output waveform as Figs. 8-8b and d, using only a single SCR and four load-carrying diodes. This circuit has the double advantage of producing the waveform shown and also the waveform of Fig. 8-8e when the motor is placed in series with the SCR between terminals x and y.

8-5.2
Variable-Voltage
Variable-Frequency
Control

As in the case of polyphase SCIMs, WRIMs, and synchronous motors, the speed of a single-phase induction or synchronous (including hysteresis) motor may be controlled using either a cycloconverter or rectifier-inverter (Section 7-2). Such devices simultaneously vary both frequency *and* voltage (in the same proportion) as excitation for the stators of induction-type or synchronous motors,* in controlling motor speed. When operated from an ac supply, the inverter presents a higher power factor to the line; it can supply higher as well as lower frequencies and does not require as many SCRs nor as complex control circuitry as the cycloconverter. For these reasons, most *single-phase* variable-voltage, variable-frequency ac solid-state single-phase motor packages are rectifier-inverters rather than cycloconverters.

Four different types of single-phase inverters are shown in Fig. 8-9, each supplying ac to a single-phase motor. At first glance it would appear that the chopper inverter, Fig. 8-9a. is most preferable since it is the simplest and uses only one SCR. The chopper, unfortunately, produces a high dc component in the output ac waveform. The remaining three configurations produce output waveforms having an average value of zero (no dc present) but each has its disadvantages.

The center-tapped load inverter shown in Fig. 8-9b requires a transformer (which increases size and weight of the solid-state package). In some designs, a center-tapped choke may be used (reducing the weight) and the motor is supplied using the choke as an autotransformer. In addition to the transformer, the blocking voltage of each SCR in Fig. 8-9b must be twice the dc supply voltage.

The inverter shown in Fig. 8-9c requires no transformer and the blocking voltage of each SCR is only half the total supply voltage. But

* See Section 7-2 for a complete discussion as to reasons why both voltage and frequency must be varied simultaneously and in the same proportion.

216

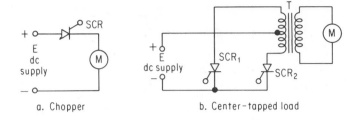

a. Chopper   b. Center-tapped load

(M) = Synchronous, hysteresis or induction motor, all single phase

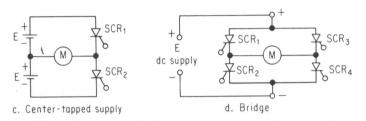

c. Center-tapped supply   d. Bridge

Figure 8-9

Single-phase inverters supplying
variable voltage and frequency.

this inverter requires a center-tapped dc source. The bridge inverter
(Fig. 8-9d) requires neither a center-tapped dc supply nor a transformer,
but it employs 4 SCRs.

A typical rectifier-inverter package for single-phase speed control of
ac single-phase synchronous (or hysteresis) motors and induction motors
is shown in Fig. 8-10. The package uses no transformers and is capable
of producing frequencies well above the line frequency (60 or 400 Hz)
and down to zero, with output voltages in proportion to frequency.
The circuit is also regenerative and may be used for dynamic braking of
the motor.

It should be noted that motor nameplates usually specify the frequency of operation. At a higher frequency, the iron friction and windage
losses (due to higher speed) are all increased despite some increased internal motor ventilation. Since voltage is also increased simultaneously with
frequency with this type of control, the motor torque is correspondingly
increased (since torque varies as the square of the stator impressed voltage). Since rated horsepower of any motor is the torque-speed product
(hp $= kTS$), care must be taken that at higher frequencies the motor
rating is not exceeded. Single phase synchronous and/or induction motors
packaged with rectifier-inverter drives should be rated for the full range
of intended speed variation.

217

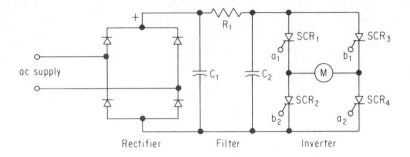

Rectifier   Filter   Inverter

Figure 8-10

Solid-state single-phase rectifier-inverter package for speed control of ac synchronous and induction single-phase motors (SCR-phase control circuits not shown).

Finally, it should also be noted that not all induction-type motors respond equally to wide ranges of frequency variation as a method of speed control. The polyphase two-phase motor gives the best performance with frequency variation similar to the three-phase motor with a cycloinverter or rotary amplifier drive. Capacitor-start motors or split-phase start motors are least suitable. These present centrifugal switch problems at lower speeds (as noted earlier).

The permanent-split capacitor motor presents some problem, as well, because the fixed capacitor (in microfarads) for ideal operation should be decreased with increased frequency (and vice versa). A design using a two-phase, three-wire, solid-state supply driving a small three-wire balanced winding, permanent-split capacitor motor over a range of 30 to 180 Hz is commercially available. But the frequency control of such a power supply is not sufficiently stable for use with a synchronous permanent split capacitor motor.

Since shaded-pole motors are inherently two-phase motors (i.e., one phase is supplied from a variable-voltage, variable-frequency supply while the other phase may be supplied from a servoamplifier of variable voltage and frequency), such motors operate reasonably well over a wide frequency range if their rating is not exceeded.

Small single-phase reluctance, hysteresis, and subsynchronous motors* also operate extremely well with variable-voltage, variable-frequency drives, because of the inherent simplicity of their design.

Variable-frequency techniques are not required for series, universal, or repulsion-type motors because these are easily speed-controlled by control of armature voltage (see Section 8-8).

---

* Kosow, *Electric Machinery and Transformers*, Secs. 8-27, 8-28, and 8-29.

8-6.
PLUGGING OF
SINGLE-PHASE
MOTORS

Plugging of single-phase motors is essentially a process of reversing the motor supply connections in such a way that motor speed drops to zero and the motor is disconnected from the supply *before* it has a chance to reverse. Thus, any single-phase motor which is *nonreversible* cannot be stopped by plugging. This is also true of *externally reversible motors.**

Section 7-15 considers plugging for both polyphase and single-phase SCIMs and synchronous motors, so no further discussion of these is required here.

Universal and ac series motors are plugged in much the same way as dc motors and these techniques were previously described (Section 6-7) and apply equally well to single-phase ac motors.

8-7.
DYNAMIC BRAKING
OF SINGLE-PHASE
MOTORS

As in the case of dynamic braking of polyphase motors (Section 7-16), the principle of dynamic braking of single-phase motors is basically the same and correspondingly that for braking dc shunt or series motors (i.e., maintaining a dc field and shorting the rotor armature), as described in Section 6-8. Just as for polyphase motors, the braking is achieved by removing ac power from the stator and replacing it with dc. The motor then becomes a dc generator whose rotor (armature) is (normally) short-circuited. This results in heavy armature (rotor) currents which set up relatively strong rotor fields (Lenz's law) in reaction against the direct stator field. The kinetic energy of the rotor (and its connected load), therefore, is consumed in generating rotor voltage and current, bringing the rotor rapidly to rest.

Since most occupancies using single-phase motors do not have dc readily available, the dc is obtained from a rectifier package, as shown in Fig. 8-11.

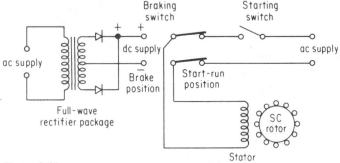

Figure 8-11

Dynamic braking of shaded pole, single-phase motor, showing rectifier package.

* See definitions, Secs. 8-1 and 8-2 on reversal of ac single-phase motors.

219

The single-phase motor shown in Fig. 8-11 to illustrate dynamic braking is a shaded-pole motor, chosen because it has (basically) only one stator winding.* The motor is started and run on ac with the DPDT braking switch in the RUN position shown in Fig. 8-11. (Note that the switch has no OFF position but is connected either to the BRAKE or RUN position.) When the DPDT switch is thrown to the BRAKE position, the motor is brought quickly to a stop when dc is applied to its stator. Since the rotor of the shaded-pole motor is identical to that of polyphase and single-phase induction-type motors, its rotor bars are normally short-circuited, and this method of braking is extremely rapid and effective.

Single-phase resistance-start and capacitor-start (split-phase) motors† are also rapidly stopped in a similar way, as shown in Fig. 8-12. Note that the dc is applied to the main or running winding to provide a strong dc field, when the DPDT switch is thrown to the BRAKE position. It might appear from Fig. 8-12 that the auxiliary or starting winding also receives dc, via the n.c. centrifugal switch CS, but there are two reasons why this does not occur. Recall that when a split-phase motor is running at rated speed the n.c. centrifugal switch, CS, is open. Furthermore, even as the speed drops due to braking action and switch CS closes, the starting capacitor C in series with the intermittently rated starting winding effectively blocks dc. Thus, over-heating of the starting winding is prevented whenever dc is applied to the main winding during the braking period.

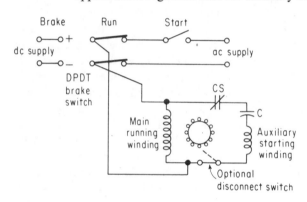

Figure 8-12

Dynamic braking of split-phase resistance-start or capacitor-start motors.

In the case of (permanent split) capacitor motors‡ both windings receive dc during the braking period, and the switching is somewhat more complex. The two identical windings may be energized in series, as shown in Fig. 8-13a or in parallel, as in Fig. 8-13b. In either case, a TPDT switch (having no OFF position) is employed to prevent capacitor C from blocking the dc. A higher dc supply voltage is required for the windings in series (Fig. 8-13a) than when connected in parallel (Fig. 8-13b),

* See Kosow, *Electric Machinery and Transformers*, Sec. 10-9, for a description of various types of shaded-pole motor constructions and methods of reversing direction.

† *Ibid.*, Secs. 10-5 and 10-6, for resistance-start and capacitor-start split-phase motors.

‡ *Ibid.*, Sec. 10-7, for single-value permanent-split capacitor motors.

220

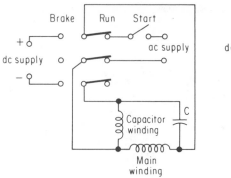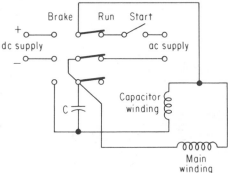

a. Series-connected windings during braking   b. Parallel-connected windings during braking

Figure 8-13

Dynamic braking of permanent split capacitor
motors using both windings either in series or
parallel connected to a dc supply.

although in either case the dc coil current should not exceed rated rms current of either winding. This method of braking using dc on the stator windings applies to both integral- and fractional-horsepower capacitor motors.

A method of braking extremely small (less than 1/500 hp) fractional-horsepower capacitor motors using no dc whatever is shown in Fig. 8-14b, but the braking efficiency decreases when attempting to use this method with larger frame sizes. This method is only used with extremely small capacitor (permanent-split) motors having identical windings (also reversible hysteresis synchronous motors) and characteristics of high-resistance, high-slip rotors. Since the two windings are identical, this motor is easily reversed by transferring the capacitor to either winding as shown in Fig. 8-14.*

The braking is accomplished by the relatively simple addition of a SPST switch shown in Fig. 8-14b. The SPST braking switch, when closed, serves two purposes, simultaneously:

a. It shorts the motor capacitor, C, thereby reducing the two-phase motor torque severely.
b. It connects both identical stator windings in parallel across the ac supply causing the motor to single phase, resulting in elliptical torque.†

The maximum torque developed by the extremely small capacitor motor (under conditions of a shorted capacitor) is less than the applied load torque and the motor is rapidly brought to a standstill.

Ibid., Sec. 10-7.
† *Ibid.*, Sec. 10-4.

221

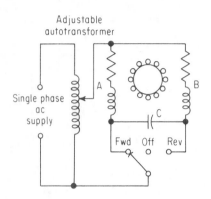

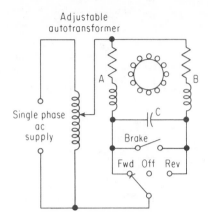

a. Connection diagram for reversing
capacitor motor with provision for
speed control by stator voltage
variation

b. Addition of switch for dynamic braking
in either direction of rotation

Figure 8-14

Dynamic braking of extremely small capacitor
motors with provision for reversing and speed
control.

It should be noted that Figs. 8-14a and b show provision for speed
control of the motors using stator voltage variation by means of an
adjustable autotransformer (or variac). Since this is a relatively expensive
device, in association with a small motor, a series resistor could provide
for control of stator voltage by stator supply voltage variation.* Furthermore, since braking involves shunting the capacitor, C, by means of a
switch (Fig. 8-14b), this shunting could also be provided by a low resistance. A relatively inexpensive method of speed control, reversing, and
dynamic braking is provided by the application of a one-turn continuous
potentiometer rheostat shown in Fig. 8-15. With the rheostat arm in the
vertical (BRAKE) position shown, a high-resistance R_3 shunts low-resistance
R_1. A corresponding equally equally high resistance R_4 shunts low-resistance R_2. Capacitor C, therefore, is shunted by the sum of low resistances
R_1 and R_2 and the motor is incapable of starting when the ON switch is
closed for reasons decribed above.

Rotating the potentiometer arm in either direction from the BRAKE
position shown in Fig. 8-15 accomplishes three purposes:

a. It increases the resistance shunting capacitor C (acting as an open
switch across the capacitor) permitting the capacitor to charge.

b. By means of series resistance, it selectively switches capacitor C in series
with either winding A or winding B.

* *Ibid.*, Sec. 10-7, Fig. 10-5b. See also Fig. 8-2a in this volume.

c. By means of series resistance it also reduces the applied voltage across the stators of both windings providing speed control by voltage variation.

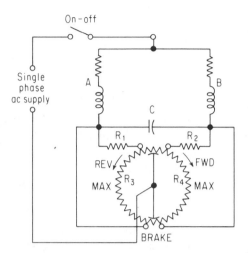

Figure 8-15

Dynamic braking, reversing and speed control of small capacitor motors.

The motor of Fig. 8-15 is capable of attaining rated speed in either direction when the potentiometer arm is rotated in either a clockwise or counterclockwise direction and left in a horizontal (maximum speed) position. Braking is accomplished by moving the potentiometer arm toward the vertical position. Note, however, that while no dc whatever is employed to achieve dynamic braking, this method applies only to *very* small single-phase permanent-split capacitor motors or hysteresis synchronous induction motors. This method of control finds its greatest application in small instrument servomotors.

A method of applying dc to the stator windings from a charged *braking capacitor* appears to be gaining in popularity. It has the advantages of reducing the number of rectifiers required to produce dc and involves less complicated switching compared to Figs. 8-11 through 8-13 previously described. The braking capacitor, C_B, may be an electrolytic (polarized) capacitor in larger integral-horsepower motors (or a "dry" type in low fractional-horsepower sizes) and due regard for capacitor polarity must be observed, as noted in Fig. 8-16. This means that the anode of the rectifier and

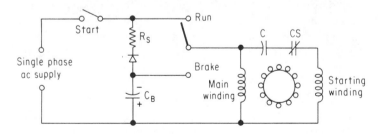

Figure 8-16

Dynamic braking of resistance-start, capacitor-start and permanent-split capacitor motors using capacitor discharge method of braking.

223

the negative terminal of the electrolytic capacitor are connected together and brought out as the BRAKE terminal, as shown in Fig. 8-16.

The method of braking shown in Fig. 8-16 using capacitor discharge to furnish dc to the main winding may be used with resistance-start split-phase motors or capacitor-start split-phase motors as well as (permanent-split) capacitor motors. A capacitor-start motor is shown in Fig. 8-16, which, when operating at rated speed, has its starting winding disconnected. Consequently, when the SPDT braking switch is thrown to the BRAKE position, only the MAIN winding receives dc, for reasons described in connection with Fig. 8-12.

Using capacitor discharge braking with Fig. 8-16, a resistance-start split-phase motor would also have a centrifugal switch, CS, but would have no starting capacitor C. Conversely, a permanent-split capacitor motor would have a starting capacitor C but no centrifugal switch, CS, using Fig. 8-16. Proper selection of the value of the braking capacitor, C_B, produces sufficient dc to ensure rapid braking without overheating the main or running winding.

In summary, the above methods of dynamic braking have two distinct advantages over mechanical braking methods: (1) There is no brake wear requiring adjustment of brake linings or drums, and (2) both the initial and maintenance costs for brake assemblies are considerably lower. On the other hand, the power for the braking action is consumed within the motor itself producing overheating of the windings (reducing winding life) and the switching is somewhat more complex in certain designs, as noted above.

8-8.
SERIES ac AND
UNIVERSAL MOTOR
SPEED CONTROL
USING SCRs

Small fractional and subfractional series ac and universal motors are extensively used in blenders, electric drills, portable tools, and small appliances requiring high-speed operation. Solid-state speed controls for such devices are less complex than for dc and polyphase motors of higher horsepower ratings. Essentially SCR-controlled* these solid-state packages usually provide manual adjustment for regulating the speed to the specific torque requirements. The drive circuits may be classified into either half-wave or full-wave and further categorized as uncompensated (nonfeedback) or feedback circuits.

8-8.1
Uncompensated
(nonfeedback) circuits

The simplest half-wave uncompensated circuit for control of a series ac or universal motor is shown in Fig. 8-17a. This circuit is the most economical in that it uses only one SCR and one diode. Since the circuit is a half-wave circuit, current flows only during the half-cycle when the SCR is positively (forward) biased and the SCR gating voltage (across R_1) is positive. Torque is produced, there-

* The SCR was introduced in Sec. 6-12.

fore, only once per cycle and only during that part of the half cycle during which SCR conduction occurs. Nevertheless, since the speed regulation of an ac series motor (or universal motor) is inherently poor, adjustment of the variable-voltage regulating resistor, R_v, provides a wide range of stable operation for a given load. The circuit shown in Fig. 8-17a operates as follows:

1. Average motor voltage across the armature is controlled by adjusting the firing point of the SCR, via the simple R-C circuit consisting of R_v in series with capacitor C.
2. The time constant of R_vC is adjusted by variable resistor, R_v, thus controlling the time required for C to charge positively to the gate turn-on voltage of the SCR.

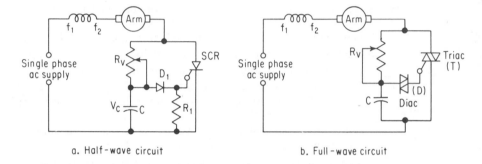

a. Half-wave circuit   b. Full-wave circuit

Figure 8-17

Half- and full-wave universal or series ac motor
control circuits without feedback.

3. During the time the SCR fires, the voltage across the capacitor, V_v, is less than the forward voltage drop across the SCR for the remainder of the positive half-cycle.
4. During the negative half-cycle of input, the diode D_1 blocks the gating pulse and the SCR goes out of conduction because it is reverse-biased.
5. When the positive half cycle begins again, capacitor C charges to provide the required gating pulse at the desired time set by the RC time constant.

A somewhat improved and relatively inexpensive circuit provides torque pulses twice per cycle using the full-wave circuit shown in Fig. 8-17b. The circuit employs a bidirectional SCR known as a *triac* (Section 8-5.1) and a bidirectional or three-layer diode known as a *diac*. Using the same motor, the full-wave circuit of Fig. 8-17b provides almost twice the torque of the half-wave circuit of Fig. 8-17a. For full conduction of both half-cycles, the motor operates as if it was connected directly across the ac line, at maximum speed. Increasing the value of R_v increases the time constant and retards the firing angle on each half-cycle, thus reducing

225

the speed. The operation of the circuit on each half-cycle is the same as that described for Fig. 8-17a, except that the voltage across C must exceed the breakdown voltage of the diac, D, before the triac, T, can be turned on. When positively biased, the triac may be triggered by a positive gate; when negatively biased, the triac is triggered by a negative gate. The triac turns off every time the input supply voltage drops to zero. A slight difference may be observed on a CRO between the positive and negative pulses across the motor armature because the triac requires more power for negative drive than positive drive.

It should be noted that neither of the circuits of Fig. 8-17 provides feedback. The torque supplied to the motor, the motor armature voltage, and the current are all manually controlled by setting R_v appropriately for a desired speed at any given load. Should the load increase, the speed decreases rapidly, as in the case of any series or universal motor.

8-8.2
Feedback circuits

The simplest possible compensated or *feedback* circuit is the Momberg half-wave circuit shown in Fig. 8-18a. This circuit is both ingenious and unique in that the voltage across the armature, V_a, is also the voltage at the gate of the SCR, controlling the firing of the SCR. The circuit shown in Fig. 8-18a operates as follows:

1. The desired (reference) voltage is set by voltage-regulating potentiometer, R_v, to provide reference or desired voltage, V_d.
2. The counter emf of the motor, V_a, taken across the motor armature, provides feedback only during the time the SCR is not conducting. This feedback is essentially inversely proportional to the applied load. (By simple dc motor theory, when the load *increases*, speed drops and counter emf *decreases*.)
3. Diode D_1 only conducts when reference voltage V_d is more positive than armature feedback voltage, V_a.
4. Assume that a motor running at a desired preset speed experiences an

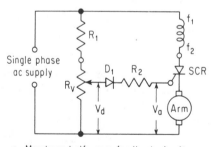

a. Momberg half-wave feedback circuit

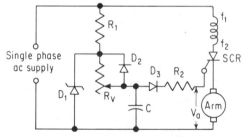

b. Gutzwiller half-wave feedback circuit

Figure 8-18

Single-phase half-wave universal or series ac motor control circuits using armature voltage feedback.

increase in load. The speed drops, decreasing V_a, due to a decreased counter emf.

5. This causes diode D_1 to conduct sooner on each positive half-cycle gating the SCR sooner each half-cycle.

6. The average motor voltage rises providing more motor torque, *tending* to return the motor speed to its original preset value.

The Momberg circuit has two serious disadvantages. First, the feedback circuit does not restore the speed exactly to its preset value whenever the load increases and, second, as the preset desired speeds are reduced to progressively lower speed values, the torque-speed curves tend to flatten out, substantially limiting the range of speed control.

The Gutzwiller half-wave feedback circuit represents an improvement over the original Momberg circuit through its addition of a reference zener diode, D_1, a blocking diode D_2, and a capacitor C.

The regulation, speed range, and stability of the Gutzwiller circuit is a decided improvement over the three previous circuits shown, but it is accomplished at the higher cost of increased numbers of solid-state devices. This circuit has formed the basis of a number of prepackaged universal motor "plug-in speed controls" (see Fig. 8-21).* Operation of this circuit (Fig. 8-18b) is essentially the same as the Momberg feedback circuit described above in which the setting of variable resistor R_v charges capacitor C, whose voltage is then compared to the counter emf across the armature, V_a. The greater the voltage difference between V_c and V_a, the earlier the firing angle of the SCR and the higher the torque and speed of the motor.

Greater torque for the same load and higher no load speeds are obtainable with the full-wave circuit shown in Fig. 8-19. This circuit employs a full-wave bridge rectifier (diodes D_1 through D_4) to provide positive pulses twice per cycle to forward bias the SCR, virtually doubling the torque of the motor. The SCR conduction angle is controlled by setting potentiometer R_2 providing a diode bias voltage, V_d, as a reference for comparison with armature counter emf, V_a. Note that Fig. 8-19 uses a diode D_5 across the series field, f_1–f_2, to provide a current path for the self-induced emf of the series field. In the absence of this diode, the energy stored in the magnetic field of the series winding would provide a positive pulse at the anode of the SCR preventing the latter from turning off, even when the rectified input pulse is at zero. In addition, a capacitor C_1 serves to maintain a relatively constant average dc voltage across potentiometer R_2, to assure a constant voltage reference value of V_d, for comparison with V_a.

It should be noted that universal and series motors exhibit extremely high no-load speeds, and correspondingly high counter emfs across the

* A. A. Adem, "Speed Controls for Universal Motors" (Auburn, N.Y.: General Electric Co., Application Note 200.47, June 1966).

SEC. 8-8 / *Series ac and Universal Motor Speed Control Using SCRs*

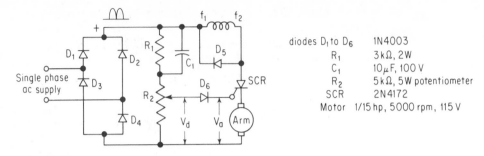

diodes D_1 to D_6	1N4003
R_1	$3 k\Omega$, 2W
C_1	$10 \mu F$, 100 V
R_2	$5 k\Omega$, 5W potentiometer
SCR	2N4172
Motor	1/15 hp, 5000 rpm, 115 V

Figure 8-19

Full-wave, feedback compensated control, for dc,
universal, or ac series-wound motors.

armature. Thus, if the load is removed from the motor with potentiometer R_2 at a low speed setting, the motor races away and V_a may always be greater than V_d in Fig. 8-19. Under these circumstances, the diode D_6 is *always* reverse-biased and the SCR cannot conduct on either half-cycle. The motor automatically slows down until V_a is less than V_d, gating the SCR once again, causing an abnormal rise in speed. This may result in a no-load instability in which speed rises and falls periodically, until the load is restored. Commercial appliances using the circuit (Fig. 8-19) usually connect the load through a gear drive which ensures some loading on the motor at all times by virtue of the gear drag, thus eliminating this instability.

The component values shown on Fig. 8-19 are for a $\frac{1}{15}$-hp, 5000-rpm, 115-V universal or series dc or ac motor. The values may be used safely up to $\frac{1}{6}$ hp without modification. For $\frac{1}{4}$ hp and above, higher-current-rating SCRs, diodes, and higher-power-rating resistors are required.

Commercial appliances, such as blenders using the circuit of Fig. 8-19, employ pushbutton switches which short out taps on resistor R_2, to select the various desired speeds. Since the blender is never operated without load, the circuit of Fig. 8-19 provides a wide selection of blender speeds, roughly from 800 to 6400 rpm.

All the circuits shown (Figs. 8-17 through 8-19) provide rotation in one direction only. If reversal as well as speed control is desired, the circuit of Fig. 8-20 may be employed. This circuit provides for reversal of current through the series field, f_1-f_2, while maintaining current in the armature always in the same direction. The theory of operation is based on triggering the SCRs (SCR_1 through SCR_4) in diagonal pairs via either transformer T_1 or T_2. When transformer T_1 is pulsed, current flows through the series field via SCR_1, f_1-f_2, and SCR_4. Alternatively, when transformer T_2 is pulsed, current flows through the series field via SCR_3, f_2-f_1, and SCR_2. Thus current flows either through f_2-f_1 or f_1-f_2, depending on whether transformer T_1 or T_2 is energized by the SPDT switch, which selects either the reverse or forward mode. The circuit shown in Fig. 8-20 operates as follows:

1. Full-wave bridge rectifier (D_1 through D_4) provides positive pulses twice each cycle, charging capacitor C_1 and providing a zener voltage to diode Z_1.

2. As capacitor C_1 charges through resistors R_1 and R_2, its voltage rises and ultimately exceeds the zener voltage rating of Z_1. Capacitor C_1 discharges through zener diode Z_1, providing a positive pulse to the gate of SCR_5.

3. When SCR_5 conducts, it provides a return path for the energy of C_1 either through the primary of T_2 or T_1, depending on the position of the SPDT switch, thus pulsing the respective secondaries in the diagonal bridge circuit.

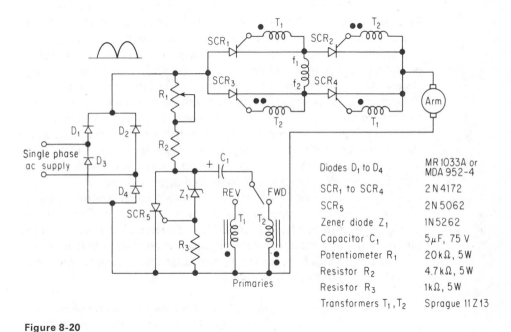

Diodes D_1 to D_4	MR 1033A or MDA 952-4
SCR_1 to SCR_4	2 N 4172
SCR_5	2 N 5062
Zener diode Z_1	1 N 5262
Capacitor C_1	$5\mu F$, 75 V
Potentiometer R_1	$20 k\Omega$, 5 W
Resistor R_2	$4.7 k\Omega$, 5 W
Resistor R_3	$1 k\Omega$, 5 W
Transformers T_1, T_2	Sprague 11 Z 13

Figure 8-20

Speed control and direction reversal for universal dc or ac series-wound motors.

4. With S_1 in the forward position shown, T_2 is energized, providing a positive polarity pulse via its secondaries to the gates of SCR_3 and SCR_2, permitting rectified dc to flow through f_2-f_1 and the series-connected SCRs and the armature.

5. When SCR_5 fires it continues to conduct for the duration of the half-cycle, dropping the voltage of C_1 to the forward volt drop of SCR_5. When the full-wave input pulse drops to zero, SCR_5 is biased off and

C_1 recharges again (as described in step 2) as the rectified input pulse starts to rise toward its maximum value.

6. The motor speed is controlled by potentiometer R_1. The higher the resistance of R_1, the greater the time constant of $(R_1 + R_2)C_1$ and the longer the time for C_1 to charge to a value above the zener breakdown voltage of Z_1. This time constant determines the conduction angle of the gates of SCR_2 and SCR_3 (or SCR_1 and SCR_4), and in turn the average motor torque, armature voltage, and motor speed.

The component values shown in Fig. 8-20, again, are those for a $\frac{1}{15}$-hp motor. In sizing components for higher horsepower ratings, the average steady-state current and the stalled rotor current of the motor determine the current and power ratings of the SCRs, diodes, transformers, etc.

It should be noted that the above types of motor speed controls also lend themselves to other types of *resistive* appliances provided that the power rating of the appliance does not exceed the power rating of the semiconductor devices used in the circuitry. These applications include photographic floodlight dimming, blenders, projection lamp dimming, electric soldering irons (resistive-element type), electric glue guns (resistive-element type), saber saws, food mixers, sewing machines, etc. As of this writing, commercial models of prepackaged speed controls* include a standard receptacle (for plugging in the particular appliance, such as an electric drill or soldering iron), replaceable fuses of the "slow-blow" type or a manually resetting circuit breaker, and a three-way switch that provides full ac power, variable and controlled dc power, and off.

A typical commercial schematic of a plug-in speed control manufactured by General Electric is shown in Fig. 8-21.† The circuit consists of a unit designed to be plugged into a 115-V ac receptacle. Located on the surface of the unit is a speed-control potentiometer, R_2, a switch, S, and a receptacle to which is connected the particular appliance to be controlled. Operation of the circuit is as follows:

1. During the positive half-cycle of the supply voltage, the arm on potentiometer R_2 taps off a fraction of the sine-wave supply voltage and compares it, in a bridge circuit, with the counter emf of the motor at the receptacle of the motor. If the counter emf of the motor exceeds or is equal to the positive potential at R_2, no current will flow to the motor.

* Some of the companies manufacturing speed controls are Bridgeport Hardware Mfg. Co., Bridgeport, Conn.; Electrotone Laboratories, Chicago, Ill.; G. E. Wiring Device Dept., Providence, R. I.; Lutron Electronics Co., Emmaus, Pa.; and Omni Enterprises, Chicago, Ill.

† Figure 8-21 is reproduced by special permission of the General Electric Co. See Application Note 200.4, *Universal Motor Speed Controls*, and Application Note 201.1, *Plug-In Speed Control for Standard Portable Tools and Appliances*, General Electric Co., Rectifier Components Department, Auburn, N.Y.

2. If the potentiometer voltage at the arm of R_2 exceeds the receptacle (motor armature) voltage, current flows through resistor R_1 and CR_1 to trigger the gate of the SCR. The SCR is triggered into conduction for a portion of the positive half-cycle to supply half-wave current to the universal motor. The motor thus operates at a speed determined by the setting of R_2.

3. Given a particular setting of R_2, assume that an increased load is applied to the motor. The motor speed and counter emf will decrease. This causes the bridge to be unbalanced earlier in the positive half-cycle. Potentiometer arm R_2 will thus trigger the gate of the SCR earlier in the cycle, and an additional increased positive half-wave component of voltage is applied to the universal motor. Thus, increases in load are compensated by increases in applied voltage, maintaining the motor speed constant, regardless of load and torque variations.

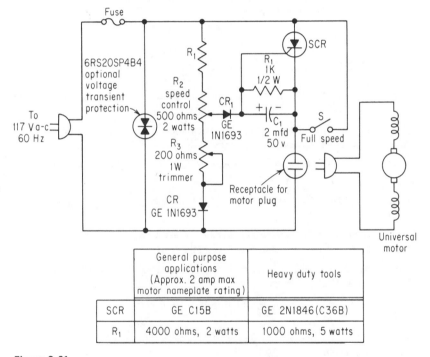

	General purpose applications (Approx. 2 amp max motor nameplate rating)	Heavy duty tools
SCR	GE C15B	GE 2N1846(C36B)
R_1	4000 ohms, 2 watts	1000 ohms, 5 watts

Figure 8-21

Plug-in speed control for standard portable tools and appliances, using SCRs.

4. For any given speed-load setting, potentiometer R_2 may be varied to provide a range of approximately 3 to 1 over the normal speed with complete stability, in addition to excellent speed regulation. Stability is achieved through the use of
   a. Rectifier CR_1, which prevents excessive reverse voltages at the gate of the SCR.

231

b. Rectifier CR_2, which similarly prevents inductive field currents from the motor (due to changes in load or speed setting) tending to produce circulating currents which may trigger the SCR or produce hunting.

c. Resistor R_3, a trimmer resistor, which is adjusted to provide minimum motor speed at which no instability may be produced.

d. Resistor R_4 and capacitor C_1, which similarly improve stability by bypassing high-frequency "transients" produced by the universal motor (commutator "hash") tending to trigger the SCR.

e. A zener diode (optional) across the supply to reduce speed changes due to line voltage fluctuations.

5. Normal operation at maximum speed may be achieved by closing switch S. This switch, external to the "black box," short-circuits the SCR and permits the universal motor to operate directly on alternating current. When operated in this manner, neither speed control nor feedback speed regulation is possible.

BIBLIOGRAPHY

Alger, P. L., and Y. H. Ku. "Speed Control of Induction Motors Using Saturable Reactors," *Electrical Engineering* (February 1957).

Baude, J. "Multi-Function Magnetic Amplifier Control for Large Motors," *Electrical Manufacturing* (October 1957).

Cockrell, W. D. *Industrial Electronics Handbook* (New York: McGraw-Hill Book Company, 1958).

Dailey, J. J. "Dual Circuit Electronic Motor Control," *Electrical Manufacturing* (July 1957).

Graphical Symbols for Electrical Diagrams (ASA Y32.2) (New York: American Standards Association), IEEE Std. No. 315-1971.

Gutzwiller, F. W. "The Silicon Controlled Rectifier," *Electrical Manufacturing* (December 1958).

————. "Universal-Motor Speed Controls," *Electro-Technology* (December 1961).

Heumann, G. W. *Magnetic Control of Industrial Motors*, 3 vols. (New York: John Wiley & Sons, Inc., 1961).

Industrial Control Equipment (Group 25) (ASA C42.25) (New York: American Standards Association).

Jones, R. W. *Electric Control Systems* (New York: John Wiley & Sons, Inc,. 1953).

Kosow, I. L. *Electric Machinery and Transformers* (Englewood Cliffs, N.J.: Prentice-Hall, Inc., 1972).

Laithwaite, E. R. "Two New Ways to Vary Induction Motor Speed," *Control Engineering* (July 1960).

232

Motor and General Standards. (MG1). New York: National Electrical Manufacturers Association.

Press, V. W. and W. R. Jones "Rugged Adjustable Speed Drives Use Magnetic Amplifiers," *Electrical Manufacturing* (November 1958).

Siskind, C. S. "A-C Motors for Adjustable-Speed Systems," *Control Engineering* (June 1959).

Uri, J. Ben: "Variable Speed Control Systems," *Electro-Technology* (March 1961).

Winsor, L. P. and E. E. Moyer. "Adjustable Speed Drives," *Electrical Manufacturing* (November 1952).

Zollinger, H. A. "Reactor Control of Induction Motors," *Electrical Manufacturing* (January 1960).

PROBLEMS AND QUESTIONS

8-1.   Classify each of the following motors as either *integral-* or *fractional-*horsepower motors:

hp	rated speed (rpm)
a. $\frac{3}{4}$	2400
b. $\frac{3}{4}$	1200
c. $\frac{1}{2}$	650
d. $\frac{1}{4}$	200
e. 2.0	4000

8-2.   a. Explain why a capacitor start motor is considered a reversible motor but not a reversing motor.
   b. Why is the capacitor motor both a reversing and reversible motor?
   c. Why are ac series or universal motors both reversing and reversible motors?
   d. Why are shaded-pole single-phase induction motors reversible but not reversing motors?

8-3.   In reversing ac series or universal motors,
   a. Why is armature reversal preferable to series-field reversal?
   b. What will happen if field and armature line connections are both reversed? Why?

8-4.   Using Figs. 7-17 and 8-2a, explain
   a. Why line voltage control lends itself to single-phase induction motors but not to polyphase SCIMs and WRIMs
   b. Why rotors of Classes C and D are more sensitive to voltage changes than other classes (A, B, or F).

8-5.   Using Fig. 8-3, explain
   a. Advantages of reactors over resistors for primary voltage control
   b. Which method produces better speed regulation?
   c. Whether the methods shown are adaptable for use with ac series or universal motors

233

d. Which single-phase ac motors cannot be speed controlled by the methods shown?

8-6. Compare saturable reactor (or magnetic amplifier) control with the methods shown in Figs. 8-3 and 8-4 for
a. Smoothness of speed control from standstill to rated
b. Efficiency
c. Speed of response
d. Ability to ensure speed regulation using closed-loop operation.

8-7. Explain why the various methods of electronic control shown in Fig. 8-8
a. Are not suited for use with synchronous or hysteresis motors
b. Possess advantages over the methods shown in Figs. 8-3, and 8-4 and list them.

8-8. a. What types of single-phase motors are speed-controlled by rectifier-inverter or cycloconverter packages?
b. Why are most single-phase, variable-frequency, variable-voltage ac solid-state motor packages usually rectifier-inverters rather than cycloconverters?
c. Compare the various inverters shown in Fig. 8-9 and discuss advantages and disadvantages of each.

8-9. Compare the magnetic amplifier of Fig. 8-6 with the electronic inverter package shown in Fig. 8-10 for controlling speed of a single-phase capacitor-start induction motor (Class A rotor) with respect to
a. Range of speeds available
b. Speed regulation
c. Ability to achieve closed-loop control of speed
d. Possibility of regeneration for braking.

8-10. a. List those ac single-phase motors which are incapable of being braked to a standstill using plugging
b. Why is plugging a simpler and more efficient means of braking than dynamic braking?
c. Is it possible to achieve dynamic braking of capacitor motors without using dc? Explain
d. What are the limitations of the method used in part c?
e. Give two advantages of dynamic braking over mechanical braking
f. Give two disadvantages of dynamic braking over mechanical braking.

8-11. For controlling speed of universal or small ac motors.
a. Compare the two circuits shown in Fig. 8-17 and discuss merits of each
b. Compare the circuits of Fig. 8-17 with those of Fig. 8-8 and discuss merits of each
c. Compare feedback circuits (Fig. 8-18) with those of Part b and discuss advantages of former.

8-12. a. List two advantages and two disadvantages of the Gutzwiller half-wave over the Momberg half-wave circuit
b. What are the advantages of the circuit of Fig. 8-19 over those in Part a?
c. What disadvantages are present in the circuit of Parts a and b?

static control

9-1.
GENERAL Coincidental with the development of electronic methods for the control of rotating machinery has been the introduction of electronic static switching devices and computer logic elements. Such static elements employ diodes, transistors, SCRs, FETs, and magnetic amplifiers as *switches*. Having neither moving parts nor mechanical contacts, such static devices possess the advantages of extremely rapid operation, long life, increased reliability, reduced size of the controlling element package (or module), and the relative ease with which they may be linked to computers for automatic process control in conjunction with tape (magnetic or paper) and punch-card programming.

The application of computer logic to the requirements of motor control tends to simplify the manner in which control circuits and systems may be designed. Indeed, all the basic functions provided by pushbuttons and relays may be duplicated by specific static-logic elements. An *element* is an electronic circuit capable of performing a given logic function.*

Five basic logic functions and their elements will be described, followed by more complex variations. These basic functions are the OR, AND, NOT, MEMORY, and DELAY functions, respectively. To further enhance the understanding of these functions, the specific relay function is first shown (relay logic), followed by two analogous static logic systems: NEMA logic† and NOR logic.‡

9-2.
OR FUNCTION

A circuit shown to be momentarily energized by means of a relay is given in Fig. 9-1a. Three remote pushbutton stations, A, B, and C, are capable of momentarily energizing an output, i.e., a load, a motor, or another relay. The analogous static switching function that accomplishes the same purpose is the OR function shown in Fig. 9-1b. The upper figure shows three semiconductor diodes connected in such a manner so that a positive voltage pulse at inputs A, B, or C will be conducted to the output. Thus, for the duration of time that a positive pulse is applied to any or all of these contacts, the output is energized.

The logic of such a condition is represented in the lower portion of Fig. 9-1b using the NEMA logic representation for the OR function. Note the use, also, of a "zero" at both input and output. The 0 indicates the absence of a signal or voltage, whereas the numeral 1 indicates either a positive or negative pulse. Thus, the implication of the NEMA logic using the "zero-one" convention is that in the *absence* of a pulse at *either A*, or B, or C, there is no (zero) output. Given a 1 at *any* of the inputs, there is a 1 at the output for the *duration of the pulse at the input*.

Referring back to the diode configuration of Fig. 9-1b, in the presence of a negative bias, the diodes CR_1, CR_2, and CR_3 are reverse-biased and act as open switches. Any positive or 1 pulse at *A or B or C* will cause that respective diode to conduct (or act as a closed switch). Since the volt drop across the diode when conducting is negligible, the positive or 1 pulse appears at the output.

NOR logic elements are essentially transistor inverter amplifiers.

---

* It is the purpose of this chapter to introduce the reader to a subject that has become one of the more significant developments in the control of electrical machinery during the latter half of this century. A detailed study of static switching, including the various logical design postulates and theorems using Boolean algebra, is beyond the scope of this volume. For more detailed information on the subject, see B. H. Arnold, *Logic and Boolean Algebra* (Englewood Cliffs, N.J.: Prentice-Hall, Inc., 1962) and G. A. Maley, *Manual of Logic Circuits* (Englewood Cliffs, N.J.: Prentice-Hall, Inc., 1970).

† See *Application Manual, Transistorized Static Control*, General Electric Bulletin GET 3551A, General Purpose Control Department, Bloomington, Ill., June 1970.

‡ See *NORPAK Solid State Logic Control*, Bulletin 8851, 2, 3, Square D Company, Mineola, N.Y.

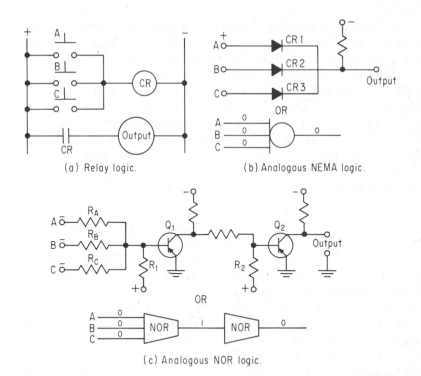

(a) Relay logic.

(b) Analogous NEMA logic.

(c) Analogous NOR logic.

Figure 9-1

OR function representation using relay,
diode, and transistor logic elements.

These are customarily mounted on printed circuit cards and are easily adapted to control functions. The upper portion of Fig. 9-1c shows two common emitter *pnp* transistors (Q_1 and Q_2) in cascade, each serving as a NOR logic element. The first NOR element, Q_1, is provided with the possibility of one or more inputs connected to its base through suitable resistors, R_A, R_B, and R_C. The base is reverse biased to a positive supply voltage through resistor R_1, thereby preventing collector-base leakage. The negative supply voltage at the collector, therefore, is cutoff with respect to the grounded emitter as long as the base is positively biased. In the absence of a negative input at A, B, or C, transistor Q_1 is cutoff and this condition may be represented as a negative output voltage (a pulse of 1) at the output of the first NOR element, as shown in the lower portion of Fig. 9-1c. This negative output applied to the input of transistor Q_2, however, causes the base of this (second) transistor to become negative with respect to its emitter. Transistor Q_2, therefore, is undergoing conduction causing both the output and its collector to be grounded.

237

In the absence of an input to the first transistor, each of the transistor amplifiers serves to *invert* the input signal. Thus, a 0 input to the first NOR element produces a 1 input to the second NOR element; the latter inverts the signal to provide a 0 output. Similarly, if a negative 1 pulse is applied to either A, B, or C, the output of the first transistor Q_1 is grounded providing a 0 output and input to Q_2. In the absence of an input signal, Q_2 has a negative or 1 output. Thus, a 1 at either A, *or B, or C* provides a 1 at the output for the duration of the input pulse. Ignoring polarity differences, the relay, the diode, and the transistor representations shown in Fig. 9-1 all perform identical OR logic in permitting a given output to be energized (only) when an input is energized.

9-3.
AND FUNCTION

Figure 9-2a shows the relay logic for a typical AND circuit. The implication of the relay logic is that contacts A, *and B, and* C must *all* be closed before control relay CR is energized, momentarily. Once CR is energized, furthermore, should A, B, or C be opened, relay CR will be deenergized.

The equivalent static switching AND function using semiconductor diodes is shown in Fig. 9-2b. In this configuration, the diodes are capable of conduction only when a negative going or zero pulse is applied to inputs A, B, and C. Assuming there is a positive pulse (of 1) applied *simultaneously* to *all* inputs, the output during this conduction period is also positive (1).

It may be considered that, except for a small (negligible) voltage drop across the diodes, the input negative voltages are connected through the diodes to the output in *coincidence*; hence, the name *coincidence circuit* for this configuration. If one of the inputs receives a zero or negative going pulse, the output voltage becomes less positive (zero) as the voltage drop across R increases. Given zero input to all terminals A, B, and C, therefore, the output is the same as the positive supply voltage (zero). The implication of the AND symbol represented by the NEMA logic shown in the lower portion of Fig. 9-2b that a (positive) pulse of 1 is required at A, *and B, and* C, simultaneously, before a (positive) 1 pulse can appear at the output.

Equivalent AND logic may be obtained using electronic tube triodes, magnetic amplifiers, or transistor amplifiers. For illustration, transistor inverter amplifiers used as AND elements are shown in Fig. 9-2c. The AND logic of this circuit is that it will produce a 1 output only when each of the inputs from A, B, and C is 1. Operation of the circuit is such that only when the positive reverse bias of each of transistors Q_A, Q_B, and Q_C is overcome by a negative pulse at the base of these transistors, simultaneously, are all three transistors driven into conduction. The base of transistor Q_2 is made sufficiently positive, therefore, to be cutoff producing a negative output of 1.

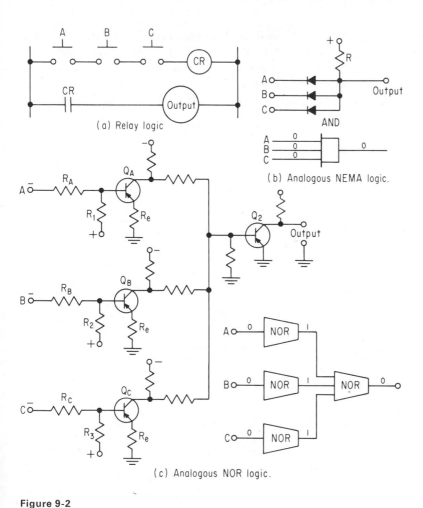

(a) Relay logic

(b) Analogous NEMA logic.

(c) Analogous NOR logic.

Figure 9-2

AND function representation using relay,
diode, and transistor logic elements.

9-4.
NOT FUNCTION

Figure 9-3 shows the relay logic for a typical NOT circuit. If momentary contact A (in the left figure) is closed, relay CR is energized, which, in turn, deenergizes the output. The same result is produced in the right part of Fig. 9-2a when the n.c. STOP is depressed. The logic of a NOT circuit is such that when energized, there is "not" an output (assuming, of course, that there was an output prior to the application of energy to the NOT function).

A fuse, circuit breaker, and overload relay are, in addition to a STOP button, examples of the application of NOT logic.

The use of electronic triodes to perform the NOT function is shown in

239

Fig. 9-3b. Triodes were used prior to transistors and other solid-state devices as static switches. A positive (1) pulse at *A* produces conduction of the triode whose plate resistance is low compared to the plate load resistor *R*. A 1 pulse at *A* causes the triode to act as a shorted switch effectively grounding the output, i.e., producing a 0 at the output. Conversely, in

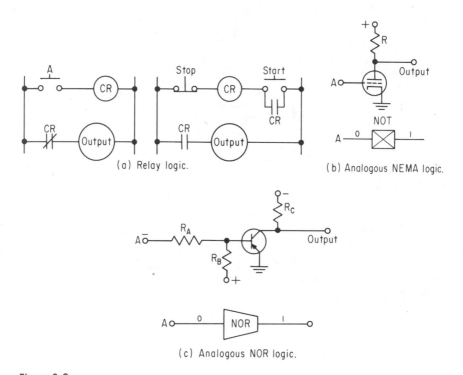

(a) Relay logic.

(b) Analogous NEMA logic.

(c) Analogous NOR logic.

Figure 9-3

NOT function representation using relay, diode and transistor logic elements.

the absence of a pulse at *A* (0 input), the triode acts as an open switch of almost infinite resistance and the output is the same as the positive plate supply voltage or 1.

The NOT function is an inversion of the input signal and the grounded-cathode triode amplifier is basically an inverter.

The NOT function and its analogous NEMA logic performed by a NOR gate is shown in the lower part of Fig. 9-3c.

The NOT function may also be performed by magnetic or transistor amplifiers, since its process is basically that of inversion. A single transistor NOT inverter is shown in the upper portion of Fig. 9-3c. The positively biased base prevents conduction and produces a negative output (or 1) as long as the transistor is cutoff. A negative 1 pulse at *A* drives the transistor into conduction, effectively grounding the output to "zero." Thus, no input (0) produces a negative output (1), whereas a negative input (1) is inverted to a 0 output.

Momentary contacts, either n.c. or n.o., were used in a relay logic of Figs. 9-1a, 9-2a, and 9-3a. The tacit assumption of such devices is that once the pushbutton is returned to its *original* (normal) state, the *logic is no longer retained*. If it is desired that the logic produced by a momentary contact *retain* its original intention, it is necessary (in the case of an "on" impulse) to shunt the pushbutton A with a n.o. CR auxiliary contact as shown in Fig. 9-4a. But if A is now provided with a "memory" to continue to energize relay CR, even when it is released (and n.o.), it is also necessary to provide an additional "erase" or "return-to-off" feature, otherwise CR will always be energized as long as the control circuit is energized. In a relay circuit, the n.c. pushbutton B provides such a reset, as shown in Fig. 9-4a.

A corresponding transistor inverter logic circuit using two NOR inverters in a back-to-back configuration provides the same "memory" and reset functions as represented in the relay logic. Parenthetically, it should be noted that magnetic amplifiers and transistorized or vacuum tube bistable (flip-flop) circuits also may be used for this purpose. The circuit shown in Fig. 9-4b is, in effect, a three-terminal network having two inputs and one output. Input A has the function of "setting" the output to 1 or to an "energized" state. Input B has the function of "resetting" the output to 0 or, in the case shown, the "deenergized" state.

As shown in Fig. 9-4b, transistor Q_1 is in a nonconducting state and transistor Q_2 is in a conducting state when there is no input on both terminals A and B. In the absence of an input signal at A, nonconducting transistor Q_1 develops a negative output at point X, which counteracts the positive base bias of transistor Q_2. The output at Q_2 is, therefore, "zero." A negative signal at A drives Q_1 into conduction and Q_2 out of conduction, causing a negative output (1). The output simultaneously provides a (negative) feedback signal to the input of Q_1, maintaining the "memory" of the input at A even when the negative impulse has been removed. The circuit can only be reset by removal of the bias power (compare to relay circuit) or by a negative reset signal at B, driving Q_2 into conduction and short-circuiting its output load resistor to ground. In the former case, restoration of power restores the positive bias required to maintain Q_1 in a nonconducting state and Q_2 in a conducting state. For this reason, the circuit is sometimes called an OFF-RETURN MEMORY because when power is OFF, the circuit is RETURNED to its original state. This action is distinguished from a RETENTIVE MEMORY which retains its state even after power has been reapplied.

In some MEMORY modules, the output at point X of Fig. 9-4b is also used for inverter or NOT function purposes.

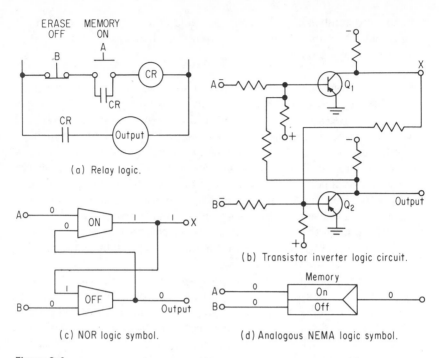

(a) Relay logic.

(b) Transistor inverter logic circuit.

(c) NOR logic symbol.

(d) Analogous NEMA logic symbol.

Figure 9-4

MEMORY function representation using transistor logic elements (OFF-RETURN MEMORY).

The NEMA analog of the NOR function logic symbol is shown in Fig. 9-4d and the NOR logic symbol is shown in Fig. 9-4c. In some MEMORY modules, the output at point X of Figs. 9-4b and (c) is also provided for inverter or NOT function purposes, whereas the NEMA analog of the NOR function logic symbol is generally represented as a three-terminal network (Fig. 9-4d).

9-6.
DELAY FUNCTION

Definite time acceleration, described in Sections 4-2 through 4-5, required some means of providing a definite *time delay* in accelerating a motor to its proper running speed. The logic of such time-delay relays is shown in Fig. 9-5a for "on-delay" purposes and in Fig. 9-5b for "off-delay" purposes. Briefly, relay *TR* is only activated at some predetermined interval after the momentary contact is maintained in a depressed state and serves either to energize or deenergize the output after an additional delay period. The customary auxiliary contact shunting pushbutton *A* is deliberately omitted because it requires a MEMORY function. The delay function may be obtained through various electronic methods including *R-C* and *R-L* time-constant circuits, neon-

CHAP. NINE / *Static Control*

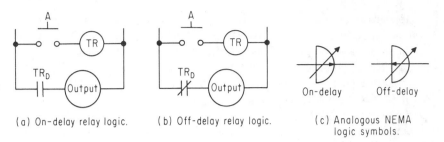

(a) On-delay relay logic.   (b) Off-delay relay logic.   (c) Analogous NEMA logic symbols.

On-delay   Off-delay

Figure 9-5

DELAY function representation using NEMA logic symbols.

tube timers, monostable multivibrator circuits, unijunction transistors, and dual magnetic-core amplifier units.* Regardless of the circuitry or techniques employed, the NEMA logic symbols for these delay devices are represented in Fig. 9-5c and Table 9-1.

9-7.
NOR AND NAND
FUNCTIONS

Two other functions which produce the inversion of the OR and AND functions covered in Secs. 9-2 and 9-3, respectively, are the NOR and NAND functions. The logic of the OR function is that a 1 at either *A or B or C* will produce a 1 at the output (Fig. 9-1b). The NOR function, shown in Fig. 9-6, produces the inversion (or opposite) of the OR function. As shown in the relay logic of Fig. 9-6a, *energizing* either *A or B or C*, in turn, *deenergizes* the output. The logic symbol is shown in Fig. 9-6b. Note that in the absence of a 1 at either *A* or *B* or *C*, there is a 1 at the output. When a 1 appears at *any* input, a 0

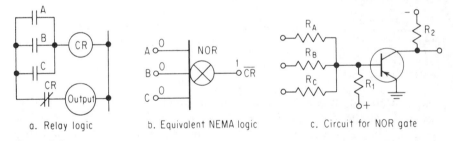

a. Relay logic   b. Equivalent NEMA logic   c. Circuit for NOR gate

Figure 9-6

NOR function logic.

    * Space, unfortunately, does not permit a discussion of the various principles used in delay elements associated with static switching. The state-of-the-art is currently advancing so rapidly that even some of the more recent techniques have been rendered obsolete by current developments. Information on such delay and other static devices may be obtained by writing to the General Electric Co., Schenectady, N.Y. (G. E. Static Controls Div.); the Square D. Company, Cleveland, Ohio (Norpak elements); the Delco Radio Co. (Delcon System); and the Westinghouse Electric Co. (Cypak magnetic control system).

243

appears at the output (compare to Fig. 9-1b). Thus, any input produces no output. Stated another way, an output is only produced when none of the inputs are energized. This is the logic of the NOR function.

The circuit for the NOR function is shown in Fig. 9-6c. Note that transistor Q_1 is at cutoff in the absence of a negative pulse (1) at A or B or C. The output at the collector of Q_1, consequently, is negative or 1 in the absence of any pulses at the input. But given a negative pulse of 1 at any input, Q_1 conducts, heavily shorting its collector to ground and producing a 0 output.

It is most important that the reader note that the NOR element shown in Fig. 9-6c is the basis of the logic elements shown in Figs. 9-1 through 9-4. By combining series of NOR gates, it is indeed possible to duplicate all the logic functions and elements previously covered (see Table 9-1). In effect, the NOR is a *generic* element which permits simplifications that allow many combinations of functions using the minimum number of active elements (transistors).* Thus, it is possible to purchase integrated circuits (ICs) having six or more individual active NOR gates for synthesizing a variety of logic functions from a single module.

Just as the NOR function produces the logical inverse of the OR, another NAND function produces the logical inverse of the AND. The logic of the AND is that an output is only produced (Fig. 9-2) when all inputs are energized. The logic of the NAND as shown in Figs. 9-7a and 9-7b is

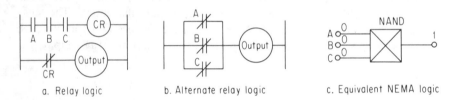

a. Relay logic   b. Alternate relay logic   c. Equivalent NEMA logic

Figure 9-7

NAND function logic.

that an output is always present in the absence of *all* the inputs. Only when *all* the inputs are energized (at a 1 state) is the output deenergized (at a 0 state). Note that the relay logic of Figs. 9-7a and b are identical. Figure 9-7c shows the NEMA symbol for the NAND function. The NAND is also sometimes called an AND NOT function because it is the inverse of the AND function, as previously described. The reader should compare the equivalent NEMA logic for the NAND function (Fig. 9-7c) with the AND function (Fig. 9-2b) and note that the NAND is the inverse of the AND.

Since the NOR and NAND functions have been shown to be the inverse

* The Square D Company has developed the NORPAK Plug-In, Solid-State Control System based on NOR elements exclusively; see *NORPAK Solid State Logic Control*, Square D Co., 138 Mineola Blvd., Mineola, N.Y.

of the OR and AND functions, the NOR and NAND outputs are readily available from those portions of the circuitry which are always the inverse of the output. For this reason they are not summarized as basic logic functions in Table 9-1.

9-8.
SUMMARY OF
BASIC FUNCTIONS

Table 9-1 shows the seven basic logic functions described in the previous sections. The first column shows the function, the second shows the relay logic for the function. The third column shows how NOR elements (transistor inverters previously shown in Figs. 9-1 through 9-4) may be used to synthesize the particular function. The fourth column shows the English logic NEMA symbol for the function with the logic written in Boolean-algebra form. The last column shows how the logic is read and what it implies.

For the AND circuit shown in the first row, the Boolean expression

$$A \cdot B \cdot C = CR \qquad (9\text{-}1)$$

is read (energizing) *A and B and C* equals *CR*.

Similarly, for the OR circuit shown in the second row, the plus (+) sign symbolizes the word "or." And the Boolean expression for the OR circuit is

$$A + B + C = CR \qquad (9\text{-}2)$$

read (energizing) *A or B or C* equals *CR*.

Note the distinction between OFF RETURN MEMORY and RETENTIVE MEMORY. The former static logic element is the same as the START and STOP configurations of conventional relay circuits. As shown in the second column for the OFF RETURN MEMORY, momentarily depressing *A* (the START) energizes *CR* closing all (n.o.) *CR* contacts and opening all (n.c.) *CR* contacts. Thus, $A = CR$. But as shown in the relay circuit, depressing *A* but not *B* is equivalent to *B* (see NOR circuit, third column) and thus

$$A \cdot \bar{\bar{B}} = A \cdot B = CR \qquad (9\text{-}3)$$

is read: *A* and not not-*B* equals *A* and *B* equals *CR*.

When the STOP ($\bar{B}$) button is depressed in the OFF RETURN MEMORY circuit, relay *CR* is deenergized, returning all contacts to their deenergized (normal) state as shown in Table 9-1, and the only output is the n.c. $\overline{CR}$. Thus, depressing $\bar{B}$ yields $\overline{CR}$. This returns the memory circuit to its original off state in which

$$\bar{A} + \bar{B} = \overline{CR} \qquad (9\text{-}4)$$

is read: not-*A* or not-*B* equals not-*CR*.

Note that the NOR circuit (column 3, Table 1) for the OFF RETURN MEMORY is also provided with a reset or OFF signal, which returns the

245

circuitry to its original state in which $\overline{CR}$ is the only output. This "off" signal may be generated by an overload transducer, a centrifugal or limit switch, or any other device that normally would be the equivalent of a n.c. contact in series with the STOP button.

In the event of power failure or generation of an OFF signal, the OFF RETURN MEMORY returns to its original state in which only $\overline{CR}$ is generated, an effect similar to the removal of power from the basic relay circuit, Table 9-1.

The effect of a latching relay (Section 5-3) is simulated by RETENTIVE MEMORY. Energizing A energizes the latching coil or relay CR_L. The armature of the latching relay, shown in the basic relay circuit (column 2, Table 9-1) will maintain n.o. CR in a closed position, even after the power has been removed and restored. Energizing B energizes the *unlatching* coil or relay CR_U. This coil opens all CR contacts and closes all $\overline{CR}$ contacts, restoring the relay to its "normal" state shown in the basic relay circuit.

The RETENTIVE MEMORY logic symbol shows the equivalent of the relay logic just described. Only A produces CR (because when A is energized, $\overline{CR}$ is opened). Only B produces $\overline{CR}$, because when B is energized, CR contacts are cleared. The logic for RETENTIVE MEMORY, as shown in the last column of Table 9-1, is

$$A = CR \qquad (9\text{-}5)$$

is read: (only) A equals CR, and

$$B = \overline{CR} \qquad (9\text{-}6)$$

is read: (only) B equals not-CR.

The equivalent of a time-delay relay having either n.o. or n.c. contacts, which operates at some adjustable (or variable) time after having been energized, is shown as TIME DELAY AFTER ENERGIZATION. As shown in the basic relay circuit (column 2, Table 9-1), when contact A closes, energizing control relay CR, *n*ormally *o*pen *t*iming contacts are *c*losed, energizing $CR(\text{NOTC})$; simultaneously, *n*ormally *c*losed *t*iming contacts are *o*pened, energizing $CR(\text{NCTO})$.

The equivalent NOR circuit and NEMA logic symbol for the TIME DELAY AFTER ENERGIZATION show that the logic of this circuit is the equation

$$A = CR_{\text{td}} \qquad (9\text{-}7)$$

is read: A equals CR after a time delay (of x seconds), and conversely,

$$A \neq \overline{CR}_{\text{td}} \qquad (9\text{-}8)$$

is read: A is not not-CR after a time delay (of x seconds).

Logic function	Basic relay circuit	NOR circuit	English logic NEMA symbol	Read logic as
AND			$A \cdot B \cdot C = CR$	A and B and C equals CR
OR			$A + B + C = CR$	Either A or B or C equals CR
NOT			$A = \overline{CR}$	A equals not CR
OFF RETURN MEMORY		Off signal	$A \cdot B = CR$ $\overline{A} + \overline{B} = \overline{CR}$	A and not-B equals CR A and B equals CR Not A or not B equals not CR
RETENTIVE MEMORY				A equals CR B equals not CR
(Variable) TIME DELAY AFTER ENERGIZATION			$A = CR_{td}$ $A \ne \overline{CR}_{td}$	A equals CR after time delay A does not equal not CR after time delay
(Variable) TIME DELAY AFTER DE-ENERGIZATION				Deenergizing A equals CR after time delay Deenergizing A does not equal not CR after time delay

TABLE 9-1. THE BASIC LOGIC FUNCTIONS, EQUIVALENT RELAY CIRCUITRY, EQUIVALENT NOR CIRCUITRY, ENGLISH LOGIC SYMBOLS AND HOW LOGIC IS READ.

The last basic logic element, shown in Table 9-1, is called TIME DELAY AFTER DEENERGIZATION. This is the relay equivalent of a resetting relay which is necessary to reset all contacts to their original condition in the event of power failure (see step 9, Section 4-4, Fig. 4-3, for this application using relay logic). The logic of this element is the same as that expressed in Eqs. (9-7) and (9-8) except that the module operates (as shown in the NEMA logic symbol) on *deenergizing A*.

9-9.
CONVERTING RELAY
LOGIC TO STATIC
LOGIC

In converting relay to static logic, it is possible to examine the relay circuit, and "talk the circuit through" using what designers sometimes call "the language of logic." Once the statement is written for the relay circuit, it is then possible to write the logic equation calling for the particular static logic elements which will duplicate the relay logic.

9-9.1
(Circuit Shown in
Fig. 9-8a)

A typical fail-safe circuit requires that four separate and remotely located contacts are all closed before a control relay, *CR* is energized. An examination of the relay circuit leads to the following statement which is followed by the logic equation equivalent to the statement: "Relay *CR* is energized only when contacts *A and B and C and D* are closed":

$$CR = A \cdot B \cdot C \cdot D \qquad (9\text{-}9)$$

Given Eq. (9-9), it is obvious that four inputs (*A, B, C, D*) must be "anded together" and the output must feed a control relay. The upper right of Fig. 9-8a shows such a configuration using a four-input AND element, but this circuit is impractical for two reasons: four-input ANDs are not usually available, and an output amplifier is also required.* Consequently, the lower right of Fig. 9-8a shows a single module containing two three-input ANDs with the second (AND 1_B) element feeding an output amplifier (Section 9-10.4) since the logic element output is insufficient to drive a control relay.

9-9.2
(Circuit Shown in
Fig. 9-8b)

This relay circuit energizes a control relay (*CR*) from *any* one of four possible remote locations. An examination of the relay circuit leads to the following statement and logic equation: "Relay *CR* is energized when contact A *or* B *or* C *or* D is closed."

$$CR = A + B + C + D \qquad (9\text{-}10)$$

---

* Using GET-3551A (as of this writing), the following types of commercial AND circuits are available on a single module: 2 two-input AND with standard and inverted outputs; 2 three-input AND with standard output (used throughout this discussion); 1 six-input AND with standard and inverted outputs; 1 seven-input AND with standard output. The balance of this chapter assumes that all logic elements, such as AND, OR, NOT, NOR and NAND, are three-input elements.

Note that the logic equivalent shown at the right of Fig. 9-8b implies that a 1 input at *B* (for example) provides a 1 input and 1 output at each of the cascaded ORs, causing the amplifier to energize control relay *CR*.

9-9.3
(Circuit Shown in Fig. 9-8c) This configuration of cascaded normally closed contacts permits a control relay (and its associated circuitry) to be *deenergized* from *any* one of four remote locations. An examination of the relay circuit leads to the following statement and logic equation: "Relay *CR* is energized when contact *A* or *B* or *C* or *D* (energized) is opened."

$$\overline{CR} = A + B + C + D \qquad (\text{or } CR = \bar{A} \cdot \bar{B} \cdot \bar{C} \cdot \bar{D}) \qquad (9\text{-}11)$$

is read: not-*CR* equals *A* or *B* or *C* or *D*.

A three-input OR in combination with a three-input NOR (see Fig. 9-6b) provides the above logic as shown at the right of Fig. 9-8c. Note that in the absence of any inputs, the NOR output to the amplifier is 1 and *CR* is energized. A 1 input at B, for example, provides a 1 output from the OR gate which produces a 0 output from NOR, deenergizing the amplifier and relay *CR*. Figure 9-8c also shows one additional symbol modification, a *signal converter* (Section 9-10.3), normally required with all pushbuttons and other switch contacts, to ensure reliability. Parenthetically, Eq. (9-11) also shows that the AND configuration could also be used to duplicate the logic.

9-9.4
(Circuit Shown in Fig. 9-8d) This configuration is the conventional start-stop pushbutton with overloads used for motor starting. An examination of the relay circuit leads to the following statements and equations:

1. "Relay *CR* is *deenergized* when contact *A* or *C* or *D* is energized (opened)."
2. "Relay *CR* is energized and remains energized when contact *B* is momentarily energized."

$$\overline{CR} = A + C + D \qquad (9\text{-}12\text{a})$$

$$CR = B \text{ (memory)} \qquad (9\text{-}12\text{b})$$

Equation (9-12b) calls for an OFF-RETURN MEMORY element, as shown in Fig. 9-8d at the right. When a 1 input appears at B, the 1 is transferred to the MEMORY [as in Eq. (9-12b)] and to the amplifier, energizing relay *CR*. Inputs *A*, *C*, and *D* are ORd simultaneously to the OFF-RETURN deenergizing the amplifier, as shown in Eq. (9-12a).

Relay logic   Static-logic

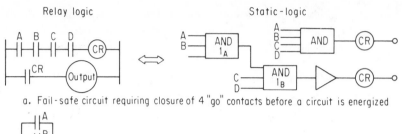

a. Fail-safe circuit requiring closure of 4 "go" contacts before a circuit is energized

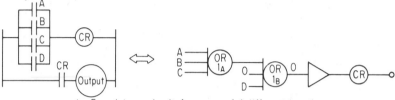

b. Energizing a circuit from any of 4 different locations

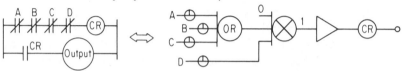

c. Deenergizing a circuit from any of 4 different locations

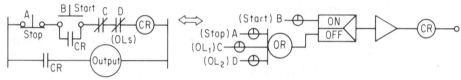

d. Conventional basic start-stop configuration with overload disconnects

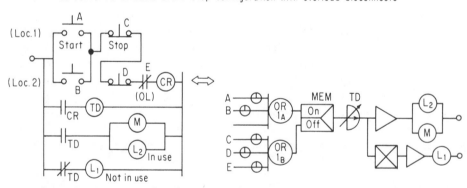

e. Conventional start-stop from 2 remote locations with time delay relay and pilot-lamps

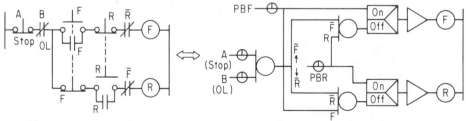

f. Conventional forward-reverse relay configuration having electrical and mechanical interlocks

Figure 9-8

Relay circuits and equivalent static-logic circuits.

The relay configuration for this circuit shows a relay, CR, started from two remote locations (A or B) and stopped from two remote locations (C or D) or an overload contact (n.c.) E in series with control relay CR. When energized, CR contacts seal (MEMORY) the start buttons and energize a time-delay relay TD. The latter's contacts energize a motor relay M and an indicator lamp (IN USE) L_2 and deenergize a second indicator lamp (NOT IN USE) L_1.

The logic statements which emerge from the above description are

1. "Relay CR is energized by A or B and once energized is sealed in memory."
2. "Relay CR must be energized before relays TD and M can be energized in sequence."
3. "When TD is energized, it energizes M and L_2 simultaneously, after a time delay."
4. "When TD is energized, it deenergizes L_1."
5. "Relay CR is deenergized by C or D or E."

The equations representing the above statements are

1. $A + B = CR$
2. $CR = TD = M$ (after DELAY)
3. $TD = L_2$ (after DELAY)
4. $TD = \bar{L}_1$
5. $\overline{CR} = C + D + E$

The static logic diagram for these five equations is shown on the right side of Fig. 9-8e. Note that in the absence of any signal, L_1 receives power from the amplifier because a 0 output from TD is inverted by a NOT to a 1, energizing L_1 through the amplifier. Inputs C, D, and E are ORd together to deenergize OFF-RETURN MEMORY, CR output, TD output and relay M and L_2 in parallel. Inputs A or B energize MEMORY providing a 1 output to a TIME DELAY AFTER ENERGIZATION element which when amplified energizes L_2 (IN USE) and motor relay M.

This is the conventional forward-reverse start-stop circuit with electrical and mechanical interlocks to prevent short circuits when both forward and reverse are energized. The logic statements which emerge from an examination of the relay circuit are:

1. Either (n.c.) STOP or (n.c.) OL deenergize F or R relays, whichever is energized.
2. Forward relay F is energized by pushbutton F and once energized is sealed in memory.

251

3. Reverse relay R is energized by pushbutton R and once energized is sealed in memory.

4. Relay F is also deenergized either by (n.c.) contact R or (n.c.) reverse pushbutton PBR.

5. Relay R is also deenergized either by (n.c.) contact F or (n.c.) forward pushbutton PBF.

The equations representing the above statements are

1. $STOP + OL = \bar{F} \cdot \bar{R}$
2. $PBF = F$, sealed in MEMORY
3. $PBR = R$, sealed in MEMORY
4. $\bar{F} = R + PBR$
5. $\bar{R} = F + PBF$

The static logic diagram for these five equations is shown on the right side of Fig. 9-8f. Note that the STOP and OL pushbuttons are ORd to deenergize both the FORWARD and REVERSE relays, by producing both F and R inputs. When the forward pushbutton PBF is energized, it simultaneously sends an OFF signal to deenergize relay R and also energize relay F. Similarly, when the reverse pushbutton PBR is energized, it simultaneously sends an OFF signal to deenergize relay F and also energize relay R.

Logic elements consisting of standard logic units or combinations of units previously described are currently available in modular or printed circuited "packages" to facilitate modification and maintenance of the control logic.

Although other chapters indicate and employ conventional relay logic, the reader should bear in mind that equivalent static-control devices may be substituted for any of the relay logic shown in the *control* circuits. Insofar as the *power* circuits energizing rotating machines are concerned, conventional relays and contactors are always employed.

9-10.
AUXILIARY DEVICES
FOR STATIC-LOGIC
SYSTEMS

Unlike electromechanical relay and control elements which are capable of handling fairly large currents and require no auxiliary power for their operation, a complete static logic system requires a number of *auxiliary devices*:

1. *DC power supplies* which are relatively ripple-free, having appropriate voltages and sufficient power to operate all the solid-state devices in the system.

2. *Sensors, transducers*, or other suitable means for receiving and accepting *pilot input* signals from sources external to the control system; such devices include thermal sensors, pressure switches, photosensitive transducers, float-level transducers and switches, etc.

3. *Signal converters* which will accept and convert ac, dc, and pulse-type waveforms to levels suitable for use as logic input signals (either a 1 or 0 level of proper magnitude).

4. *Power amplifiers* to operate various output relays, solenoids, steppers, or even small servomotors, in accordance with the decision-making logic (or information proceessing) performed by the static-control system.

Each of the above auxiliary devices vary in design in accordance with the type of static logic system provided by a particular manufacturer. In specifying and designing a complete static logic system, it is necessary that all components are *compatible*. In this case, the components described below are those available for the G.E. transistorized static-control system.*

9-10.1
Power Supply and Incoming ac Line Filter

A typical 5-A dc logic element *power supply* is shown in Fig. 9-9a and the incoming ac line *filter choke* and *capacitor* are shown in Fig. 9-9b. The schematic of both the filter and power supply is shown in Fig. 9-9c. The purpose of the input filter is to suppress harmonics other than the fundamental 60-Hz input at 115-V ac.

The outputs of the power supply shown in Fig. 9-9c provide the following voltages with respect to the COMMON terminal, ground:

a. 12-V ac supply for logic elements such as the variable delay (8 to 300 sec) and auxiliary devices (such as sine-wave to square-wave converter).

b. −12 V, dc, which is the basic power required for all logic modules, associated pilot lamps (G. E. No. 345), and output amplifier triggers. The full-wave rectifiers and the transformer is rated at 5 A, which is sufficient power to drive 600 load units (logic functions rated at approximately 8.3 mA).

c. −4 V, dc, which is a reference voltage to be used for troubleshooting purposes by applying this voltage to specific test points and observing results of output, using a dc 20,000-Ω/V voltmeter.

d. −125 V, dc, which is called the *original input power*, is used in conjunction with various input transducers or pilot devices such as limit switches, pushbuttons, pressure switches, etc. Rated at 200 mA, this supply can simultaneously feed the equivalent of 40 transducers at 5 mA each (see Sec. 9-10.2).

9-10.2
Input (Pilot) Devices

The various input *transducers* (limit switches, pressure switches, etc.) usually selected for operation with static-control elements require 125-V dc for their operation. Ultimately, because of the nature of static logic systems, these are digital (rather than analog) devices of the on-off (1 or 0) type. The electrical contacts of such devices

* G. E. Bulletin GET 3551A, *op. cit.*

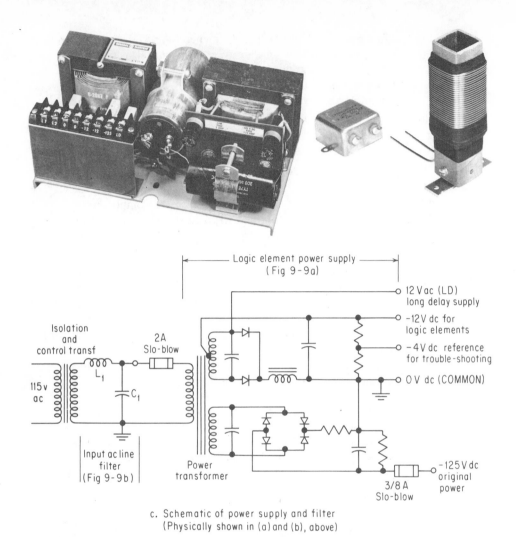

c. Schematic of power supply and filter
(Physically shown in (a) and (b), above)

Figure 9-9

Logic element power supply (5% A rating).

do not exhibit either the contact fidelity or the reliable circuit continuity at the low voltage (12-V dc) required for logic elements. At 125-V dc, good contact fidelity is assured because this high voltage burns away any oxidizing film on the contacts of the input transducers (which may remain idle or unused for long periods). In some applications, because of the design and nature of the transducer, 115-V ac is used (e.g., an E-core transformer used as a linear displacement transducer) as an input to the static-control system.*

* For a complete discussion of transducers see Norton, H.N. *Handbook* of *Transducers* for *Electronic Measuring Systems* (Englewood Cliffs, N.J.: Prentice-Hall, 1969).

Figure 9-10a shows a photocell (essentially an analog device) whose voltage output is converted to a digital logic output using a resistive sensitive converter. The photocell's resistance decreases as more light strikes the cell. The resistive sensitive converter is a device which is capable of using input resistance as a switching means (Section 9-10.3), converting the analog resistance change into a digital 0 or 1 input into the logic circuitry.

The symbol for such an optical input, using a one-line logic diagram, is also shown in Fig. 9-10a. Note that the variable arrow indicates that the trip-point adjustment is variable so that a given illumination level can be adjusted to a 1 input to the logic. Below this illumination intensity, the input to the logic is 0.

A *proximity* transducer input is shown in Fig. 9-10b. The proximity switch input serves as a detector of metals (either ferrous or nonferrous) without any physical contact and without any moving parts. In association with its sensor amplifier and power supply, this transducer supplies a 1 signal to the logic system when the sensing head detects the presence of metal in the proximity of the transducer, or conversely, a 0 level in the absence of metal.

The use of switches (microswitches, limit switches, pushbuttons, or relay contacts) in conjunction with the 125-V dc input is shown in Fig. 9-10c. When the contacts of such a switch is open, the signal to logic is at a 0 level (monitor light OFF). When the switch is closed, the signal to the logic is at a 1 level. Note that the transducer switch is connected to the -125-V dc supply described in Section 9-10.1d.

Transducers requiring ac for their operation require converters (ac to dc) for their operation. Such a combination is shown in Fig. 9-10d.

9-10.3
Converters and Signal
Conversion Circuitry

Converters, as shown in Section 9-10.2, are necessary in conjunction with various transducers to convert the transducer output to an appropriate level and type of logic signal which is acceptable to the static logic element.

Thus, when using analog-type transducers, as shown in Fig. 9-10a, some type of analog-digital converter is required to provide an appropriate signal to the static logic circuitry.

This is also true for the case of the proximity switch transducer shown in Fig. 9-10b. Here the converter consists of a sensor amplifier and a proximity switch power supply. The final output of the converter is that it yields a 1-level signal when metal is detected by the sensor.

The most common type of converter is the dc-dc converter shown in Fig. 9-10c. This converter is designed to operate from a -125 V (or alternatively -24 V) input, and uses two transistors (one *npn* and the other *pnp*) in complementary symmetry. The description of operation is as follows*:

* G. E. Bulletin GET 3551A, *op. cit.*, p. 2–6.

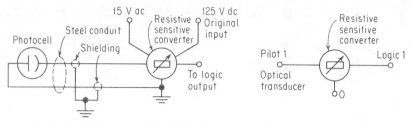

a. Voltage generating optical transducer with converter

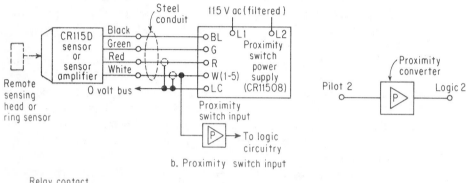

b. Proximity switch input

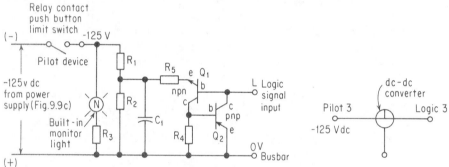

c. Limit switch (or relay contact, push button, etc) converter

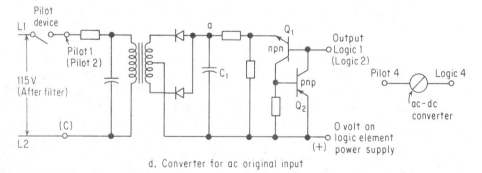

d. Converter for ac original input

Figure 9-10

Transducers and converter inputs.

1. With output terminal of converter, L, connected to the input terminal of a logic element, -4 V is always present at L when the transducer pilot device (PILOT 1) is in an OFF (0) condition. Most important, however, is that capacitor C_1, when connected to the input of a logic element, charges to a negative voltage of -4 V when the logic element is OFF and the converter input is OFF.

2. Transistor Q_1 cannot conduct however, when the pilot device is open (OFF) without a current path to the -125-V supply and its base is more negative than its emitter. The logic output at L is, therefore, well below the 0 V or short-circuit required for a 1 input to the static logic.

3. When the pilot device switches to ON (a 1 input), -125 V is applied to the voltage divider (R_1, R_2) and across the neon monitor lamp, N. The capacitor C_1 begins charging to a more negative voltage and when it rises to more than -4 V, transistor Q_1 conducts heavily, since its emitter is more negative than its base.

4. When transistor Q_1 goes into conduction, it acts as a shorted switch, setting the base of Q_2 at a high negative potential with respect to its emitter. Transistor Q_2 also goes into full saturation, resulting in the base of Q_1 becoming even more positive than its emitter. With both transistors in full conduction, the emitter voltage of Q_1 is less than -0.5 V. Thus, the short of the pilot device (a 1 condition) is duplicated by a short at the input of the logic device (also a 1 condition).

5. With the emitter of Q_1 at practically 0 V, capacitor C_1 discharges its negative charge completely. This acts as a reset so that if the pilot signal opens, C_1 is in a discharged state and ready to charge negatively once more, as indicated in step 3. This technique makes the converter relatively insensitive to electrical noise and line disturbances and increases the reliability of the system. In effect, a 1 output is only possible when accompanied by a valid shorting of the pilot device (a 1 input).

The same principle is also used in the ac to dc converter shown in Fig. 9-10d. The essential difference in the ac-dc converter is the provision of a full-wave rectifier which is only energized when the pilot device is shorted. As in the case of the dc-dc converter, capacitor C_1 is negatively charged by virtue of its connection to the logic circuitry (see step 1). Shorting the pilot device makes point a more negative, causing conduction of Q_1 (and Q_1 in turn as described above) and discharge of C_1, as previously described.

Symbols for the various types of converters are shown in Fig. 9-11. The dc-to-dc converter of Fig. 9-11c is used to furnish a compatible output for inputs of -125-V dc, -24-V dc, or -12-V dc. The sine-wave to square-wave converter, shown in Fig. 9-11e, is used to convert the 12-V ac (available at the LD terminal of Fig. 9-9) to a square wave of the same frequency. With a 60-Hz input, the square wave provides an ON signal for approximately 8 ms and an OFF signal for approximately 8 ms at output terminal 4. An inverse or NOT output is also available at terminal 3.

257

a. Resistive sensitive converter

b. Proximity converter

c. dc to dc converter

d. ac to dc converter

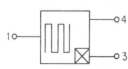

e. Sine to square wave converter

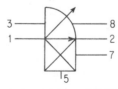

f. Long delay unit (LD)

Figure 9-11

Converter symbols.

The symbol for the DELAY unit is shown in Fig. 9-11f. Time ranges are variable from 6 ms up to 300 s. The input terminal 1 is designed to receive the ON signal and provide a delayed signal at output terminals 7 and 8. Terminal 8 is a (n.o.) time-closing contact while terminal 7 is a (n.c.) time-opening contact. Thus, an ON input may be delayed as much as 300 s before an ON input appears at terminal 8. Terminal 2 of Fig. 9-11f is an inverter output of terminal 1. When an ON appears at 1, an OFF instantaneously appears at 2. Terminal 3 is a connection to a variable potentiometer for producing the variable delay. Terminal 5 is a connection to a RESET input which can override operation at any time and restore the DELAY to its original state. Each output of the DELAY (terminals 7 and 8) has a *fanout capability* of 12 other static-logic inputs.

9-10.4
Output Amplifiers

The output of any static logic system is *insufficient* to drive or operate commercial solenoids, relays or stepping switches, or even to switch the power feeding relatively large loads. It is customary, therefore, to amplify the static logic system output. Depending on the amount of power controlled, output amplifiers are made available in a variety of voltage and current ratings. Commonly used amplifier voltage ratings are 24-V dc (6 W and 36 W) and 115-V ac (1 A, 4 A, and 10 A).

A number of output amplifiers are shown in Fig. 9-12a. It should be noted that these amplifiers cannot be paralleled (if it is desired to increase rating) but a smaller amplifier may be *cascaded* to drive a larger amplifier. A typical first-stage dc amplifier, capable of operating a 24-V dc relay (or driving a larger dc amplifier) is shown in Fig. 9-12b, the schematic for two complete and independent 6-W, 24-V, 0.25-A dc amplifiers on a single plug-in module. The amplifier shown in Fig. 9-12b operates in the following manner:

1. Given an OFF signal (-4 V) at terminal 1, both Q_1 and Q_2 are in conduction and full saturation. The collector voltage of Q_1 (acting as a shorted switch) is at 0-V dc. Therefore, both the base and emitter of Q_2 are at the same potential and Q_2 is at cutoff.

2. With Q_3 nonconducting, the power transistor, Q_4, is also cutoff, with

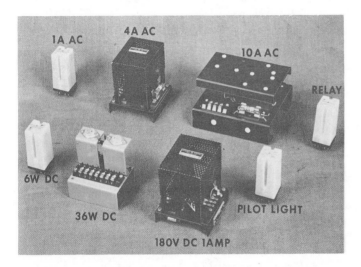

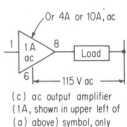

(c) ac output amplifier (1A, shown in upper left of (a) above) symbol, only

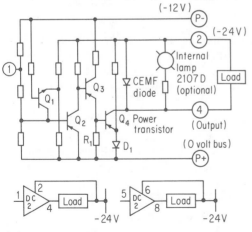

(b) Schematic for 6 W dc, 24 V, 0.25 A, dc, amplifier

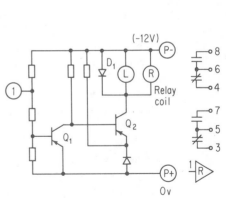

(d) Relay output amplifier

Figure 9-12

Output amplifiers, dc and ac.

diode, D_1, causing the emitter to be more negative (with respect to the zero bus voltage) than the volt drop across the power transistor base resistor, R_1.

3. Given an ON signal (0 volts) at terminal 1 (or a short from 1 to the 0-V bus), both Q_1 and Q_2 are driven into cutoff. With Q_2 no longer shorting the base of Q_3 to ground and the emitter of Q_3 more positive than its base, Q_3 is driven into full saturation. With Q_3 shorted, the base of Q_3 is more negative than its emitter, causing full saturation of the power transistor, Q_4.

4. With power transistor Q_4 acting as a closed switch, current now flows from the 0-V bus through the diode, D_1, and transistor, Q_4, to output terminal 4 through the load and to the -24-V bus. In effect, 0 V (and/or a short from terminal 1 to ground) connects the load across -24 V and the ground (0 V) bus.

The amplifier of Fig. 9-12b also has a number of fail-safe features, notably:

1. Transistor Q_1 ensures that whenever the -24-V supply is present and the -12-V supply is cut off (or vice versa), transistor Q_2 conducts, thus cutting off the power transistor. Thus, in the event of loss of either -12 V or -24 V power, there is no possibility that the load is energized by the amplifier.

2. The CEMF diode across terminals 2 and 4 serves as a surge-protection short circuit to the counter emf generated by a relay or inductive load whenever the amplifier is suddenly switched to the OFF position.

3. Lamp 2107-D in parallel with the load provides visual indication of the ON state of the amplifier. As a quick check of proper amplifier operation, a short between terminal 1 and the 0-V bus should light this lamp.

The symbol for the 6-W dc output amplifier is shown directly below the schematic (Fig. 9-12b). Note that since two complete and independent 6-W amplifiers are packaged on a single plug-in module, selection of terminal 5 for the input dictates automatically the use of terminals 6 and 8 for the load.

In addition to dc output amplifiers, similar amplifiers are available for use with 115-V ac relays, steppers, or solenoids. Only the symbol is shown for these ac amplifiers in Fig. 9-12c. These amplifiers are available in 1-A, 4-A, and 10-A holding current with inrush currents of approximately four times this value. As in the case of the dc output amplifier described above (Fig. 9-12b), shorting terminal 1 to the ground bus (a 1-level input) connects the load through the amplifier to the 115-V ac line.

Another type of amplifier provides a built-in (dc) relay whose contacts are capable of switching *remote* 115-V ac circuits. The relay output amplifier schematic shown in Fig. 9-12d is a self-contained two-transistor amplifier and relay whose contacts are rated 3 A, continuous, and 12-A inrush, with a rated-load life of more than 10^7 switching cycles. The relay output amplifier (Fig. 9-12d) operates in the following manner:

1. An ON signal at terminal 1 (0 V or a short to line P+) cuts off transistor Q_1, placing the collector of Q_1 at −12 V. With its base at −12 V, transistor Q_2 goes into saturation, placing relay coil R (as well as lamp L and the reverse-biased diode) across the −12-V and 0-V buses.

2. When relay coil R is energized and lamp L is lit, all n.c. relay contacts (across 3-5 and 4-6) are opened and all n.o. relay contacts (across 5-7 and 6-8) are closed.

3. Conversely, with no signal (OFF) applied to terminal 1, Q_1 conducts placing the base of Q_2 at 0 potential and causing cutoff of Q_2. This results in an open circuit between the relay coil and ground potential resulting in switching the relay to its OFF state and bringing all contacts to their normally deenergized state shown in Fig. 9-12d.

4. As in the amplifier shown in Fig. 9-12b, diode D_1 serves to dissipate the counter emf generated by the relay coil when it is deenergized.

The relay output amplifier of Fig. 9-12d has a pickup time of approximately 10 to 15 ms and a dropout time of approximately 30 to 45 ms.

9-11.
COMPLETE STATIC
LOGIC CONTROL
SYSTEM

Having treated all the elements of a static logic control system, let us combine them to determine how a complete system operates. The block diagram of such a system is shown in the upper part of Fig. 9-13a and the schematic diagram is shown in Fig. 9-13b.

A simplified description of each of the blocks in Fig. 9-13a follows:

A. Transducer, *FS*

The sensor (transducer) for this particular system is a float switch, *FS*, having two stable states:

1. The ON state occurs when the liquid level is *below* a specified *minimum*. When transducer *FS* is in the ON state, its output level is 1.
2. The OFF state occurs when the liquid level is *above* a specified *maximum*. When transducer *FS* is in the OFF state, its output level is 0.

B. Signal Conversion
Circuitry

Depending on the nature of the transducer employed, some signal conversion is usually required to match the transducer output signal to the input of the static switching circuitry, as noted in Section 9-10.3. As noted from the symbol in Fig. 9-13a, the signal conversion is dc to dc. The components which perform the signal conversion consist of resistors R_1 through R_4 lamp L_1 and reverse-biased diode D_1, (see Fig. 9-13b). The 1 and 0 levels are provided in the following way:

1. OFF state

a. When transducer *FS* is OFF (open), no voltage drop occurs across lamp L_1, due to the open circuit. With *FS* at 0 level and L_1, respectively,

261

in the OFF state, the voltage at point a is at a negative potential, somewhere between -6 V and ground.

b. The voltage at point a is determined by the parallel combination of R_1 and R_2 in series with R_3 and R_4. The values of resistors R_1 through R_4 are selected so that the voltage at point a is -5 V, and the voltage at point b is -4 V.

c. Thus, the signal-conversion circuitry converts a 0-level transducer output to a -4-V input to the base of transistor Q_1, representing the 0-level output from the signal converter. Lamp L_1 is OFF indicating a 0-level condition.

2. ON state

a. When the transducer FS is ON (closed), there is now a total of 54 V across the series combination of L_1 and R_1, and lamp L_1 lights indicating the 1 level or ON state.

b. The potential at point a tends to go positive, but this causes diode D_1 to conduct heavily, bringing point a to ground potential. With a at ground potential there is 48 V across L_1 while transducer FS is in the ON state.

c. With point a at ground potential, point b is also grounded. The signal conversion circuitry converts a 1-level transducer output to a 0-volt input at the base of transistor Q_1, representing the 1-level output from the signal converter. Lamp L_1 is ON indicating a 1-level condition.

3. Summary

The results of the signal conversion described above, yields the following summary:

(FS) Transducer	Transducer Level	Potential at Point b (V)	Lamp L_1
OFF	0	-4	OFF
ON	1	0	ON

C. Static Switch

The static switching is accomplished by transistors Q_1 and Q_2 and their associated bias resistors R_2 through R_6. Both transistors are identical in CE configuration with emitter grounded. The biasing ensures that the transistors are either in saturation or cutoff (and never in the active region of their emitter characteristic). Let us consider the action of transistors Q_1 and Q_2 given a 0- and 1-level input at point b, respectively, in turn.

1. 0 Level input at point b $(-4\ V)$

a. Transistor Q_1 with -4 V applied to its base, conducts heavily, and is fully SATURATED. In saturation, it acts as a closed switch, placing its collector, point c at ground potential. Thus, with transducer FS in the OFF state, transistor Q_1 is ON. For this reason, Q_1 is regarded as an inverter, inverting a 0-level input to a 1 level output (ground potential at point c).

b. Transistor Q_2 with 0 V applied to its base is in a nonconducting state of cutoff. When cut off, the potential at its collector, point d, is -6 V.

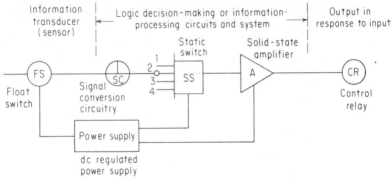

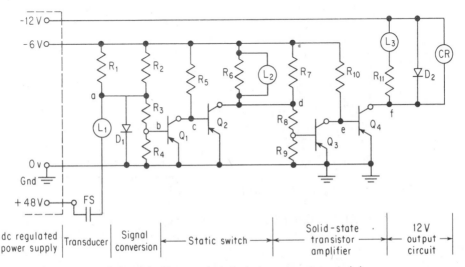

a. Block diagram of a static logic system in elementary form

b. Schematic diagram of static logic system shown in (a)

Figure 9-13

A typical static-switching control system.

This results in no voltage drop across L_2 and the lamp L_2 is OFF. Note that transistor Q_2 inverts the logic of Q_1, restoring the output of the static switch to the original input condition.

c. Summary: Q_1 inverts a 0-level input to a 1 output, Q_1 is ON.

Q_2 inverts a 1 input to a 0 output, Q_2 is OFF.

L_2 is OFF when the input to Q_1 is 0 level and L_1 is also OFF.

2. 1-Level input at point b (0 volts)

a. Transistor Q_1 is biased OFF, producing a -6-V, 1-level signal at its collector.

b. Transistor Q_2 is fully saturated, ON, acts as a shorted switch, applying 6 V across L_2. Lamp L_2 is ON, reflecting a 1-level input at point b.

263

c. Summary: Q_1 inverts a 1-level input to a 0 output, Q_1 is OFF.

Q_2 inverts a 0 input to a 1 output, Q_2 is ON.

L_2 is ON when the input to Q_1 is a 1-level and L_1 is also ON.

From the above it may be seen that when the transducer output is 0, the static switch output is 0, and when the transducer output is 1, the static switch output is also 1. A double signal inversion is performed by transistors Q_1 and Q_2 to make this possible. Neither Q_1 nor Q_2 are in conduction or cutoff at the same time (see Table 9-2 below).

D. Transistor Amplifier The output of switching transistor Q_2 is insufficient to drive a commercial control relay, CR. The amplifier portion consists of power transistors Q_3 and Q_4, biased in such a way that they produce both current and voltage amplification (a characteristic of CE transistors) and produce at the same time signal inversion. Thus, Q_3 is on when Q_1 is on, and Q_2 and Q_4 are correspondingly off whenever an OFF signal is indicated by lamp L_1.

In the presence of an ON from transducer FS the following action occurs:

1. Lamp L_1 is ON, transistor Q_1 is OFF, and transistor Q_2 is ON (1 level).
2. Transistor Q_3 inverts the logic of Q_2 and is cut off because point d (Fig. 9-13b) is grounded.
3. Transistor Q_4 inverts the logic of Q_3 and turns ON, placing collector f at ground potential and providing an output voltage of 12 V across L_3 and control relay CR.

In effect, when FS is at a 1 level (ON), the following are all ON or 1: Q_2; Q_4; lamps L_1, L_2, L_3; and control relay CR.

E. Output Circuit The above-described operation ensures that the control system may be used to refill a tank to a given liquid level. In response to a low liquid level, transducer FS is switched ON, energizing all lamps, Q_2, Q_4, and control relay CR. This relay starts the motor that drives the pump to refill the tank. Relay CR remains energized until the liquid level rises sufficiently to return float switch FS to its OFF (0 level) position. The response to a 0 transducer input turns all lamps off, deenergizes relay CR, and stops the motor and pump.

Table 9-2 summarizes the operation of the static-switching control system shown in Figs. 9-13a and b. In this table the 0 denotes the OFF state and the 1 denotes the ON state of each device.

From the above description of operation and from Table 9-2 we may summarize the following generalizations and interesting properties of static devices:

1. Transistor elements in both the static switch (Q_1 and Q_2) and the solid-state amplifier (Q_3 and Q_4) perform a signal inversion of the digital

TABLE 9-2 SUMMARY OF STATES IN STATIC-SWITCHING CONTROL SYSTEM

	TRANSDUCER	LAMP		TRANSISTORS			LAMP	LAMP	RELAY
	FS	L_1	Q_1	Q_2	Q_3	Q_4	L_2	L_3	CR
OFF	0	0	1	0	1	0	0	0	0
ON	1	1	0	1	0	1	1	1	1

logic. A 1 input applied to the device results in a 0 output and vice versa. It has been shown that a number of logic elements (such as NOR and NOT devices) perform this function. (See Sections 9-2 through 9-7.)

2. The lamps reveal (at any instant) the existing states of the logic elements and serve as both a means of ensuring normal operation and troubleshooting of the circuitry. For example, *all* lamps must be at the *same state* at *all* times. The failure of any lamp to light or go out is an indication of the nature and cause of the malfunction of the system. For example, if L_1 is OFF but L_2 and L_3 remain ON, it indicates a shorted transistor Q_2, an open transistor Q_1, a shorted resistor R_5, etc. (See Fig. 9-13b.)

3. In both the static switch (Q_1, Q_2) and the solid-state transistor amplifier (Q_3, Q_4), one of the transistors is in conduction while its companion transistor is cut off. At no time are both transistors either at a 1 or 0 state, simultaneously. (See Table 9-2.)

4. Unlike conventional linear transistor amplifiers, the transistors in the static switch (Q_1, Q_2) and the power amplifier (Q_3, Q_4) are biased to operate either at cutoff or in a heavily saturated condition. At no time are the transistors operating on the linear portion of their emitter characteristics.

9-12.
DESIGNING STATIC-LOGIC SYSTEMS

The approach of this text has been to present conventional relay circuits first, followed by their static logic equivalents. In designing simple static logic control circuits, it is usually simpler to draw the conventional relay circuit first and then attempt a conversion to static logic. In highly complex systems, this procedure may be more time consuming, and designers experienced with logic elements may work directly in terms of functional specifications performed by the system. In designing static logic systems, the following steps are useful, particularly with relatively less complex systems:

1. Make a list of the various functions performed by the system, including relay pickups, contactor energizations, and relay dropouts.

2. Write each functional step in the sequence in the same order in which the proper system performance occurs.

3. Write equations for each functional step in Boolean-algebra form.

4. Substitute static logic elements for the equations written, and draw logic diagrams for each.

265

5. Combine the various static logic elements into a single, composite diagram.

6. Reexamine the composite diagram to eliminate redundancies, simplify circuits, and conform to commercial modules.

The above procedure is illustrated in Exs. 9-1 and 9-2, in which a static logic system is designed to operate a commercial three-speed polyphase squirrel-cage induction motor (SCIM). The low and high speeds of the SCIM are provided by one three-phase delta-connected winding providing 12 poles and 6 poles, respectively, when the phase belts are connected in parallel and series, respectively. The medium speed is provided by a second, wye-connected winding producing 8 poles. At a frequency of 60 Hz, since speed, $S = 120f/P$, the low and high speeds are 600 and 1200 rpm, respectively. The medium speed is 900 rpm. The line contactors which provide such operation are shown in Fig. 9-14. Note that this arrangement is but one step more complex than the arrangement shown in Fig. 7-10a, which also has two windings to produce a two-speed SCIM using pole changing. The difference is that the SCIM of Fig. 9-14 connects each phase winding in series and parallel, using the method described in Sec. 7-3. (Actually, four speeds are possible if this is done to both windings.)

The starting specifications for such a motor are as follows:

1. The motor may be *started* in *any* of its three speeds: low, medium, or high.

2. If the motor is running at a low or medium speed, it may be advanced to its next higher speed (i.e., from low to medium *or* from medium to high) without requiring that the motor stop or slow down.

3. If the motor is operating at a higher speed it is necessary to press the STOP contact so that the motor slows down before energizing it at a lower speed. This prevents a direct change from a higher to a lower speed which might damage high-inertia equipment in the event of an abrupt speed transition.

4. Any overload (or short circuiting of line contacts due to failure to clear) will prevent operation of the motor in any speed.

The various functions which must be performed are as follows:

1. Three control relays (CRs) are required to energize the main line contacts, H, M, and L, respectively in Fig. 9-14, designated as CR_1, CR_2 and CR_3.

2. CR_1 will energize line contactor, L, closing all low-speed contacts.

3. CR_2 will energize line contactor, M, closing all medium-speed contacts.

4. CR_3 will energize line contactor, H, closing all high-speed contacts.

5. CR_1 is energized (relay pickup conditions) under the following conditions:

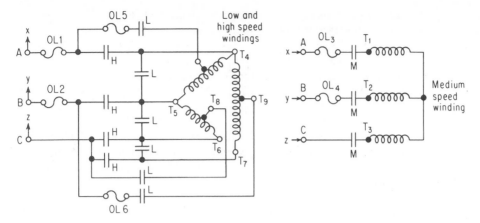

Figure 9-14

High, medium, and low line contacts for 3-speed
SCIM used in Exs. 9-1 and 9-2.

    a. All overload contacts are closed.
    b. The STOP button is closed (not depressed).
    c. The LOW button is depressed (energized).
    d. Control relay CR_2 is not energized (nor contactor M).
    e. Control relay CR_3 is not energized (nor contactor H).

6. CR_2 is energized under the following conditions:
    a. All overload contacts are closed.
    b. STOP is not depressed.
    c. The MEDIUM button is depressed (energized).
    d. Control relay CR_3 is not energized (nor contactor H).

7. CR_3 is energized under the following conditions:
    a. All overload contacts are closed.
    b. STOP is not depressed.
    c. The HIGH button is depressed (energized).

8. Line contactor L is energized under the following conditions:
    a. CR_1 is energized (see limitations 5).
    b. CR_2 and CR_3 are not energized.

9. Line contactor M is energized under the following conditions:
    a. CR_2 is energized (see limitations 6).
    b. CR_3 is not energized.

10. Line contactor H is energized under the following condition:
    a. CR_3 is energized (see limitations 7).

11. CR_1 is deenergized under the following conditions:
    a. Any overload relay in the main line clears the relay circuit.
    b. STOP is depressed.
    c. Relay CR_2 is energized.
    d. Relay CR_3 is energized.

12. CR_2 is deenergized under the following conditions:
   a. Any overload relay is energized.
   b. STOP is depressed.
   c. Relay CR_3 is energized.
13. CR_3 is deenergized under the following conditions:
   a. Any overload relay is energized.
   b. STOP is depressed.

Note that items 1 through 7 deal with the manner in which control relays are picked up. Items 8, 9, and 10 cover the manner in which line contactors are energized. Items 11 through 13 cover the manner in which control relays are deenergized.

EXAMPLE 9-1

a. Write the Boolean equations to summarize the above statements.
b. Draw the relay circuit to accomplish the above equations, using relay logic.

Solution

a. 1. Relay pickup equations

$$CR_1 = \overline{OLs} \cdot \bar{S} \cdot \overline{CR_2} \cdot \overline{CR_3} \cdot L$$
$$CR_2 = \overline{OLs} \cdot \bar{S} \cdot \overline{CR_3} \cdot M$$
$$CR_3 = \overline{OLs} \cdot \bar{S} \cdot H$$

2. Contactor equations:

$$L = CR_1 \cdot \overline{OLs} \cdot \bar{S} \cdot \overline{CR_2} \cdot \overline{CR_3}$$
$$M = CR_2 \cdot \overline{OLs} \cdot \bar{S} \cdot \overline{CR_3}$$
$$H = CR_3 \cdot \overline{OLs} \cdot \bar{S}$$

3. Relay dropout equations (note that *either* of the following may cause dropout):

$$\overline{CR_1} = OLs + S + CR_2 + CR_3$$
$$\overline{CR_2} = OLs + S + CR_3$$
$$\overline{CR_3} = OLs + S$$

b. The relay circuit for the above equations is shown in Fig. 9-15. Note that for CR_1 to pick up (as shown in Ex. Solution a.1), all conditions stated must be met.

The procedure for developing static logic from the relay equations stated in Ex. 9-1 follows the procedure outlined in the beginning of Section 9-12. Example 9-2 shows how static logic equations may be developed using the relay logic equations.

EXAMPLE 9-2

Using the relay logic equations developed in Ex. 9-1,
a. Write static logic equations in Boolean form

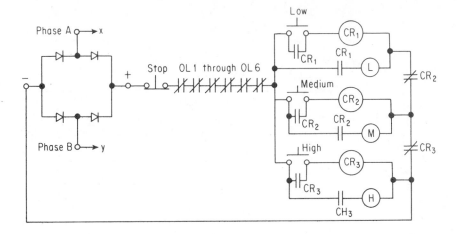

Figure 9-15

Relay circuit for design specifications
of Ex. 9-1.

b. Draw the static logic circuit to perform the functions of Fig. 9-15.

Solution

a. 1. To start the motor at any of the three speeds,

$$CR_1 = L \cdot \bar{S} \cdot \overline{CR}_2 \cdot \overline{CR}_3$$
$$CR_2 = M \cdot \bar{S} \cdot \overline{CR}_3$$
$$CR_3 = H \cdot \bar{S}$$

2. Once the motor is running in either low (L), medium (M), or high (H), the following conditions exist for the line contactors:

$$L = CR_1 \cdot \overline{OLs} \cdot \bar{S} \cdot \overline{CR}_2 \cdot \overline{CR}_3$$
$$M = CR_2 \cdot \overline{OLs} \cdot \bar{S} \cdot \overline{CR}_3$$
$$H = CR_3 \cdot \overline{OLs} \cdot \bar{S}$$

3. While the motor is running in LOW, MEDIUM, or HIGH, one of the three relays is energized or

$$CR_1 = (\bar{S} \cdot \overline{OLs} \cdot \overline{CR}_2 \cdot \overline{CR}_3)(CR_1 + L)$$
$$CR_2 = (\bar{S} \cdot \overline{OLs} \cdot \overline{CR}_3)(CR_2 + M)$$
$$CR_3 = (\bar{S} \cdot \overline{OLs})(H + CR_3)$$

4. To stop the motor, the relay dropout equations (Ex. 9-1a.3) may be used, or

$$\overline{CR}_1 = S + OLs + CR_2 + CR_3$$
$$\overline{CR}_2 = OLs + S + CR_3$$
$$\overline{CR}_3 = OLs + S$$

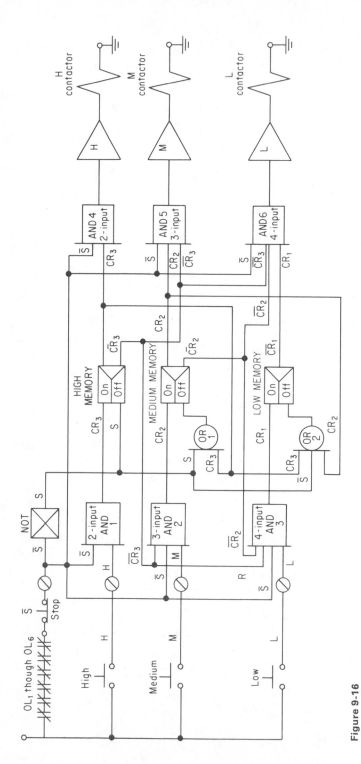

Figure 9-16

Static-logic circuit for design specifications of Ex. 9-2.

5. Some simplification is realized in that overload contacts OLs and STOP are always acting together. Therefore, these may be conditioned from the same signal source, $\bar{S}$.

b. The static logic circuit for the above equations is shown in Fig. 9-16.

The use of OR and MEMORY functions in Fig. 9-16 in conjunction with AND functions implements the logic of the equations stated in Ex. 9-2a. Thus, the motor may be stopped in any speed by generating either $\overline{CR}_1$, or $\overline{CR}_2$, or $\overline{CR}_3$, as noted in Ex. 9-2a4. The OR2 turns off the LOW MEMORY by *either S or CR_3 or CR_2* signals. Similarly, OR1 turns off MEDIUM MEMORY by either S or CR_3.

Note also the necessity for signal conversion circuitry prior to the input to static logic in Fig. 9-16 and the necessity for amplifiers to drive the various line contactors, as described in previous sections.

BIBLIOGRAPHY

Arnold, B. H. *Logic and Boolean Algebra* (Englewood Cliffs, N.J.: Prentice-Hall, Inc., 1962.

Baker, J. P. "Cypak Systems, An Application of Logic Functions," *Westinghouse Engineer* (July 1956).

Duffy, J. J., and Spatz, S., "Static Switching Today," *Electrical Manufacturing* (May 1959).

General Electric, *Application Manual, Transistorized Static Control*, Bulletin GET 3551A, General Purpose Control Department, Bloomington, Ill., June 1970.

Harwood, P. B. *Control of Electric Motors*, 4th ed. (New York: John Wiley & Sons, Inc., 1970).

Heumann, G. W. *Magnetic Control of Industrial Motors*, 3 vols. (New York: John Wiley & Sons, Inc., 1961).

Jones, R. W. *Electric Control Systems* (New York: John Wiley & Sons, Inc., 1953).

Kosow, I. L. *Electric Machinery and Transformers* (Englewood Cliffs. N. J.; Prentice-Hall, Inc., 1972).

Maley, G. A., *Manual of Logic Circuits* (Englewood Cliffs, N.J.: Prentice-Hall, Inc., 1970).

Norpak Solid State Logic Control, Bulletin 8851, 2, 3, Square D Company, Mineola, N.Y.

Norton, H. N. *Handbook of Transducers for Electronic Measuring Systems* (Englewood Cliffs, N. J.: Prentice-Hall, Inc., 1969).

QUESTIONS AND PROBLEMS

9-1. Define
   a. Element
   b. Function
   c. Relay logic
   d. NEMA logic
   e. NOR logic.

9-2. a. Draw a three-input OR symbol using NEMA logic. Show the effect of a 1 input at terminal A on the output
   b. Draw a three-input OR circuit using NOR logic. Show the effect of a 1 input at terminal A on the output
   c. Using the two *pnp* transistors shown in Fig. 9-1c, explain why a negative input pulse of 1 produces a negative output pulse at collector of Q_2.

9-3. a. Using the diode AND circuit of Fig. 9-2b, explain why a positive (1) pulse must be applied to all inputs in order that the output is positive (1)
   b. Explain why a 0 input at any AND circuit produces a 0 output
   c. Explain why a negative input of 1 is required simultaneously at all three transistor inputs of Fig. 9-2c to produce a negative output of 1 at the collector of Q_2.

9-4. Explain why the NOT function is performed by any active device capable of phase inversion.

9-5. a. Distinguish between OFF-RETURN MEMORY and RETENTIVE MEMORY in terms of purpose, function, and application
   b. Using the NOR logic of Fig. 9-4c, explain why a 1 pulse at A produces a 1 pulse at the output
   c. Using the transistor circuit of Fig. 9-4b, explain why a negative pulse at A produces a negative pulse at the output, even when the pulse is removed.

9-6. Using a simple NOT inverter, show how it is possible to convert
   a. An OR to a NOR, using diodes and a transistor inverter
   b. An AND to a NAND, using diodes and a transistor inverter
   c. Explain the operation of the circuits drawn in parts a and b, in detail.

9-7. a. Explain how ON-DELAY (TIME DELAY AFTER ENERGIZATION) differs from OFF-DELAY (TIME DELAY AFTER DEENERGIZATION)
   b. Give one application of each of the above functions in control circuits.

9-8. Draw from memory the English logic NEMA symbol, and write the Boolean equation expressing the logic, for each of the following functions:
   a. AND function
   b. OR function
   c. NOT function
   d. OFF-RETURN MEMORY
   e. RETENTIVE MEMORY
   f. TIME DELAY AFTER ENERGIZATION
   g. TIME DELAY AFTER DEENERGIZATION.

272

9-9.   Using NEMA logic symbols, draw a static logic control system to replace the control circuitry shown in Fig. 2-2.

9-10. Using NEMA logic symbols, draw a static logic control system to replace the control circuitry shown in Fig. 2-3.

9-11. Repeat Problem 9-10 for the wound-rotor starter of Fig. 5-5.

automatic feedback control systems

10-1.
GENERAL

The closed-loop technique, as a means of providing feedback information for the automatic control of speed, was first introduced in Sections 4-6 through 4-10, 8-4, and 8-7. In summary, these (previously described) devices provide an automatic control (or regulation) of voltage, current, and speed by comparing a *reference* signal (representing the *desired* condition) with the actual output signal (voltage, current, or speed) fed back to rotary or static amplifier windings.

Some rotary amplifiers* use a differential flux principle, so that the net output flux controlling the exciter emf represents the *difference* between the *desired* output and the *actual* output. The output of such exciters responds to (1) any change in the reference signal input (i.e., the desired output), (2) a change in the load or output conditions producing a change in the feedback signal, and (3) a change in the prime-mover speeds of either

* See I. L. Kosow, *Electric Machinery and Transformers* (Englewood Cliffs, N.J.: Prentice-Hall, Inc., 1972), Chap. 11.

274

the exciter or the generator (or both) producing a change in output, reference, or feedback signals. The output of the static or rotary amplifier, therefore, is a function of the difference between (and change in) the opposing reference and feedback signals. When both these signals are constant, the output is constant. When either of these signals varies, the amplifier output will vary in such a manner as to return the system to a (desired) constant output or *null* (zero rate of change) condition.

> A *closed-loop system* in which the output (or a portion of it) is *fed back* for purposes of *comparison* with the input (and the difference between the two is amplified and fed to an actuator) so that the *output follows variations of the input*, is called a *servomechanism.** The illustrations used in previous chapters were concerned with the control of such electrical quantities as voltage and current (in addition to speed). Using servomechanisms, it is currently possible to control, automatically, a vast number of physical quantities such as are indicated in Section 10-8. The ability to control an almost unlimited variety of physical quantities has been made possible through a reduction of the problem of automatic control to its basic or fundamental elements, known as the *generalized servomechanism.*

10-2.
THE GENERALIZED
SERVOMECHANISM

The closed-loop circuits previously introduced and discussed may be considered as having certain *common* characteristics in that they all possess certain *common elements*. The recognition of an insight into these common characteristics is a relatively recent development. Automatic devices for lifting and transferring grain were known to the Egyptians. The governor developed by James Watt to maintain a constant steam-engine speed (by adjusting the throttle of the steam engine using the centrifugal force of a flyball) is a typical servomechanism. A thermostat, used to control the temperature of a room or a liquid bath, is another example.

Nicholas Minorsky† is credited with the first insights that led to a contemporary servomechanism theory when he published his analysis of automatically steered bodies in 1922. This perception was followed by that of Harry Nyquist† (Bell Telephone Laboratories), whose analysis of feedback in electronic amplifiers (1932) provided general insight into the dynamic characteristics of control systems. Prior to and during World War II, the demand for automatic control in radar-tracking and gun-fire control systems increased the general dissemination and study of the properties of these systems.

---

* The subject of servomechanisms, from either a theoretical or a practical point of view, is an extensive study. This chapter is intended as an introduction to the subject, merely covering broad applications of rotating machines as used in servomechanisms for the closed-loop control of industrial processes. For a more detailed study of the subject, see E. R. Johnson, *Servomechanisms* (Englewood Cliffs, N.J.: Prentice-Hall Inc., 1963), and references cited at the end of this chapter.

† See references at the end of this chapter.

275

Figure 10-1 shows the similarity between the feedback amplifier and the generalized servomechanism. An amplifier with feedback is shown in Fig. 10-1(a). The output voltage, e_{out}, is fed through a feedback network, having a feedback factor, β. The output of the feedback network, $\beta(e_{out})$, is compared with the input signal in such a way that the phase of the feedback voltage is opposed to that of the input voltage. The difference or error signal, $e_{in} - \beta(e_{out})$, is fed to the amplifier having a gain, G. The output of the amplifier is, therefore, $G(e_{in} - \beta e_{out})$. The amplifier output

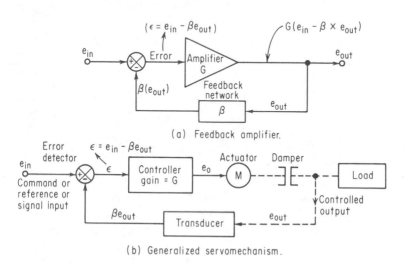

(a) Feedback amplifier.

(b) Generalized servomechanism.

Figure 10-1

Analogy between feedback amplifier and generalized servomechanism.

has both direction and magnitude, two properties that make for maximum amplifier stability. For example, an increase in output voltage tends to increase the feedback voltage and to reduce the error-signal input to the amplifier, thereby reducing the output voltage to its desired (normal) value. The ratio of output to input for the entire system, e_{out}/e_{in}, is the closed-loop gain of the system.

The feedback amplifier of Fig. 10-1a is a specific case of the generalized servomechanism, shown in Fig. 10-1b. The primary function of the feedback amplifier is to maintain a constant and desired output voltage for a given input voltage, regardless of such variables as internal temperature changes and external load changes. This is accomplished by means of negative feedback, described above.

Other advantages of negative feedback in addition to constant output are improved frequency response, reduced distortion, and reduced noise.

The generalized servomechanism, as indicated earlier, may be used to control a number of physical quantities, such as displacement, pressure,

CHAP. TEN / *Automatic Feedback Control Systems*

velocity, force, temperature, etc., in addition to electrical quantities such as current and voltage. This method of control involves the use of a transducer or sensor which converts the controlled output quantity to an electric voltage. A similar transducer may be used for providing the input signal, as well. Both of these electric signals are fed in opposition to each other (negative feedback) to an error detector, whose output signal represents the difference between the reference and the feedback signal. The output error signal, ϵ, is amplified by a controller having a gain, G. The controller amplifies the error signal to produce sufficient power to drive an actuator, as shown in Fig. 10-1b. The load, as represented by any of the above physical quantities, is controlled through displacement or suitable modification of the physical quantity in such a manner that the actuator tends to reduce the error between the (transduced) command signal, e_{in}, and the (transduced) output signal, βe_{out}. As in the case of the feedback amplifier, the closed-loop system gain H is

$$H, \text{ closed-loop system gain} = \frac{\text{output signal}}{\text{input signal}} = \frac{e_{out}}{e_{in}} \quad (10\text{-}1)$$

and the controller gain is

$$G = \frac{\text{controller output}}{\text{controller input}} = \frac{e_o}{\text{error}} = \frac{e_o}{e_{in} - \beta e_{out}} \quad (10\text{-}2)$$

where e_o = voltage output (or its analog equivalent) of the controller amplifier
e_{in} = transduced command signal input voltage
βe_{out} = transduced output of the physical quantity being controlled

Equation (10-2) may be solved for e_{out} in terms of the other quantities, yielding

$$e_{out} = e_{in} \frac{G}{1 + \beta G} \quad (10\text{-}2a)$$

and when expressed as defined in Eq. (10-1), the closed-loop system gain H, becomes

$$\text{Closed-loop system gain} = \frac{e_{out}}{e_{in}} = \frac{G}{1 + \beta G} = G\frac{1}{1 + \beta G} = H \quad (10\text{-}3)$$

where all the terms have been defined previously.

Note that since G and β are both ratios (dimensionless), Eq. (10-3) expressing the overall system gain is also dimensionless. The product, βG, in the denominator of Eq. (10-3) is called the open-loop gain, or the *loop gain*, for short. The open-loop gain is the resultant signal output-input ratio produced by a given input signal which passes through the

277

controller, the actuator, and the transducer without any feedback into the error detector. If the open-loop gain, βG in Eq. (10-3), is significantly less than unity, the overall closed-loop system gain is essentially the same as that of the controller amplifier, G. Conversely, when the open-loop gain is significantly larger than unity, the overall system closed-loop gain is approximately $1/\beta$. If the transducer feedback factor, β, is approximately unity when $\beta G \gg 1$, the overall closed-loop gain is approximately unity, since it equals $G/(1 + \beta G)$. Thus, a high open-loop gain βG does not necessarily imply a high closed-loop gain. The (open-) loop gain, therefore, is a factor which determines the sensitivity of the overall closed-loop system to internal or external changes.

The polarity of the (negative) feedback signal in Fig. 10-1b is such that it will tend to *reduce* the system error and to stabilize when the controller output is zero. If, on the other hand, the feedback signal is *positive*, such that it tends to *add* to the input signal and to *increase* the error, *instability* results. A *positive feedback signal* will cause the loop gain βG in Eq. (10-3) to become *negative*.

A loop gain of between 0 and -1 results in an overall system gain which exceeds the controller amplifier gain. When the (open) loop gain is -1, the overall system gain approaches infinity,* producing marked instability.† In summary, therefore, the static characteristics of the generalized servomechanism are:

1. The factor $1/(1 + \beta G)$ in Eq. (10-3) determines the difference between the command signal and the feedback signal from the controlled output. Thus, any changes in either the external or internal voltages of the closed-loop system will produce an effect of $1/(1 + \beta G)$ on the controlled output quantity.
2. Negative feedback $(+\beta G)$ results in a reduction in gain but an improvement in overall system stability, linearity of response, and bandwidth.
3. Positive feedback $(-\beta G)$ results in increased gain, marked instability, and nonlinearity of response as the controller amplifier approaches saturation.

10-3.
SERVOMECHANISM
ELEMENTS

"*Afficionados*" notwithstanding, there is little difficulty in assembling a working high-fidelity system, because the individual components are clearly established and recognizable. The recognition of the function of each unit and its method of application in the overall system is essential. The individual components (*amplifier, speakers*, and inputs such as *record player, tuner, tape recorder*, and

---

* This assumes a constant controller amplifier gain G. As the overall system gain increases, however, saturation of the controller amplifier is produced, thus tending to limit the system gain.

† Note the similarity of the magnetic amplifier feedback effects (Section 8-4) to those of the generalized servomechanism.

microphone) may be assembled using various and interchangeable units, *regardless of origin and manufacturer*, within limits of impedance matching.

A complete servomechanism may be assembled in a somewhat similar manner, using various individual components or elements supplied by various manufacturers to control a variety of physical quantities automatically. Before these elements can be specified, however, it is necessary to identify each recognizable component in terms of its function, as in the case of the high-fidelity system. Recognition of the function of each of the units and its method of application in the overall system is, therefore, the first step for proper specification of a working servomechanism.

The elements and function(s) of the various units comprising a servomechanism are the following:

1. Error detector. Three functions are performed by this unit:
a. Measurement of the variations of the controlled physical quantity using suitable transducers (see 5) to convert various measurable physical quantities into electrical quantities.
b. Comparison between the relative magnitudes and directions (phase) of two input quantities; these are the command signal (reference input) and the controlled output.
c. Error generation of an error signal of such magnitude and direction (phase) to represent the difference between the two inputs (see preceding) at any instant.

2. Controller. This unit is a power amplifier which serves to amplify the error signal produced by the error detector. It may be an electronic, magnetic, rotary, hydraulic, or pneumatic amplifier.

3. Actuator. The driven element which responds to the amplified signal produced by the controller. The actuator drives the output mechanism and, in so doing, produces a change in the variable or physical quantity to be controlled in response to a command signal.

4. Damper. (See Section 10-5.) A damper is required to reduce oscillation or instability in the response of the servomechanism to command signals.

5. Feedback loop. Contains elements which convert the variations in output (controlled variable) to suitable quantities for input to and comparison within the error detector. It is customary to sense the output by means of transducers (Section 10-8) which convert this output variation into an electric signal of proper form for input and comparison within the error detector.

Unlike a high-fidelity system, however, the identification of the above five servomechanism elements as separate and individual units is quite difficult. In some instances these functions may be combined in a single piece of apparatus. In the case of the amplidyne, for example, three of the five elements are contained within the multifield exciter itself.* Acting

* See Kosow, *Electric Machinery and Transformers*, Sec. 11-16.

through its differential flux principle, the amplidyne is an error detector. As a cross-field rotary power amplifier, it serves as a controller. Damping is achieved through appropriate antihunt networks used in the commercial amplidyne. Only the actuator (i.e., the dc motor driven by the amplidyne) and the feedback network representing the load output are external to the single apparatus which incorporates the three functions described above.

Identification, therefore, of discrete and separate servomechanism elements is difficult, particularly when one considers all of the possible variations and techniques employed in closed-loop control. Despite differences in appearances, however deceiving, *all* servomechanisms possess specific similarities:

1. Sensing devices to detect the magnitude and direction of variation in physical quantities (commanded and controlled).
2. A means of effecting comparisons and evaluating the differences between the reference input (command) signal and the output or actual physical condition to be controlled.
3. A means of amplifying this difference and providing sufficient and proper power to actuate.
4. A mechanism capable of responding to the difference in such a manner that
5. A null is produced, i.e., a condition of no difference between the desired and the actual physical quantities at the output.

A complete servomechanism, therefore, *detects* and *compares* input against output, *generates an error signal* of such amplification in power as may be required and sufficient to correct the error, and in this way *produces a null*, i.e., a correspondence between input and output. It goes without saying that the servomechanism must be capable of doing this accurately, automatically, continuously, and instantaneously.

**10-4.
SERVOMECHANISM
ACCURACY**

For the precise automatic control of any process or physical variable, a servomechanism should respond perfectly to *any and all* changes of commanded input, however slight. Each of the factors which contribute to servomechanism error, therefore, must be identified and reduced to a minimum. Some of the factors included under possible sources of servomechanism error are: dead space, electrical noise, instrumental or transducer error, lost motion in mechanical linkages (as well as other mechanical devices, including dampers), velocity lag, acceleration lag, and maximum controlled servomechanism rate.* A brief description of each of these error factors is given below.

---

* A complete theoretical discussion of servomechanism error considerations is beyond the scope of this volume. For a more thorough treatment, see Johnson, *Servomechanisms* pp. 28–68.

1. Dead space. An actuator, when driven by a controller in response to an error, may cause the load to be driven *beyond* the proper null position (*overshoot*) because of the inertia of the load. A contact type (on-off) transducer in an error detector may not recognize the existence of such error if the overshoot is so small that the contacts fail to close. Similarly, using proportional-type transducers, a very small error signal resulting from overshoot (or undershoot) may not be large enough (even when amplified) to enable the actuator to overcome the starting friction of the load (or the parameter to be controlled). In the same way, a slight change in the input signal may also reveal this error, because the actuator may not develop sufficient torque to respond to the command. Thus, the servomechanism error detector and the overall servosystem is said to have a *dead space* on either side of its proper null position.

2. Electrical noise. Electronic, magnetic, or rotary amplifiers, as used in controllers, are essentially electric devices. These amplifiers tend to produce and generate small spurious electric signals called "electrical noise." The transducers and error detectors associated with the complete servomechanism likewise are subject to this error. If the error signal to be sensed is a very small change in input signal, electrical noise may *mask* or counteract this signal. Hydraulic or pneumatic controllers do not suffer from this type of error. When a servomechanism must respond to exceedingly small signals, some method must be employed to filter aperiodic noise from the periodic error signal.

3. Instrumental or transducer error. Transducers (Section 10-8) are used to convert various physical to electric quantities. The electric signals fed to and generated by the error detector depend on the sensitivity and accuracy of the transducers employed. Errors which fall into this class include the accuracy and linearity of the error detector and amplifiers of the controller, in addition to the transducers which supply signals to them.

4. Lost motion. Whenever mechanical shafts, differentials, gears, friction dampers, or discs are used, there is always the possibility of error resulting from backlash or lost motion. Given a small input signal, therefore, no change in output may be produced until the lost motion is taken up. Thus, it is possible that small input signals may fail to produce a change in the output.

5. Velocity lag. Inertia, friction, and dampers all tend to cause a lag in the controlled output whenever a change is introduced in the input command signal. This error varies with the first derivative of the error signal. When it is required that the load respond almost instantaneously with rapid changes in input, this error may be of serious consequence.

6. Acceleration lag. During periods of rapid acceleration or deceleration of input signals, the controlled output may lag behind the input. This error varies with the second derivative of the error signal.

7. Maximum controlled rate of the servomechanism. If the velocity of the input signal is continuously accelerated, a rate may be reached where the servomechanism output can no longer keep pace with the input. This results in an ever-increasing error between the output and the input. In

281

addition to acceleration lag, the maximum controlled rate of the servo-mechanism is a factor tending to produce error whenever input changes occur at an "excessive" velocity.

"A chain is only as strong as its weakest link." An excessively large error in *any* of the factors above seriously reduces the overall accuracy of the servomechanism under certain conditions of operation. In selecting, specifying, and designing servomechanism elements, attempts are made to reduce the preceding (and other) error factors to a minimum, within limits of practicability.

10-5.
SERVOMECHANISM
INSTABILITY

Let us assume, for the moment, that, as a result of analyzing the error factors and reducing the errors to a minimum, a servo-mechanism is produced which is virtually both frictionless and errorless! Such a servomechanism responds instantly to all changes of input with a minimum of lag and error. The response of such a typical servomechanism used in *position* control (to control the angular position of a particular device) is shown in Fig. 10-2a. At time t_0, an input signal voltage is obtained by rotating an input potentiometer to angle θ_i. Assuming that such a servomechanism is "errorless," it responds instantly (no lost motion, no velocity or acceleration lag errors) since θ_i exceeds θ_o. Its actuator output, θ_o, rotates the mechanical load through the same angle in zero time. But since an errorless servo has no frictional error loss and a correspondingly high controller and closed-loop gain, the output continues to overshoot beyond the commanded input angle, θ_i. The servo-mechanism error detector now responds to an error in which θ_o exceeds θ_i, causing the actuator to be driven in the *reverse* direction, as shown in Fig. 10-2a. Thus, in a virtually "errorless" and undamped servomechanism, it would be relatively impossible for the mechanism to stabilize at a null position where θ_i and θ_o are equal and their difference is zero (zero error).

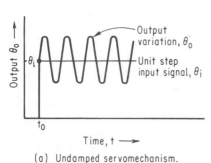

(a) Undamped servomechanism.

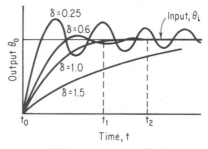

(b) Damped servomechanism.

Figure 10-2

Effect of damping on servomechanism response.

As shown in Fig. 10-2a, the result is called successive overshooting, oscillation, instability, or hunting.

It should be noted that the process of error reduction, in which the various error factors have been *reduced* to a minimum, has *increased* the possibility and tendency toward *instability*. This situation is not as unfortunate as it might appear, however, since it indicates immediately that it is unnecessary to reduce all errors to zero (a completely hopeless task). The very presence or existence of small errors cause the output θ_o to lag the input θ_i. The rate of change of output, $d\theta_o{}_{dt}$, is now less than that of a given step input (infinite rate of change). The reduced slope in the output of such a servomechanism having some error (say due to friction and velocity lag) is equivalent to the addition of *damping*.

10-6.
THE DAMPED
SERVOMECHANISM

The effect of damping on the speed of response and stability of a servomechanism is shown in Fig. 10-2b for a unit step input signal, θ_i, applied at time t_0. With practically no damping whatever, except that necessitated by some friction and velocity lag (say, $\delta = 0.25$), instability and oscillation still occur. But the successive overshoots are now smaller than without damping (Fig. 10-2a), and the oscillation gradually dies away.

Increasing the amount of damping ($\delta = 0.6$) again decreases the slope and the speed of servo response, but it also reduces the instability. For such a higher damping factor [see Eq. (10-7)], the oscillation dies out in time t_2. Further increases in damping ($\delta = 1$) cause the speed of servo response to be reduced still further, as shown by the rate of rise and the slope in Fig. 10-2b. No oscillation is produced at this degree of damping and the servomechanism reaches a null ($\theta_o = \theta_i$) in the *shortest possible* time, t_1.

A further increase in damping ($\delta = 1.5$) reduces the rate of response severely, as shown in Fig. 10-2b, and the time to reach a null is again increased.

The curves of Fig. 10-2 indicate that a certain (*critical*) amount of damping is necessary (1) to provide stability, and (2) to produce a null in the shortest possible time, with (3) the "maximum" rate of response. Increasing the amount of damping (**overdamping**) *reduces* the rate of response and increases the time required to produce a null. Similarly, if the servomechanism is damped by *less* than a critical amount (**underdamping**), the rate of response is also increased but oscillation is produced and the response time to produce a null is increased.

The degree or amount of damping required to achieve critical damping is determined by the **damping factor, δ,** of the servomechanism. Evidently, this factor varies with the nature of the dampers employed, as well as the various elements comprising the servomechanism. It is possible to generalize the factors affecting stability and transient behavior of any

283

servomechanism in terms of its inertial and frictional elements in a general equation*

$$K\theta_i = J\frac{d^2\theta_o}{dt^2} + F\frac{d\theta_o}{dt} + K\theta_o \qquad (10\text{-}4)$$

where K = output torque per unit error angle
θ_i = input angle in radians
θ_o = output angle in radians
J = output moment of inertia in slug-ft² or kg-m²
F = friction torque per unit output speed in $\dfrac{\text{lb-ft}}{\text{rad/s}}$ or $\dfrac{\text{N-m}}{\text{rad/s}}$

For a given step input, $\theta_i(t)$, as shown in Fig. 10-2b, it is evident that the output will ultimately correspond to the input and that θ_o must equal θ_i under steady-state conditions regardless of the amount of friction torque, F, or inertia, J. Eliminating the steady-state portion (equated to zero), and extracting the transient portion, yields a second-order differential equation in the form of

$$J\frac{d^2}{dt^2} + F\frac{d}{dt} + K = 0 \qquad (10\text{-}5)$$

But Eq. (10-5) is in the form of a simple quadratic equation having two roots, r_1 and r_2, whose solution may be expressed as

$$r_1, r_2 = \frac{-F \pm \sqrt{F^2 - 4JK}}{2J} = \frac{F}{2J} \pm \sqrt{\left(\frac{F}{2J}\right)^2 - \frac{K}{J}} \qquad (10\text{-}6)$$

Three possibilities emerge from Eq. (10-6) for stable systems, i.e., those whose oscillations are ultimately reduced to zero. Roots r_1 and r_2 may be:

1. Conjugate and complex, with negative real parts. This is evident when $(F/2J)^2 < K/J$, or $\delta = \dfrac{F}{2\sqrt{JK}} < 1$. This is the case of the *under-damped* servomechanism, where $\delta < 1$, and the oscillations that are produced ultimately attenuate to zero.

2. Negative, real, and equal. This occurs when $(F/2J)^2 = K/J$, or $\delta = \dfrac{F}{2\sqrt{JK}} = 1$. This is the *critically damped* case, in which the servomechanism null is produced in the shortest possible time and with the maximum rate of response, which will not produce oscillation.

3. Negative, real, and unequal. This case occurs when $(F/2J)^2 > K/J$, or $\delta = \dfrac{F}{2\sqrt{JK}} > 1$. This is the case of the *overdamped* servomecha-

---

* H. Lauer, R.N. Lesnick, and L.E. Matson, *Servomechanism Fundamentals*, 2nd ed. (New York: McGraw-Hill Book Company, 1960), p. 81.

nism, where $\delta > 1$ and a null is produced in a time exceeding the critical damping. The rate of response is also considerably reduced, but no oscillation occurs.

Examination of the three possibilities indicates that *instabilities* occur when the roots are *complex* and that *stability* occurs when the roots are *real*. It is further evident that critical damping, F_c, is obtained when $F_c = 2\sqrt{JK}$ and the ratio of these terms is unity. Given a system in which the damping may be varied, the ratio of *actual* damping, F, to *critical* damping, F_c, is the *damping factor*, or

$$\delta = \frac{F}{F_c} = \frac{F}{2\sqrt{JK}} \tag{10-7}$$

The reader should reexamine Fig. 10-2b in the light of Eq. (10-7) for the underdamped ($\delta < 1$), critically damped ($\delta = 1$), and overdamped ($\delta > 1$), servomechanism.

It was previously stated that attempts at error reduction (designed to improve servomechanism accuracy) result in a reduction of damping and a corresponding tendency toward instability. Consequently, some form of additional damping must be added to a servomechanism if null is to be obtained without instability in the shortest possible time with the maximum possible rate of response.

The situation is not quite that simple, however. If attempts are made to increase the rate of response of a servomechanism by increasing either the controller gain or the closed-loop gain, the output torque factor, K, in Eq. (10-7), is correspondingly increased. This results in a decreased damping factor δ, with resultant instability. But for any given load or output to be controlled, the inertia, J, is fairly constant. The only way to increase the damping factor (restoring stability) is to increase the viscous drag or friction, F, of the system, as indicated in Eq. (10-7).

Thus, it would appear ridiculous to use high-gain amplifiers to improve the rate of response and simultaneously add corresponding frictional drag designed deliberately to reduce the useful energy output of the servomechanism. Some type of damping is needed, obviously, which achieves stability and simultaneously has a "viscous" property equivalent to frictional damping without excessive loss of energy or lag between output and input shaft (steady-state error). A consideration of various types of damping indicates whether this is possible.

10-7.
TYPES OF DAMPING
There are two large classes of damping methods employed for stabilization of servomechanisms: viscous damping and error-rate damping.

Viscous damping consists of applying a retarding or counterforce to a servosystem which is directly proportional to the velocity (rate of change)

285

or motion of the system. Whenever a change of input occurs, therefore, viscous damping performs the useful function of reducing the amplitude and duration of the transient oscillation produced on the system. While viscous damping effectively reduces transient oscillation, it produces simultaneously a *steady-state error*. The undesirable steady-state error results from the very fact that the output is changing and that the damping is opposed to and varies in magnitude with such change. Viscous damping, therefore, suppresses oscillation but also *retards* output with respect to input. This may be observed by the various degrees of damping shown in Fig. 10-2b. In some servomechanism applications, this lag or steady-state error between input and output is of little consequence. When it is intolerable, however, it becomes necessary to resort either to error-rate damping or a combination of viscous and error-rate damping.

Error-rate damping, as will be shown, consists of mechanisms applying a retarding or counterforce to the system only in the presence of an oscillatory or transient error. Error-rate damping, unlike viscous damping, has no effect whatever on the steady-state motion of the system.

Viscous Damping   As shown in Fig. 10-1b, viscous dampers are usually located between the actuator output shaft and the load. Some of the more common types of viscous dampers include:

1. Mechanical friction devices. Fluid and pneumatic dashpots, friction discs, brakes, and gear-type devices of a mechanical nature, fall into this class of dampers.

2. Electromagnetic friction devices. Magnetic brakes, eddy-current brakes, and fluids containing magnetic particles, fall into this class. They have the advantages of permitting *electric* adjustment of the degree of viscous damping (i.e., control of the damping factor) by variation of the dc excitation applied to the magnetic field winding.

3. Frictionless dampers. A major disadvantage of the two previous types of dampers is the power loss and excessive heat created, representing a waste of useful energy and requiring larger controllers. A more simple and effective form of viscous damping may be had by developing the damping signal *electrically*. An electric signal may be obtained from a suitable transducer at the output. This signal is then fed back into the closed loop in such a manner as to (a) make an *increasing* error appear *larger*, and (b) make a *decreasing* error appear *smaller*. All frictionless viscous dampers which operate in this manner produce the same effect as viscous damping and, at the same time, are so much more efficient that a larger controller is not required. Typical frictionless dampers are:

a. Tachometer-generator damping. Shown in Fig. 10-3a, a velocity signal may be obtained from a tachometer generator coupled to the output shaft of the actuator. This signal, proportional to the rate of change of output, is fed to the error amplifier in opposition to and along with the normal error signal between the output θ_o and input θ_i. The amplifier thus responds to *two* signals acting in opposition: a position error signal

286

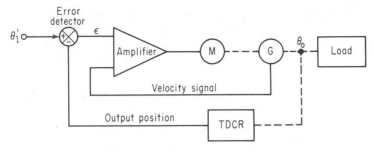

(a) Tachometer-generator damping.

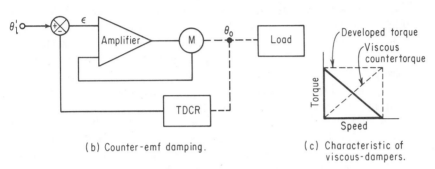

(b) Counter-emf damping.

(c) Characteristic of viscous-dampers.

Figure 10-3

Frictionless dampers.

and a speed or velocity voltage signal. The speed signal effectively damps the error signal in the same manner as a friction damper.

b. Counter-emf damping. Shown in Fig. 10-3b, the counter emf generated by a dc servomotor armature may be used directly as a measure of the speed change. Since counter emf is proportional to speed, there is no necessity for a tachometer generator whenever dc servomotors* are used as actuators. The generated counter emf is fed directly to the amplifier, along with and in opposition to the position error signal.

c. Motor-characteristic damping. Viscous damping is defined as an oppositional force proportional to the rate of change of the system. Its effect is greater at high than at low speeds. Consequently, a servosystem using a *constant-torque actuator* in combination with a viscous damper would experience a *reduction* in torque with an increase in speed of the system, as shown in Fig. 10-3c. But such a drooping speed-torque characteristic occurs inherently in certain types of motors: series, repulsion, universal, and, particularly, in induction motors having high rotor resistance (Table 5-1). When these motors are used as actuators, therefore, the need for dampers is virtually eliminated and, needless to say, such damping is frictionless.

* See Kosow, *Electric Machinery and Transformers,* Sec. 11-13, for a discussion of types of servomotors.

287

Where it is *absolutely necessary* that the steady-state error between θ_o and θ_i be reduced to a minimum, viscous damping *cannot* be used. As previously indicated, all viscous damping is accompanied by a steady-state lag, even when the variation of input command signal occurs at a fixed rate.

Error-rate damping involves the use of components or devices whose output varies as the rate of change of error (i.e., the *acceleration* or *deceleration* of input versus output). It is possible to use tachometer generators or specific components whose output voltages are proportional to the rate of change of error. It is more customary, however, to use four-terminal networks whose outputs are proportional to the rate of change of input. The advantages of such networks is that they are smaller, more easily modified, and less expensive than corresponding electromechanical components. These networks are the customary R-C or R-L differentiator networks. When the servomechanism is of a type requiring dc error signals, the network is called a dc phase-lead network (the output across any differentiator network leads its input voltage).

A typical dc *phase-lead* network in combination with a mechanical differential is shown in Fig. 10-4a for use as an error-rate-damped servomechanism. The difference between the input and output shaft positions, θ_i and θ_o, causes the output gear of the differential to drive the arm of potentiometer R_3. If the input-output error is constant, the potentiometer arm on R_3 will remain fixed, corresponding to a particular error between θ_i and θ_o, as represented by the dc voltage, e_i, at a given potentiometer setting. Voltage e_i will cause a current to flow through R_2–R_1, charging C_1 to a particular value as represented by the ratio R_2/R_1. Should the error between θ_i and θ_o change to some other value, however, corresponding changes will occur in the differential output shaft to move the slider to a new position on R_3, producing a change in input voltage, e_i, and corresponding current variations in R_2 and R_1. At the same time, depending on the variation in voltage across R_2, capacitor C_1 will either charge or discharge to meet the voltage drop across R_2. The resulting capacitor discharge or charge is proportional to the rate of change of the capacitor voltage and the voltage across R_2 and, in turn, the range of change of the differential input voltage, e_i. The voltage across R_1, fed to the dc amplifier, is proportional, therefore, to (1) the error between input and output, and (2) the time rate of change of the error between input and output.

A qualitative explanation of how damping is accomplished may be understood by considering a change of error such that e_i increases in a positive direction, causing C_1 to charge. The voltage drop across R_1 is increased, excessively, as a result of this charging current, increasing the voltage input to the dc amplifier. The dc motor, M, turns at a faster rate than necessary, causing θ_o to approach θ_i rapidly. The mechanical differential responds to the reduced difference between θ_i and θ_o by moving

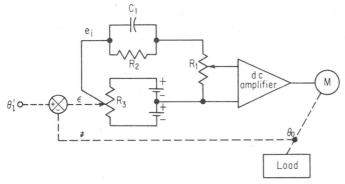

(a) Mechanical differential error-rate damped servomechanism.

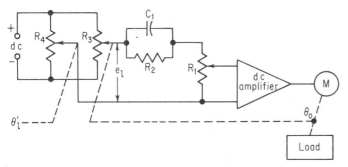

(b) Bridge potentiometer error-rate damped servomechanism.

Figure 10-4

Error-rate differentiator (dc phase
lead) network damping.

the potentiometer arm in such a direction as to reduce voltage e_i. As
the output voltage decreases, capacitor C_1 discharges when its voltage
exceeds R_2. The discharge of C_1 is in such a direction as to reduce the voltage across R_1, reducing the input to the amplifier in anticipation of the
null, thereby preventing overshooting.

A similar error-rate servomechanism, using two multiturn potentiometers in a dc bridge circuit, is shown in Fig. 10-4b, eliminating the
need for the mechanical differential as an error detector. The lead network
and its action are identical to the circuit of Fig. 10-4a, previously described.
Both of these circuits have a disadvantage, however, in requiring a *mechanical* feedback linkage between the output at the load and its input to
the error detector. When the reference input is located remotely from
the load and its actuator, the presence of long mechanical linkages creates
many problems not easily overcome. Figure 10-5 shows an ac error-rate
damping circuit using·a synchro transmitter and synchro control trans-

former in combination*. The feedback signal is electrically provided from the output of the synchro transmitter to the control transformer stator. Thus, it is possible to transmit this signal over long distances, and even to correct or adjust its phasing by using intermediate differential transmitters. Another advantage of this configuration is that the ac supply for the actuator (a two-phase servomotor) also supplies the rotor of the synchro transmitter in close proximity to the load. No power supply is required at the error detector or at the synchro control transformer.

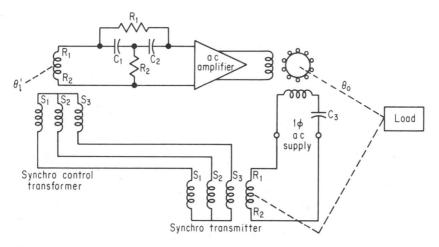

Figure 10-5

Error-rate damping using ac differentiator network.

Lastly, by virtue of its high-resistance rotor, the ac induction motor also provides viscous damping in addition to the error-rate damping provided by the differentiator network. The combination and configuration of servo elements shown in Fig. 10-5 is one of the more popular and extensively used servo systems where combined viscous and error-rate damping is required.

10-8.
TRANSDUCERS

The examples illustrating various servomechanism principles, in this as well as in previous chapters, were confined either to closed-loop voltage, position, or speed control. It was pointed out earlier, however, that a variety of physical quantities may be controlled automatically (Section 10-1) by using transducers. A *transducer* may be defined as a device by means of which (1) energy is converted from one form to another, or (2) one energy system is related to another. For purposes of electronic instrumentation and feedback control systems, we are

* See Kosow, *Electric Machinery and Transformers*, Sec. 11-11, for a discussion of synchros.

interested primarily in that class of transducers which lend themselves to the translation of a nonelectric quantity into an electric signal. Each of the four energy forms (light, heat, mechanical energy, and chemical energy) possesses various physical quantities which may be "sensed" by appropriate transducer devices. Thus, such quantities as acceleration, displacement, flow, force, color, temperature, viscosity, etc., are sensed and converted by means of transducers into electric signals whose variations are proportional to the original quantity to be measured. The advantages of converting these nonelectric parameters into some suitable electric quantity are evident when one considers (1) the possibilities of telemetering by using long-distance radio or microwave transmission; (2) the wide variety of electric and electronic amplifiers and actuators capable of responding to an electric signal; (3) the accuracy, speed of response, and extremely high sensitivity of associated electric and electronic devices; and (4) the ease with which electric information may be stored for later readout purposes.

A survey of current literature indicates that new and ingenious techniques are continuously emerging to provide electric output transducers for various areas of scientific study: medical, chemical, biological, mechanical, thermal, optical, acoustical, magnetic, etc. Indeed, it would appear that almost *any* physical quantity hitherto measured by nonelectric means is capable of being converted into an electric analog. For purposes of this work it will be assumed, therefore, that transducers are available possessing the necessary characteristics (frequency response, sensitivity, linearity, accuracy, phase sensitivity, etc.) to translate any physical parameter into an electric quantity.* For purposes of the following presentation, transducers will be represented as two-port blocks in which the input signal will be, for the most part, a nonelectric quantity and the output signal an electric one of suitable form for amplification.

Various open-loop transducer instrumentation applications are shown in Fig. 10-6. A nonelectric variable is sensed by means of a suitable transducer and converted to an electric output, which is in turn amplified and fed to some type of output device. The output device may be an electric meter having a moving pointer and a stationary scale which is calibrated directly in the units of the nonelectric parameter (force or weight in grams, pressure in millimeters of mercury, etc.) or a digital readout display representing a similar conversion. Figure 10-6 shows some of the more frequently used types of output devices associated with (open-loop) transducer systems used in instrumentation.†

* Space unfortunately does not permit a discussion of transducer construction or principles here. A complete exposition of transducer types will be found in H. Norton, *Handbook of Transducers for Electronic Measuring Systems* (Englewood Cliffs, N.J.: Prentice-Hall, Inc., 1969).

† For a complete discussion of display devices, see Thomas, H. *Handbook of Digital Display Devices and Systems* (Englewood Cliffs, N. J.: Prentice-Hall, Inc., to be published Jan. 1974).

If the input of Fig. 10-6 represented a particular chemical or physical process, it would be continuously measured and monitored by the method shown in the figure. If it is desired to *control* the process *automatically,* in accordance with some predetermined steps, or to hold it constant at some predetermined standard input, then the closed-loop system shown in Fig.

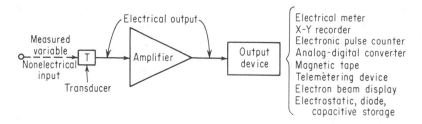

Figure 10-6

Open-loop transducer-instrumentation
systems.

10-7 is used. Instrumentation of the process is still possible, as shown in Fig. 10-7, simultaneously with its automatic control. The nonelectric output of the controlled process is converted to an electric signal for purposes of automation and is compared to a reference electric signal at an error detector or summing point in a differential amplifier (Fig. 10-10b).

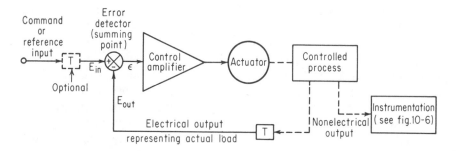

Figure 10-7

Closed-loop automatic-feedback control
system.

The general block diagram of Fig. 10-7 is, perhaps, the most frequently employed scheme used in automatic process control to ensure a uniform and automated product. Given the process, suitable transducers and control elements as well as instrumentation output devices may be selected for each of the elements represented in block form in Fig. 10-7.

292

10-9.
MODULATORS

In many instances, the output of the transducers used to sense a particular process is direct current. When used with a rotary amplifier or a dc differential amplifier (see Fig. 10-10), this is not a problem, because such amplifiers require dc input voltages. Where, however, it is desired to use an ac amplifier and an ac servomotor as an actuator for process control, it is necessary to convert the respective dc reference and output signals to an ac error signal. Such a conversion may be accomplished by means of a contact modulator, shown in Fig. 10-8a. This modulator is a mechanical vibrator (quite similar to that used in automobile radios) which alternately and rapidly samples each of the two dc input signals. The modulator output represents the difference between the two signals. Capacitor C passes the alternating current component to which it has a low impedance. When the two dc signals are equal in magnitude, the capacitor blocks the direct current and the output across R is zero. The output voltage of the contact modulator is essentially a square wave.

Numerous types of electronic modulators may be used, whose ac output magnitude and polarity depend on the magnitude and polarity of the direct current applied as an input. These may vary from the simple diode circuit, shown in Fig. 10-8b, to more sophisticated types involving transistors in bridge (or so-called ring modulator) configurations. A *pnp* and *npn* transistor in complementary symmetry may be used as a carrier-suppressed modulator to convert a dc signal to alternating current of corresponding magnitude and polarity.* The simple modulator shown in Fig. 10-8b operates as follows:

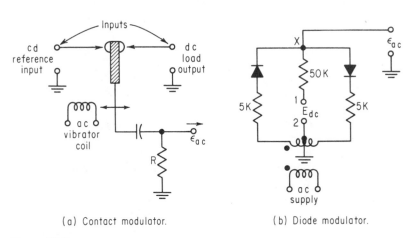

(a) Contact modulator.   (b) Diode modulator.

Figure 10-8

Modulators.

* A complete and thorough analysis of magnetic and electronic modulators is presented in W. R. Ahrendt and C. J. Savant, Jr., *Servomechanism Practice*, 2nd ed. (New York: McGraw-Hill Book Company, 1969), pp. 133–144.

1. The two dc inputs are applied to terminals 1 and 2. When these (the dc reference input and the dc load output) are equal, there is no difference of potential between points 1 and 2.

2. Assume that the instantaneous ac supply polarity is as shown in Fig. 10-8. When the diodes are conducting, the polarity at point X is the same as point 2 (ground potential), since each diode conducts equally.

3. On the reverse half-cycle, the diodes do not conduct, and point X is unaffected by conduction since the diodes act as open circuits. The ac supply, therefore, causes the voltage at X to oscillate from ground to a potential in synchronism with the supply voltage.

4. Assume a positive polarity at point 1 and a negative (less positive) polarity at point 2. Point X now alternates from a positive value to zero at the same frequency as the ac supply. Similarly, if point 1 becomes more *negative*, point X will be alternately negative and zero. The greater the difference of potential between points 1 and 2, the greater the average value of the direct current in the ac output wave.

10-10.
MODULATOR-
AMPLIFIERS

Certain types of amplifiers inherently operate as modulators; that is, they will accept a dc error signal and produce an amplified ac output signal proportional and commensurate to the input. Figure 10-9a shows a bridge-type magnetic amplifier (Section 8-4) as a modulator-amplifier. The control winding receives a reference command signal, E_{in}, and a dc transducer load (process) output signal, E_{out}. The dc bias windings of the mag-amp are connected series-aiding, whereas the control windings are connected series-opposing. Assume that the process output signal is negative (less positive) than the command signal. The circuit of Fig. 10-9a operates as follows:

1. The control winding tends to desaturate winding 1, since it *opposes* the bias winding tending to saturate winding 1. Conversely, the control winding error signal tends to saturate winding 2, since its mmf *aids* the bias winding. The reactance of winding 1 increases, and simultaneously the reactance of winding 2 decreases.

2. Windings 3 and 4 have equal reactance as the secondaries of the input transformer connected to the ac supply. Thus, with variation of the ac input, a difference of potential appears at the "bridge output," points A and B, exciting one phase of a two-phase induction motor.

3. The motor turns, regulating the process, changing the transducer output until its potential corresponds to the dc input potential. When these are equal, the control voltage is zero and no current flows in the control winding.

4. With only the dc bias winding saturating the magnetic amplifier, coils 1 and 2 are equally saturated and have the same impedance as coils 3 and 4, respectively. The potential at points A and B is, therefore, zero, and the actuator motor stops.

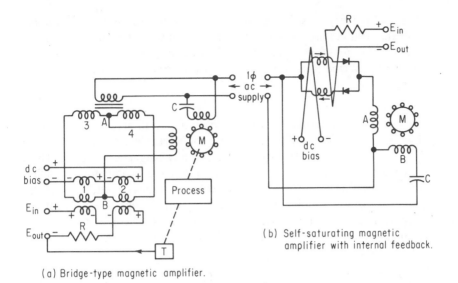

(a) Bridge-type magnetic amplifier.

(b) Self-saturating magnetic amplifier with internal feedback.

Figure 10-9

Magnetic amplifiers as modulator amplifiers.

Reversing the polarity of the error signal (E_{out} more positive than E_{in}), causes a reversal of winding saturation and reverses the phase of the alternating current applied to the motor winding. The modulator amplifier is, thus, a phase-sensitive error detector which converts a dc error signal into an amplified ac control signal. The greater the dc error signal, the greater the degree of saturation and the magnitude, respectively, of the ac control signal developed by the amplifier.

Diodes may be used in lieu of the bridge configuration to produce a self-saturating magnetic amplifier with internal feedback, as described in Sec. 8-4, 10-9b, as shown now in Fig. 10-9b. The feedback loop from the output process to be controlled has been omitted for the sake of simplification. Thus, the control winding flux always aids the core winding flux, tending to saturate both windings of the magnetic amplifier and decreasing the impedance in series with phase A of the two-phase motor. The function of the dc bias winding is the same as that described in Section 8-4. With zero error signal, the bias is adjusted so that the mag-amp windings are unsaturated, and the total supply voltage is necessary to overcome the comparatively high impedance of the magnetic amplifier. A small error signal will initiate saturation, reducing the series impedance and increasing the amplifier output voltage to phase A. The circuit of Fig. 10-9b, unfortunately, does not have the phase-sensitive characteristics of the bridge-type magnetic-amplifier shown in Fig. 10-9a. Thus, if the error signal reverses in Fig. 10-9b it will not produce reversed rotation of the ac motor.

295

External polarity-sensitive relays may be used to reverse the bias winding and the field connections of the ac motor, but this introduces a serious time lag in amplifier response. For this reason, the circuit shown in Fig. 10-9a is preferred to that of Fig. 10-9b for servosystem control application.

10-11.
dc AMPLIFIERS

If it is desired to drive a dc motor as an actuator for process control, it is a simple matter to insert a half-wave or full-wave rectifier in the output circuit of any magnetic amplifier. Thus, any magnetic amplifier may be used as a dc-to-ac (modulator) amplifier, or as a dc-to-dc amplifier with rectifiers in its output circuit.*

The rotary amplifiers (amplidyne, Ward-Leonard, Rototrol, or Regulex)† are all, inherently, dc-to-dc amplifiers. These lend themselves to high- and medium-power applications, as shown in Fig. 10-10a.

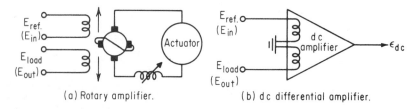

(a) Rotary amplifier. (b) dc differential amplifier.

Figure 10-10

DC to dc amplifiers.

Where a small amount of power is required, extremely sensitive electronic transistorized dc differential amplifiers may be employed. These amplifiers (sometimes called dc null detectors) are extremely sensitive *electrometers* capable of detecting relatively small differences between the dc reference and load output potentials and amplifying these differences using dc amplifiers. The extremely small dc voltage output of such a dc amplifier, as shown in Fig. 10-10b, may either (1) be fed a modulator and further amplified using ac amplifiers, or (2) be fed to a dc-to-dc amplifier of the rotary or magnetic amplifier type.

In all of the foregoing discussion, it should be noted that the nature of the process to be controlled and the transducers available to sense this process dictate, to a large degree, the type of amplifiers and actuators to be used.

* See D. L. McMurtrie, "Magnetic Amplifier Output Circuits," *Electrical Manufacturing* (August 1960).
† See Kosow, *Electric Machinery and Transformers*, Secs. 11-16 and 11-17.

10-12.

**EXAMPLES OF
AUTOMATIC
PROCESS CONTROL**

The basic diagram for process control shown in Fig. 10-7 is adaptable to any number of applications. Two examples are given below.

The first example is shown in Fig. 10-11. Two liquid chemicals of different density are to be mixed in a certain proportion. One chemical has a higher viscosity than the other. The *exact* proportion (representing the desired viscosity) will produce a given output voltage from a transducer, in this case a *viscometer*.* As shown in Fig. 10-11,

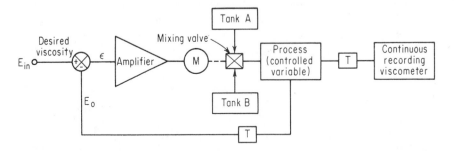

Figure 10-11

Automatic control of liquid mixture by sensing viscosity.

the desired viscosity is provided from a calibrated voltage source. The actual viscosity is provided from the tank in which the liquids have been mixed. The difference in viscosities is fed as an electric error signal to an amplifier which drives an actuating motor. The motor controls a two-way mixing valve such that rotation of the motor in one direction will reduce the flow from tank A and increase the flow from tank B. When the voltage from the output (controlled process) transducer equals the desired viscosity, the error is zero and the amplifier output is at a null. The actuator and the mixing valve are stationary when the desired mixture is obtained. The system remains at a null until a change in viscosity is produced. The proportion of the mixture is continuously recorded by means of instrumentation, as shown in Fig. 10-11.

A second example is shown in Fig. 10-12. The intent of this system is to control the temperature of a steel ingot as it leaves a furnace prior to being shaped by rollers or drawn through various dies. If the temperature is either too high or too low, the resultant product will not be uniform.

---

* Various types of sensors may be used to measure viscosity. The most common type uses a motor (constant-speed synchronous type) to drive a shaft (through a flexible spring coupling) to a paddle immersed in the liquid. The torsion on the spring is an indication of the viscosity, and the torsion may be sensed by either resistive or capacitive measurements. See Norton, H. H., *op. cit.*

297

Temperature may be sensed by any one of a variety of temperature transducers (thermistors, bolometers, optical pyrometers, or thermocouples). The desired temperature is represented as a voltage fed to a suitable error detector for comparison with the output of the temperature transducer. The error is amplified, and the voltage is used to control a motor which either opens or closes a fuel valve. Thus, if the temperature is excessive, the error will drive the motor to adjust the fuel valve in such a direction as to reduce the fuel fed to the furnace and thus lower temperature of the furnace. Conversely, if the temperature is too low, the error is of the reverse polarity and the valve is driven to increase the fuel input. A continuous-recording pyrometer monitors and measures the variation in temperature of the process to ensure that the temperature remains within the limits of variation permitted for the process.

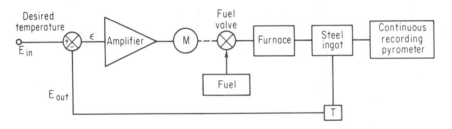

Figure 10-12

Automatic control of steel ingot by sensing temperature.

In both of these cases, it should be noted that the process is controlled by means of control of a convenient variable. In Fig. 10-11, the proportion of the mixture is controlled by controlling the viscosity of the mixture. The proportion might also have been controlled by controlling the color density of the mixture. It might also have been controlled by sensing the motor current required to drive a pump which pumps the final mixture (a higher viscosity requires a higher pump power and more load is placed on the motor).

Similarly, in the case of the steel ingot, the primary purpose of the automatic control process is to ensure a uniform shape of the steel emerging from either rollers or dies. Sensors might have been used to measure the thickness of the steel as it emerges or the tension required to pull it through dies of a given thickness. Thus, the selection of the physical quantity to be measured as an analog of the process to be controlled is but one phase of a highly complex process known as *control systems engineering*.*

* See the references at the end of this chapter.

The controller gain or open-loop gain, G, in Eqs. (10-2), (10-2a), and (10-3) may be used to determine the steady-state frequency response of a closed-loop system using the *open-loop transfer function*. A transfer function, defined as the *ratio* of the *output* signal to the *input* signal, is nothing more than a *multiplying* factor to convert input quantities into output quantities. Transfer functions enable the various components of a servosystem to be manipulated in block form, for purposes of simplication and reduction.

Certain conventional symbols are used in connection with the block diagram as shown in Fig. 10-13. The symbol for an input variable is shown in Fig. 10-13a. If this variable A is fed to an amplifier having gain G, its output is B. The operation as shown in Fig. 10-13b yields $B = GA$ and the transfer function is B/A, or simply the gain of the block.

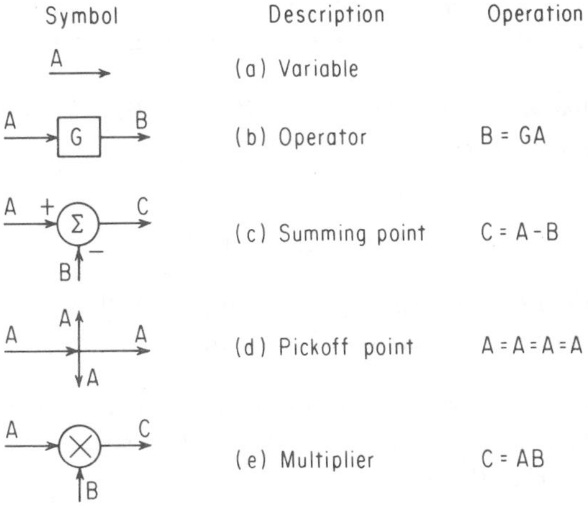

Symbol	Description	Operation
$A \longrightarrow$	(a) Variable	
$A \rightarrow \boxed{G} \rightarrow B$	(b) Operator	$B = GA$
$A + \bigcirc\!\!\Sigma \rightarrow C$, $B -$	(c) Summing point	$C = A - B$
$A \rightarrow$ (pickoff) A, A	(d) Pickoff point	$A = A = A = A$
$A \rightarrow \bigotimes \rightarrow C$, B	(e) Multiplier	$C = AB$

Figure 10-13

Transfer function block diagram symbols.

When two inputs A and B are connected to a summing point, Σ (as in the case of an error detector) such that the polarity of one is opposite the polarity of another, as shown in Fig. 10-13c, the output C is the difference of the two inputs or $C = A - B$, as shown.

When more than one output is "picked off" from a point in the block diagram, as shown in Fig. 10-13d, without amplification, attenuation, summing, or multiplication, the outputs are the same as the input, or $A = A = A$.

Given two inputs which are to be *multiplied*, shown in Fig. 10-13e, the product $C = AB$. Note the symbol for the multiplier function as distinguished from the summing function (Fig. 10-13c).

The transfer function of any block (or system) may be considered an *operator*. A system transfer function operates on an input to produce an output. The concept of such operation is shown in Fig. 10-14, where more complex operations are simplified. Thus, if two blocks A and B are cascaded (Fig. 10-14a), the overall transfer function (the ratio of output to input signal) is $\frac{xAB}{x}$ or AB. This may be represented by a single block having gain AB. This is an important rule in servomechanism operation:

> **The gain of series cascaded blocks is the product of the individual gains.**   Rule (10-1)

299

In the same way, the pickoff points and summing junctions may be eliminated in Fig. 10-14b, whose transfer function is $\frac{x(A + B)}{x}$ or $A + B$, as shown.

The classical *negative feedback* representation (Fig. 10-14c) may be simplified as shown by Eq. (10-3) to a single block having the transfer function $\frac{A}{1 + AB}$. Note that the simplification has eliminated the pickoff points and the summing point, retaining the transfer functions of each block. Figure 10-14c thus gives rise to another important servomechanism concept:

> ***Any closed-loop system may be replaced by an equivalent open-loop system.***   Rule (10-2)

If a summing point is shifted with respect to a system, the effect is shown in Figs. 10-14d and e. The rule to be inferred from Fig. 10-14d is

> ***Shifting a summing point ahead of a block having gain A requires insertion of 1/A in the gain of the variable added.***   Rule (10-3)

Conversely, the rule to be inferred from Fig. 10-14e is

> ***Shifting a summing point after (behind) a block having gain A requires insertion of gain A in the variable added.***   Rule (10-4)

Pickoff points may also be shifted but these are treated in a reciprocal manner to summing points as shown in Figs. 10-14f and g, respectively. The rule to be inferred from Fig. 10-14f is

> ***Shifting a pickoff point after (behind) a block having gain A requires insertion of gain 1/A in the variable picked off.***   Rule (10-5)

Conversely, the rule to be inferred from Fig. 10-14g is

> ***Shifting a pickoff point ahead of a block having gain A requires insertion of gain A in the variable picked off.***   Rule (10-6)

The last rule required for our purposes is illustrated in Fig. 10-14h. It may be stated as

> ***In the absence of gain blocks, the sequence of either pickoffs or order of summing does not affect the outputs or sums in any way.***   Rule (10-7)

The implication of Rule (10-7) is that sequences of sums or pickoffs

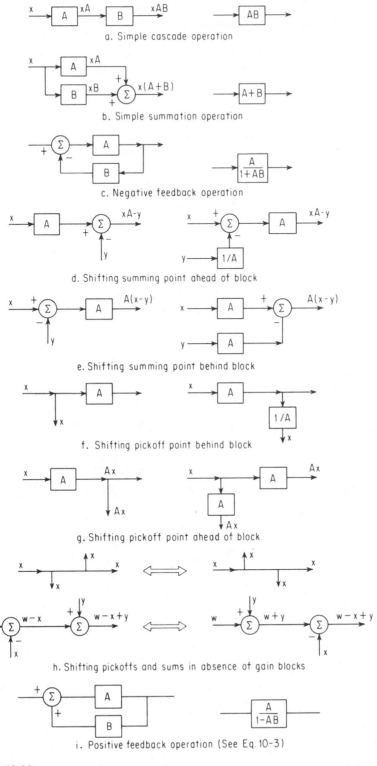

Figure 10-14

Block diagram operation and reduction techniques.

may be reversed, commutated, or even expanded, as long as gain blocks are *not* involved in the process.

Examples 10-1 and 10-2 illustrate how the above rules are applied to simplify control systems having major and minor feedback loops and cascaded blocks.

EXAMPLE 10-1

Simplify the feedback system shown in Fig. 10-15a to an open-loop system containing a single gain block.

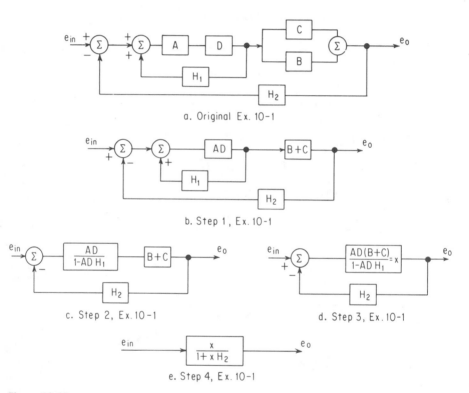

a. Original Ex. 10-1

b. Step 1, Ex. 10-1

c. Step 2, Ex. 10-1

d. Step 3, Ex. 10-1

e. Step 4, Ex. 10-1

Figure 10-15

Example 10-1 and solution.

Solution

(see Figs. 10-15b through e):

Step 1. a. Since A and D are cascaded, they may be combined into a single block AD, using Rule (10-1).

b. Since gain blocks B and C are summed, they may be combined into a single gain block $(B + C)$.

c. The result of these reductions is shown in Fig. 10-15b.

Step 2. a. Positive feedback loop H_1 may be reduced to a single equivalent open-loop system, Rule (10-2), whose transfer function, using Eq. (10-3), is $\dfrac{AD}{1 - ADH_1}$. Note that a positive feedback system produces a minus sign in the denominator.

    b. The result of these reductions is shown in Fig. 10-15c.

Step 3. a. The two cascaded blocks in Fig. 10-15c may be combined into a single transfer function, x; using Rule (10-1),

$$x = \frac{AD(B + C)}{1 - ADH}$$

    b. The result of this reduction is shown in Fig. 10-15d.

Step 4. a. The negative feedback loop shown in Fig. 10-15d may be reduced to an equivalent open-loop system, using Rule (10-2) and Eq. (10-3), where the open-loop gain is $\dfrac{x}{1 + xH_2}$ and x is defined in step 3.

    b. The equivalent transfer function for the system of Fig. 10-15a is shown in Fig. 10-15e. Note the $+$ sign in the denominator for negative feedback.

Example 10-1 was solved without requiring shifting of summing or pickoff points. This technique is illustrated in Ex. 10-2.

EXAMPLE 10-2

Simplify the feedback system shown in Fig. 10-16a to an open-loop system containing a single gain block.

Solution

(see Figs. 10-16b through e):

Step 1. a. The pickoff points between gain blocks B and C are shifted behind block C, as shown in Fig. 10-16b.

    b. Using Rule (10-5), the pickoff points are multiplied by a factor $1/C$ when shifted behind block C.

    c. The result of this reduction is shown in Fig. 10-16b.

Step 2. a. The summing point of block H/C is shifted ahead of block A so that both feedback loops have a common summing point.

    b. Using Rule (10-3), shifting a summing point ahead of block having gain H/C requires multiplication by factor $1/A$, yielding a feedback factor of H/AC.

    c. The above steps also permit cascaded blocks A, B, and C to be combined, using Rule (10-1), into a single block ABC.

    d. The result of these reductions in shown in Fig. 10-16c.

Step 3. a. The positive and negative feedback loops have the same summing points and pickoff points. The two loops may be combined into a single loop having the transfer function

$$\frac{H}{AC} - \frac{1}{C}$$

    b. The above expression may be simplified algebraically to

$$\frac{H}{AC} - \frac{1}{C} = \frac{1}{C}\left(\frac{H}{A} - 1\right) = \frac{1}{C}\frac{H - A}{A} = \frac{H - A}{CA}$$

    c. This reduction is shown in Fig. 10-16d.

303

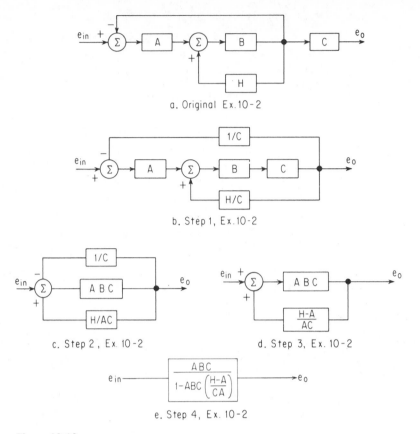

a. Original Ex.10-2

b. Step 1, Ex.10-2

c. Step 2, Ex. 10-2

d. Step 3, Ex. 10-2

e. Step 4, Ex. 10-2

Figure 10-16

Example 10-2 and solution.

Step 4. a. The positive feedback loop may be reduced to an equivalent open-loop system, Rule (10-2), having the value shown in Fig. 10-16e. Note the negative sign in the denominator because the feedback is positive. The open-loop transfer function may be simplified to

$$\frac{ABC}{1 - ABC[(H - A)/CA]} = \frac{ABC}{1 - B(H - A)}$$

The above examples show that the transfer function of cascaded elements is the product of their individual transfer functions, and that ideal elements are assumed in the individual blocks (i.e., those having constant gain regardless of frequency, and no phase shift, regardless of frequency). Let us examine the gain blocks more closely, beginning with a resistive linear block, having a gain of less than unity, as shown in Fig. 10-17a. The transfer function, H, for this circuit is independent of fre-

CHAP. TEN / *Automatic Feedback Control Systems*

quency and constantly of the value

$$H = \frac{R_2}{R_1 + R_2} = \frac{e_o}{e_{in}} \tag{10-8}$$

Passive elements such as capacitors and inductors, however, produce transfer functions whose gains are frequency-dependent and which pro-

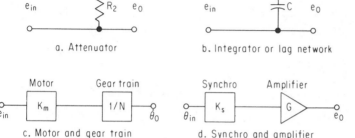

a. Attenuator

b. Integrator or lag network

c. Motor and gear train

d. Synchro and amplifier

Figure 10-17

Typical elements having transfer functions.

duce phase shifts. Consider, for example, the simple integrator shown in Fig. 10-17b. The transfer function for such a circuit is derived as follows:

$$e_o = e_{in} \frac{-jX_c}{R - jX_c}$$

and

$$H = \frac{e_o}{e_{in}} = \frac{1/\omega C}{R + 1/\omega C} = \frac{1/sC}{R + 1/sC} = \frac{1/RC}{s + 1/RC} = \frac{1/\tau}{s + 1/\tau}$$
$$= \frac{1}{1 + \tau s} \tag{10-9}$$

where $s = \pm j\omega$ or the Laplace operator
$\tau = RC$

If the transfer function is plotted such that its gain, H, is represented on the ordinate, using a linear scale, expressed in decibels (dB), versus the frequency ω, in radians per second, on a logarithmic decade scale, we obtain a *Bode diagram* for the network. The Bode diagram plots both the gain of the transfer function, H, and the phase, θ, of the transfer function separately in a very special way. Since the transfer function is a mea-

305

sure of the output amplitude divided by the input amplitude at any instant and the bel is a ratio of output to input powers, then $H = \log_{10} \dfrac{p_0}{p_{in}}$ in bels and

$$H = \log_{10} \frac{e_0^2/R_0}{e_{in}^2/R_{in}} \text{ but assuming } R_0 = R_{in}{}^*, \text{ then}$$

$$H = \log_{10} \frac{e_0^2}{e_{in}^2} = \log_{10}\left(\frac{e_o}{e_{in}}\right)^2 = 2 \log_{10}\frac{e_o}{e_{in}} \qquad \textbf{bels}$$

but since 10 dB = 1 bel(B) or a conversion ratio of 10 dB/B exists, then

$$H = 20 \log_{10}\frac{e_o}{e_{in}} \qquad \text{dB} \tag{10-10}$$

where e_o = output amplitude at any given frequency
e_{in} = input amplitude at any given frequency

The vertical axis of a Bode diagram is plotted in linear units of dB gain. In the case of passive networks (containing no amplifying current or voltage sources) shown in Fig. 10-17, the gain is always less than unity or less than 0 dB.

The horizontal axis of a Bode diagram is the frequency ω in rad/s plotted on a logarithmic decade scale. A *decade* is defined as a separation of two frequencies by a ratio of 10/1 or 1/10.

Figure 10-18 shows the Bode plot of the network of Fig. 10-17b. At extremely low frequencies, the output across the capacitor is almost the same in magnitude and in phase with the input voltage. Therefore, the gain, H, is 0 dB and the phase shift, θ, between input and output is small. As the frequency increases, the capacitive reactance decreases and the gain is less than unity (or a negative dB value). The phase shift between output and input, θ, also increases, approaching $-90°$, since the output lags input at all times.

Note that the abscissa or horizontal axis of Fig. 10-18 is plotted on a logarithmic decade scale. At $\omega = 1$ rad/s, the gain H is -3 dB. Since the -3 dB point for this circuit occurs where the phase shift is $-45°$ and $\omega = 1/\tau$, the time constant for this circuit is 1 s. Note also that the gain falls off at a slope of -12 dB per decade or -6 dB per octave. An octave is defined as a separation of two frequencies in the ratio 2:1 or 1:2.

It is sometimes convenient to determine relations between slopes expressed in decades and octaves, as shown in Ex. 10-3.

---

* Although the derivation of Eq. (10-10) assumes $R_0 = R_{in}$, conventional use permits expression of *relative* current or voltage gains in *dB* even when $R_0 \neq R_{in}$.

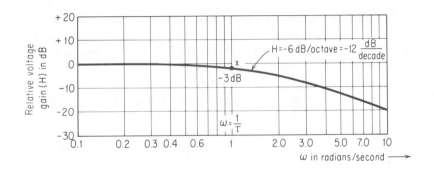

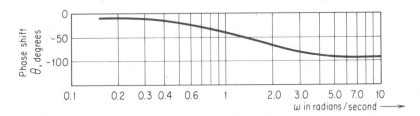

Figure 10-18

Bode plot showing and phase frequency
response of phase-lag network.

EXAMPLE 10-3

At a frequency ω_1, a Bode plot has a slope of -20 on a Bode plot. At a higher
frequency, $\omega_2 = 2\omega_1$, the slope of the Bode plot is -40. Calculate
a. The change in dB per octave for the -20 slope.
b. The change in dB per octave for the -40 slope.
c. The change in dB per decade for the -20 slope.
d. The change in dB per decade for the -40 slope.
e. The relation between dB/octave and dB/decade for each slope.

Solution

a. dB change $= -20 \log_{10} \dfrac{\omega_2}{\omega_1} = -20 \log_{10} \dfrac{2\omega_1}{\omega_1} = -20 \log_{10} 2$

    $= $ **-6 dB/octave**

b. dB change $= -40 \log_{10} \dfrac{\omega_2}{\omega_1} = -40 \log_{10} 2 = $ **-12 dB/octave**

c. dB change $= -20 \log_{10} \dfrac{\omega_2}{\omega_1} = -20 \log_{10} \dfrac{10\omega_1}{\omega_1} = -20 \log_{10} 10$

    $= $ **-20 dB/decade**

d. dB change $= -40 \log_{10} \dfrac{\omega_2}{\omega_1} = -40 \log_{10} \dfrac{10\omega_1}{\omega_1} = -40 \log_{10}10$

    $= $ **-40 dB/decade**

e. -6 dB/octave equals **-20 dB/decade;** -12 dB/octave equals
    -40 dB/decade

307

Example 10-3 reveals that a slope of -18 dB/octave corresponds to a slope of -60 dB/decade. Similarly, a slope of -24 dB/octave corresponds to a slope of -80 dB per decade, and so on. The same relations would also hold for positive slopes.

Expressing gain in terms of falloff or rise in dB/octave is sometimes extremely useful because it provides a means of estimating the phase angle, which is often difficult to calculate from a given expression for the transfer function. The following rule of thumb serves as a guide for such estimation: For a slope of $\pm 6n$ dB/octave, the phase angle θ tends to approach

$$\pm \frac{n\pi}{2} \text{ radians} \qquad (10\text{-}11)$$

where n = whole-number integers (1, 2, 3, 4, etc.) or fractional integers ($\frac{1}{2}$, $\frac{1}{4}$, etc.)

EXAMPLE 10-4

For the Bode plot shown in Fig. 10-18, estimate the phase angle when
a. The gain slope is 0 dB/octave.
b. The gain slope is 6 dB/octave.

Solution

a. $\theta = -n\pi/2$ radians when the slope is $-6n$ dB/octave
   $n = 0$ for a slope of 0 dB/octave
   Hence $\theta = -0\pi/2 = \boldsymbol{0 \ radians}$
Thus θ approaches 0 radians or 0° when the slope is 0 dB/octave, as shown in Fig. 10-18b.
b. $n = 1$ for a slope of -6 dB/octave and $\theta = -n\pi/2 = -\pi/2$ radians $= \boldsymbol{-90°}$.

Thus, θ approaches $-90°$ when the slope is -6 dB/octave, as shown in Fig. 10-18b.

The ordinate of a Bode plot is expressed in dB but measurements on passive or active networks are made using either an electronic voltmeter (EVM) or a cathode-ray oscilloscope (CRO) as the frequency of the input signal is varied. It is unnecessary to maintain the input magnitude constant because the transfer function is the ratio of the output to input amplitudes over a predetermined frequency range. The curve of Fig. 10-19 shows a convenient method for converting the ratio of output/input amplitudes into dB.*

The curve of Fig. 10-19 is extremely useful for finding either dB values or gain ratios directly for the range of values given in the figure. It is possible, however, to even use the chart of Fig. 10-19 to find either gain

---

* The reader may construct his own chart conveniently using three- or four-cycle semilogarithmic paper and connecting the following points:

Gain ratio	0.1	1.0	10	100	1000
dB	-20	0	20	40	60

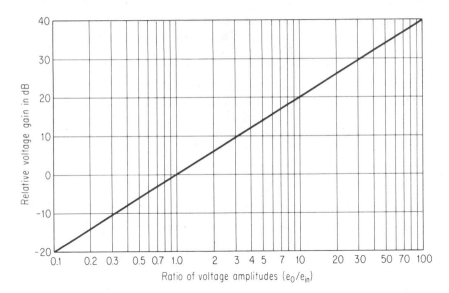

Relative voltage gain in dB

Ratio of voltage amplitudes (e_0/e_{in})

Figure 10-19

Chart for converting ratio of amplitude into
dB and vice-versa.

ratios or dB values which are not given on the chart, as shown by Exs.
10-5 and 10-6.

EXAMPLE 10-5

An amplifier has successive gains, measured over a frequency band, of
a. 60 dB
b. 54 dB
c. 48 dB
d. 42 dB
Calculate gain ratios for each of the above values.

Solution

The highest value given on the chart (Fig. 10-19) is 40 dB, or gain ratio = 100.
a. Step 1. Subtract this value (40 dB) from the given value (60 dB), yielding
    a ratio value given on the chart.

$$dB = 60\ dB - 40\ dB = 20\ dB \text{ (excess value)}$$

  2. Read the excess value from the chart, 20 dB = gain ratio of 10
  3. Multiply the two ratios obtained or $100 \times 10 = \mathbf{1000}$
     Then a gain of 60 dB corresponds to ratio of **1000**

Note: This is to be expected since 20, 40, 60, 80, and 100 dB correspond,
respectively, to gain ratios of 10, 100, 1000, 10,000, and 100,000.
b. Given 54 dB, 1. 54 dB − 40 dB = 14 dB
       2. From chart (Fig. 10-19), 14 dB = a gain ratio of 5.0
       3.   40 dB = 100
          $\underline{+\ 14\ dB =\quad 5}$
             54 dB = 5 × 100 = **500** (gain ratio)

309

c. Given 48 dB,   1. 48 dB − 40 dB = 8 dB
                    2. From chart (Fig. 10-19), 8 dB = a gain ratio of 2.5
                    3.   40 dB = 100

                        + 8 dB =   2.5
                        48 dB = 2.5 × 100 = **250**

d. Given 42 dB,   1. 42 dB − 40 dB = 2 dB
                    2. From chart (Fig. 10-19), 2 dB = a gain ratio of 1.25
                    3.   40 dB = 100

                        + 2 dB =   1.25
                        42 dB = 1.25 × 100 = **125**

EXAMPLE 10-6

Convert the following gain ratios to dB:
a. 1600
b. 800
c. 400
d. 200

Solution

Step 1. Write the gain ratio as a number between 1 and 10 times an appropriate power of 10.
      2. Find the dB value of the gain ratio of the number between 1 and 10.
      3. Add the dB value to 20 times the power of 10 found in Step 1.

a. (Step 1) $1600 = 1.6 \times 10^3$
   (Step 2) a gain ratio of 1.6 = 4 dB (from Fig. 10-19)
   (Step 3) 4 dB + 20 × 3 dB = **64 dB**
b. (Step 1) $800 = 8.0 = 8.0 \times 10^2$

TABLE 10-1   GAIN AND PHASE RESPONSES OF COMMON
TRANSFER-FUNCTION TERMS

TRANSFER FUNCTION	SLOPE OF LOG GAIN		PHASE ANGLE, θ
	(dB/decade)	(dB/octave)	(deg)
s or $j\omega$	+20	+6	+90°
s^2 or $(j\omega)^2$	+40	+12	+180°
s^3 or $(j\omega)^3$	+60	+18	+270°
$\dfrac{1}{s}$ or $\dfrac{1}{j\omega}$	−20	−6	−90°
$\dfrac{1}{s^2}$ or $\dfrac{1}{(j\omega)^2}$	−40	−12	−180°
$\dfrac{1}{s^3}$ or $\dfrac{1}{(j\omega)^3}$	−60	−18	−270°
s^n or $(j\omega)^n$	$(+20)n$	$(+6)n$	$n(+90°)$
$\dfrac{1}{s^n}$ or $\dfrac{1}{(j\omega)^n}$	$(−20)n$	$(−6)n$	$n(−90°)$
$K = 1$ (gain ratio)	0 dB slope at 0 dB		0°
$K = 10$ (gain ratio)	0 dB slope at 20 dB		0°
$K = 0.1$ (gain ratio)	0 dB slope at −20 dB		180°

(Step 2) a gain ratio of 8 = 18 dB (from Fig. 10-19)
(Step 3) 18 dB + 20 × 2 dB = **58 dB**
c. (Step 1) 400 = 4 × 10²
(Step 2) a gain ratio of 4 = 12 dB (from Fig. 10-19)
(Step 3) 12 dB + 20 × 2 dB = **52 dB**
d. (Step 1) 200 = 2 × 10²
(Step 2) a gain ratio of 2 = 6 dB (from Fig. 10-19)
(Step 3) 6 dB + 20 × 2 dB = **46 dB**

Certain values appear frequently in transfer functions, such as the expressions $j\omega$, $1/j\omega$, $j\omega^2$, etc. In approximating the gain and phase response of a transfer function, it is convenient to know the responses of these functions. Table 10-1 lists some of the more common terms found in transfer-function equations, as well as the slope in dB/decade, dB/octave, and phase angle.

10-14.
PREDICTIONS FROM
OPEN-LOOP
TRANSFER
FUNCTIONS
REGARDING
CLOSED-LOOP
STABILITY

If a Bode plot is made showing the transfer-function gain and phase curves of a complete system (input transducer, servo-amplifier, and motor) under open-loop conditions (without feedback), it is possible to predict in advance whether such a system, when connected in a closed-loop mode, is stable or unstable. As described in Sections 10-5 and 10-6, an unstable closed-loop system has a damping factor of much less than 1.0 and requires additional damping. The damping may be provided in the form of a network added either in the feedback loop (feedback compensation, Section 10-17) or cascaded in series with the amplifier (cascade compensation, Section 10-16) to provide the required stability. Let us first consider the two measures of relative stability which are obtained from a Bode plot.

The transfer-function open-loop system gain and phase responses for a cascaded input transducer, servoamplifier, and motor are shown in Fig. 10-20. At lower frequencies (from 0.1 to 0.8 rad/s), the transfer function falls off at a slope of −6 dB/octave. At middle frequencies (from 0.8 to 3.2 rad/s), the transfer function falls off at a slope of −12 dB/octave. At higher frequencies, the open-loop gain falls off at a slope of −18 dB/octave until its gain ratio is unity, or 0 dB. This 0-dB point where the gain ratio is unity is known as the *gain crossover* (Fig. 10-20).

The phase angle response of the open-loop system is also shown in Fig. 10-20. At low frequencies the phase lag is −90°. This is anticipated in accordance with Eq. (10-11) and verifies that relation for a slope of −6 dB/octave. At middle frequencies the phase lag drops from −90° to −180°. Again, this verifies Eq. (10-11) for a slope of −12 dB/octave. At the gain-crossover point, the phase lag approaches 270°, again verifying Eq. (10-11) for a slope of −18 dB/octave.

311

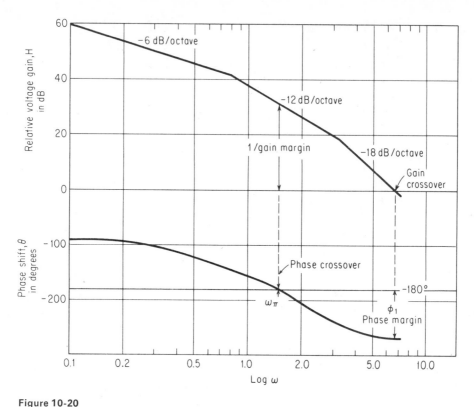

Figure 10-20

Open-loop system gain and phase responses,
showing gain and phase margins, and gain
and phase crossovers.

The point at which the phase response crosses $-180°$ is called the *phase crossover*. This point corresponds to $-\pi$ radians. The frequency at which this point occurs is sometimes designated ω_π, as shown in Fig. 10-20.

If both the phase and gain response are plotted on a common frequency axis, as shown in Fig. 10-20, it is possible to predict whether the open-loop system will produce a stable response when connected in the closed-loop mode. The *relative stability* of the system is indicated by two measures, known as the *gain margin* and the *phase margin*.

10-14.1
Gain Margin

The gain margin is defined as the absolute magnitude of the reciprocal of the open-loop transfer function, obtained at the frequency ω_π where the phase angle of open-loop system lag is $-180°$, or

$$\text{Gain margin} = \left| \frac{1}{GH(j\omega_\pi)} \right| \qquad (10\text{-}12)$$

As may be seen from Fig. 10-20, the reciprocal of the gain margin is approximately $GH(j\omega_n) = +31$ dB. It is frequently common to use the *reciprocal*, expressed in dB, as a measure of gain margin because it is so easily obtained from a Bode plot. As shown in Fig. 10-20, the gain margin is a positive value. It can be shown, however, that a closed-loop system is *unstable* when its open-loop gain margin exceeds a unity gain ratio. In effect, the *gain margin* is a measure which expresses the amount by which any closed-loop system *differs* from a *stable* system.

A system is stable when its gain margin is negative (i.e., it has a gain ratio of less than 1). As shown in Fig. 10-20, therefore, with a gain margin of $+31$ dB, the open-loop system would go into oscillation if connected in a closed loop.

<table>
<tr><td>10-14.2
Phase Margin</td><td>The *second* measure of relative stability is called the *phase* margin ϕ_1. The phase margin is defined as 180° plus the phase angle θ of the open-loop transfer function taken at a gain of *unity* or 0 dB, i.e., at the gain-crossover point, or</td></tr>
</table>

$$\phi_1 = \theta + 180° \qquad (10\text{-}13)$$

As shown in Fig. 10-20, the phase margin at the gain-crossover point is

$$\phi_1 = \theta + 180° = -270° + 180° = -90°$$

The phase margin in this case is negative, and it again serves as a measure of the angle by which the phase of the open-loop ratio of a stable system differs from $\pm 180°$ at the gain crossover.

A system is stable when its phase margin is positive, i.e., at the point where the open-loop transfer function has a gain ratio of unity (0 dB), the phase shift is less than 180°. As shown in Fig. 10-20, therefore, with a phase margin of $-90°$, the open-loop system would go into oscillation if connected in a closed loop.

10-14.3
Stability
Considerations in
Terms of Gain
Margin and Phase
Margin

The preceding discussions showed for an open-loop system:

a. *If the gain margin is positive, the system is unstable* (Sec. 10-14.1), if connected closed loop.

b. *If the phase margin is negative, the system is unstable* (Sec. 10-14.2), if connected closed loop.

c. *Both* a negative gain margin and a positive phase margin are required for a stable closed-loop system.

Note that *both* criteria must be met for stability to occur. Thus, a positive phase margin accompanied by a positive gain margin produces a situation where the closed-loop servo is somewhat "jittery" or "lively" and breaks into oscillations occasionally.

The two stability factors, phase margin and gain margin, are somewhat related to each other. A positive phase margin of approximately 0 to +30° is sometimes accompanied by positive values of gain margin, tending toward instability. *It is generally good design practice* to seek positive phase margins in the range +40° to +60° since these are accompanied usually by negative gain margins.

Increasing the value of phase margin (by damping) to a positive value of more than +60° provides too much damping, resulting in a *wide deadband* and greater servo error, as shown in Fig. 10-2b. Such a servo is said to be "sluggish" or heavily overdamped. Increasing the open-loop amplifier gain results in a less negative gain margin and also decreases the phase margin to a less positive value (see Sec. 10-16).

In general, it is good servo practice to design a servo system so that its open-loop transfer function provides an underdamped response. In this way, desired damping may be added to provide the necessary amount of stability to assure unity damping factor in Eq. (10-7). At a damping factor of unity, the system reaches a null in the shortest possible time whenever an error is introduced (Fig. 10-2b). Further, the underdamped response ensures that servo errors are reduced to a minimum (Section 10-4).

**10-15.
CASCADE
COMPENSATION**

Rule (10-1) of Section 10-13 states that the gain of series-cascaded blocks is the product of their individual gains, where the gains are expressed as a ratio of individual output divided by input. Thus, for two cascaded blocks, having gains of G_1 and G_2, respectively, the overall open-loop gain is

$$G = G_1 \times G_2 = \log G_1 + \log G_2 \qquad (10\text{-}14)$$

The obvious advantage of a logarithmic plot of the open-loop system transfer function of Fig. 10-20, for example, is that if a network is used in cascade compensation, its response may be *graphically added* to improve the overall system response, using Eq. (10-14). Thus, the *resultant* of the two plotted gain response curves precisely describes the overall open-loop and closed-loop behavior of the cascaded system.

Examining Fig. 10-20, an unstable system which has a positive gain margin and a negative phase margin, we might ask ourselves what type of network may be cascaded with this system to provide a negative gain margin and positive phase margin, thus producing stability. Such a network should have a transfer function whose gain curve is essentially less than unity (negative values of dB) so that the overall (combined) response produces a negative gain margin. Similarly, the phase response of such a network should have positive (positive-going slopes) or *leading* phase shifts to produce a positive phase margin, tending toward stability. As pointed out earlier, a positive phase margin from 40° to 60° is desirable. Three types of networks are discussed which satisfy these conditions.*

10-15.1
Phase-Lead Networks

Phase-lead networks may be inserted in series between the amplifier and error detector of a servomechanism of Fig. 10-1 to improve the closed-loop performance. Such networks introduce phase leads to compensate for the phase lag in open-loop performance of error transducers, amplifiers, and motors of a servomechanism. A typical phase-lead network is shown in Fig. 10-21a, with its Bode plot in Fig. 10-21b. Since the network consists of passive elements, its gain is always less than 1, or 0 dB. Its phase response, however, is always positive or leading. The advantages of such phase-lead compensation include the following:

1. Increased stability, providing positive phase margins with resulting increased accuracy.
2. Transient performance is not adversely affected.
3. Generally increased bandwidth of system.
4. Relatively low cost in comparison to mechanical or electromechanical dampers or compensating devices.
5. Negligible attenuation at higher frequencies (see Fig. 10-21b).
6. Several lead networks may be cascaded if a large phase lead is required.
7. Excellent operation in conjunction with dc electronic and rotary amplifiers.

The disadvantages of the phase-lead network include the following:

1. Phase lead is provided primarily in the low- to medium-frequency range.
2. In the low range, attenuation of output signal may require further amplification.
3. These networks operate on dc systems primarily. If an ac amplifier and ac error detector are used, these networks require a demodulator between the error detector and the network, and, conversely, a modulator between the network and the ac amplifier (see Fig. 10-21c).

The transfer-function equation for the lead network shown in Fig. 10-21a is

$$H = \frac{a/b(1 + j\omega/a)}{1 + j\omega/b} = \alpha \frac{1 + j\omega\tau}{1 + j\alpha\omega\tau} \tag{10-15}$$

where a/b represents various lead ratios from 0.1 to 0.5, depending on the values of R_1, R_2, and C (Fig. 10-21a).

* Other networks may be employed as well. See Appendix Table A-10 for a listing of transfer functions and gain response curves of some of the more common networks.

315

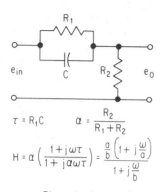

$$\tau = R_1 C \qquad \alpha = \frac{R_2}{R_1 + R_2}$$

$$H = \alpha \left(\frac{1 + j\omega\tau}{1 + j\alpha\omega\tau} \right) = \frac{\frac{a}{b}\left(1 + j\frac{\omega}{a}\right)}{1 + j\frac{\omega}{b}}$$

a. Phase lead network

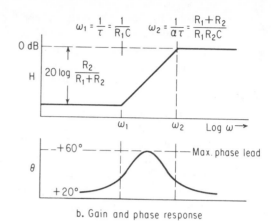

$$\omega_1 = \frac{1}{\tau} = \frac{1}{R_1 C} \qquad \omega_2 = \frac{1}{\alpha\tau} = \frac{R_1 + R_2}{R_1 R_2 C}$$

b. Gain and phase response

Phase lead network

c. Location of network in an ac servo error loop

Figure 10-21

Phase-lead network response and attenuation using cascade compensation.

       In general, the lower the ratio a/b, the greater the value of the maximum positive phase angle in Fig. 10-21b. For extremely small values, below 0.1, the maximum positive angle is $+72°$. For $a/b = 0.5$, the value is $+14°$.

       Conversely, however, small values of a/b produce greater attenuation of low frequencies. A value of 0.1 attenuates low frequencies by -20 dB while a value of 0.5 produces only -6 dB attenuation.

       Figure 10-22 shows the effect of cascading a phase-lead network (having a lead ratio of $a/b = 0.1$ and a maximum lead angle of $+55°$) in series with the unstable open-loop system H_0 of Fig. 10-20. Stability is effected in the following way:

1. The transfer function of the network [see Eq. (10-15)] has a gain-crossover frequency of approximately 3 rad/s and is represented on Fig. 10-22 as H_{PL}, where the subscripts indicated "phase lead."

2. The transfer functions are added graphically [in accordance with Eq. (10-14)] such that $H_{PL} + H_0$ provides a resultant transfer function H_R, which crosses the unity gain or 0-dB axis at approximately 3 rad/s.

316

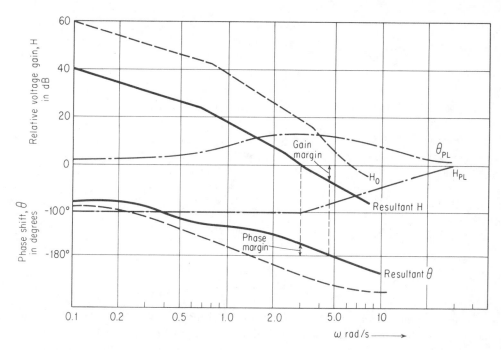

Figure 10-22

Cascade compensation using phase-lead
network to provide stability.

3. Note that the original transfer function, H_0, crosses the 0-dB axis at about 5 rad/s while the resultant, H_R, falls off about 3 rad/s. It is this earlier falloff of amplifier gain which provides the negative gain margin, ensuring stability, in part.

4. The phase response of the phase-lead network, θ_{PL}, is also added to the original phase response, θ_0, to provide increased bandwidth of the servo and the positive gain margin (approximately $+25°$), ensuring stability.

10-15.2
Phase-Lag Networks

The lead network discussed in the previous section caused the resultant open-loop transfer function to approach unity gain at a lower frequency and at a higher rate of falloff. This ensures that phase margin is positive and gain margin is negative. In effect, the open-loop gain must fall below unity before the phase lag approaches $-180°$, particularly in the low-frequency region. This implies that any network whose characteristic ensures a decrease in low-frequency gain might also serve to stabilize a servomechanism.

A typical phase-lag network is shown in Fig. 10-23a, with its controlling equations for time constant, τ, and transfer function, H. The gain and phase response of a phase-lag network are shown in Fig. 10-23b. Note that at *very* low frequencies the passive network has 0-dB gain but

317

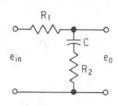

$\tau = (R_1 + R_2)C$

$\alpha = \dfrac{R_2}{R_1 + R_2}$

$H = \dfrac{1 + s\alpha\tau}{1 + s\tau} = \dfrac{1 + j\dfrac{\omega}{b}}{1 + j\dfrac{\omega}{a}}$

a. Phase lag network

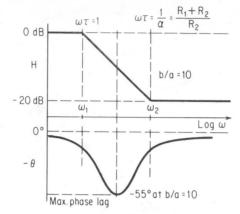

b. Gain and phase response

Figure 10-23

Phase-lag network.

rapidly falls off at a breakpoint frequency of $\omega_1 = 1/\tau$. The attenuation for frequencies above $\omega_2 = 1/\alpha\tau$ may be as much as -20 dB. Phase-lag networks are also placed in direct cascade with amplifiers to stabilize a servomechanism. Their *advantages* include

1. Increased stability, providing increased accuracy.
2. Relative insensitivity to noise or other high-frequency components affecting performance.
3. Relatively low cost in comparison to mechanical or electromechanical damping devices.
4. Improved low-frequency gain of the overall system.

The *disadvantages* of phase-lag networks include the following:

1. Generally decreased bandwidth of system.
2. Severe attenuation at the higher frequencies (Fig. 10-23b).
3. Like phase-lead networks, demodulators and modulators are required for their use in conjunction with ac amplifiers.
4. Because of 2, such compensation produces sluggish operation in response to rapidly changing inputs.

Figure 10-24 shows how a lag network may be used to produce negative gain margins and stable systems by causing the resultant open-loop transfer function to approach unity gain at a more rapid rate of falloff. The original response H_0 of a given amplifier falls off at a rate of -6 dB/octave. A lag network which has a maximum attenuation of -18

318

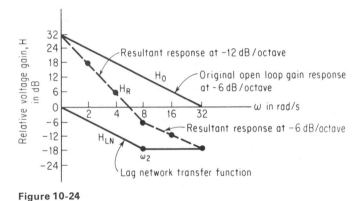

Figure 10-24

Use of phase-lag network to
improve closed-loop stability.

dB (ratio of $b/a = 8$) is added in cascade, having the transfer function H_{LN} shown in Fig. 10-24. The resultant response, H_R, now drops off at a faster rate of -12 dB/octave, producing a negative gain margin at $\omega = 8$ and unity gain at $\omega = 6$. (Originally, unity gain occurred at $\omega = 32$, as shown in Fig. 10-24.) Since phase shift of most amplifiers becomes increasingly negative with frequency, the lag network provides early gain crossover and relatively positive phase margins (angles of lag less than 180°) with negative gain margins, thus producing stability.

10-15.3
Lag-lead
Compensation

The two previous subsections showed that either a lead network or a lag network may improve the open-loop transfer characteristic, resulting in stability when used in a closed-loop system.

Phase lead is provided by the phase-lead network in the vicinity of the gain-crossover point, while the phase-lag network attenuates gain to unity rapidly before the phase lag of the system exceeds 180°. It should be possible, therefore, to cascade a specific phase-lead and phase-lag network in series to produce the desired compensation and stability. While this is theoretically possible, it is more convenient to use a single network, shown in Fig. 10-25a, called a *lag-lead* network. The network is readily derived from its gain and phase response, shown in Fig. 10-25b. A lagging phase angle is first produced as well as attenuation at low frequencies, typical of a lag network. At higher frequencies, a leading phase angle is produced as well as the positively rising slope characteristic of a lead network, with relatively little attenuation at higher frequencies.

The manner in which a dc lag-lead network may be cascaded to a servo system to provide closed-loop stability is shown in Fig. 10-26 and explained in Exs. 10-7 and 10-8.

319

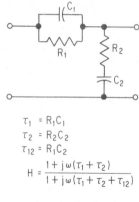

$\tau_1 = R_1 C_1$

$\tau_2 = R_2 C_2$

$\tau_{12} = R_1 C_2$

$H = \dfrac{1 + j\omega(\tau_1 + \tau_2)}{1 + j\omega(\tau_1 + \tau_2 + \tau_{12})}$

a. Lag-lead network

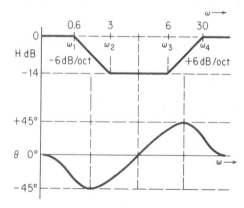

b. Gain and phase response

Figure 10-25

Lag-lead network; transfer function and Bode plots.

EXAMPLE 10-7

Using the original system open-loop response and original system phase response shown in Fig. 10-26, determine

a. Frequency at which gain is unity (0 dB) in rad/s.
b. Frequency at which phase shift is 180° in rad/s.
c. Original system gain margin in dB.
d. Original system phase margin in degrees.
e. Whether the system is stable.

Solution

(see Fig. 10-26):

a. Unity gain occurs at **20 rad/s**.
b. Phase shift is 180° at **9 rad/s**.
c. Gain margin (at 9 rad/s) is **+21 dB**.
d. Phase margin (at 20 rad/s) is **−225°**.
e. System is **unstable** without compensation because phase margin is negative and gain margin is positive.

EXAMPLE 10-8

A lag-lead network is cascaded to the original system of Ex. 10-7 and produces the resultant transfer-function open-loop gain and resultant phase response shown in Fig. 10-26. For the resultant responses determine

a. Frequency at which the gain is unity.
b. Frequency at which phase shift is 180°.
c. Gain margin.
d. Phase margin.
e. Whether the system is stable.

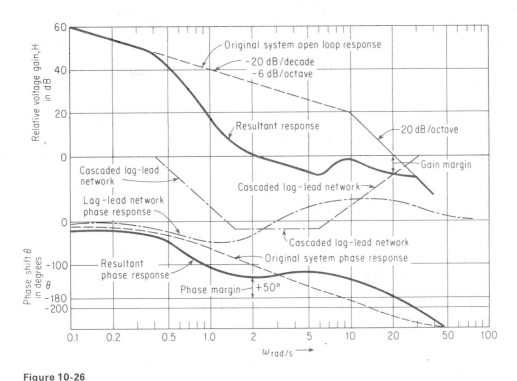

Figure 10-26

Cascade compensation of lag-lead network to
provide stability. See Exs. 10-7 and 10-8.

Solution

(see Fig. 10-26):
a. Unity gain occurs at **2 rad/s**.
b. Phase shift is 180° at **20 rad/s**.
c. Gain margin (at 20 rad/s) is **−9 dB**.
d. Phase margin (at 2 rad/s) is **+50°**.
e. System is *stable* because gain margin is negative and phase margin is positive.

10-16.
GAIN-FACTOR
COMPENSATION
The stability of a servomechanism is determined by its responses
to inputs or external disturbances at the load. Ideally, a stable
servomechanism is one that remains at rest (or at a null) until
excited by some input or external source and which returns to
rest (to a null) in response to such excitations. As pointed out previously,
the design of a servomechanism is a compromise, at best, between such
factors as stability, speed of response, and accuracy. The previous sections
showed how networks may be used to provide the required damping to
ensure stability of a servo system when connected in the closed-loop
mode. The assumption made in the previous section is that the original

321

system is underdamped. Let us now consider the various components cascaded in the error loop of a servosystem (Fig. 10-27a) to determine how they affect stability.

A complete closed-loop system is shown in Fig. 10-27a. The error detector for the system is a synchro control transformer, shown in Fig. 10-27b. A mechanical error input θ_{in} generates an output voltage e_{01}. The transfer function representing the output per unit input is K_s, expressed in V/rad, essentially the voltage output per unit error angle of input.

The electrical output of the error detector, e_{01}, serves as the input to an ac amplifier having gain G and electrical output e_{02}. The transfer function of the amplifier is G (or e_{02}/e_{01}), having no units since it is a numerical ratio, as shown in Fig. 10-27c.

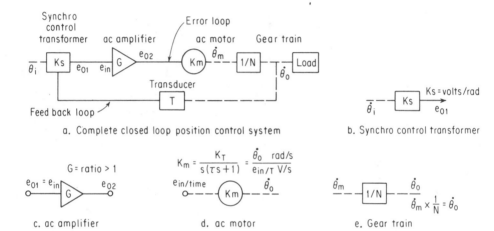

a. Complete closed loop position control system

b. Synchro control transformer

c. ac amplifier   d. ac motor   e. Gear train

Figure 10-27

Transfer functions of the various cascaded elements in the error loop of a servomechanism.

The electrical output of the amplifier is applied to an ac motor whose speed (in rad/s) is a function of the error input (in V/s). Thus, the transfer function of the motor, K_m, is a ratio of θ/e_{in} per unit time and is expressed in $\dfrac{\text{rad/s}}{\text{volts/s}}$ or rad/V, as shown in Fig. 10-27d.

The motor output is coupled through a gear train, having a transfer function $1/N$, to a load, shown in Fig. 10-27e.

The overall open-loop transfer function, H, for the various cascaded elements in the error loop between the disturbing input and the load is

$$H = \frac{K_s G K_m}{N} \tag{10-16}$$

where all terms have been defined above.

EXAMPLE 10-9

The servosystem shown in Fig. 10-27a has the following specifications: K_s = 23 V/rad, G = 2500, K_m = 5.415/s(0.0217s + 1) rad/s/V/s, and 1/N = 1/2000. Write the complete expression for the open-loop transfer function.

Solution

$$H = \frac{K_sGK_m}{N} = 23\frac{V}{rad} \times 2500\left[\frac{5.415}{s(0.0217s + 1)}\right]\frac{rad/s}{V/s} \times \frac{1}{2000}$$

$$= \frac{156.7}{s(0.0217s + 1)} = \frac{156.7}{j\omega(0.0217j\omega + 1)} = \frac{156.7}{0.0217j^2\omega^2 + j\omega}$$

$$= \frac{156.7}{-0.0217\omega^2 + j\omega} = \frac{156.7}{\omega(-0.0217\omega + 1j)}$$

The transfer function obtained in Ex. 10-9 is plotted in Fig. 10-28 showing the gain and phase response for the particular function.* The plot may be verified, however, in the following way:

1. At $\omega = 1$, the transfer function has the approximate value of 156.7, which corresponds to $1.567 \times 10^2 = 4$ dB + 40 dB = 44 dB (see Ex. 10-6).

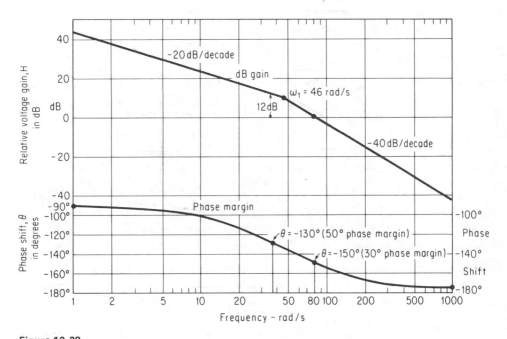

Figure 10-28

Bode plot for transfer function of Ex. 10-9.

> * The method of plotting gain and phase for various transfer functions is not covered in this brief chapter. For details of this technique, see B. E. De Roy, *Automatic Control Theory* (New York: John Wiley & Sons, Inc., 1966), Chap. 3.

2. At low frequencies, ω is small and ω_2 even smaller. The transfer function has the form of $1/j\omega$ because the ω^2 term may be neglected. The transfer function of $1/j\omega$ has a negative slope of -20 dB/decade or -6 dB/octave.

3. Using Eq. (10-11), a slope of -6 dB/octave approaches a phase angle of $-90°$.

4. Thus, at $\omega = 1$, $H = 44$ dB at an angle of $-90°$.

5. The gain continues to fall off at -6 dB/octave until ω_1, the frequency at which the second-order term $-0.0217\omega^2$ becomes significant. Breakpoint ω_1 occurs at approximately $1/0.0217$ or 46 rad/s (see Fig. 10-28).

6. At $\omega = 46$ rad/s, the gain falls off at a rate of -12 dB/octave or -40 dB/decade because of the influence of the slope of $1/j\omega^2$.

7. At the same time, using Eq. (10-11), the phase shift approaches 180°.

8. The phase angle at which the transfer-function expression crosses the 0-dB line is found by equating the solution of Ex. 10-9 to unity and substituting $\omega = 80$ rad/s (see Fig. 10-28). This yields 80 $(0.0217 \times 80 + j)$ or $-139 + j80$, which appears in the third quadrant at a lagging angle of $-150°$.

As may be seen from Fig. 10-28, the system if operated closed loop would be relatively stable because the phase margin is positive ($+30°$) while the gain margin is negative (-42 dB). As indicated earlier, however, a phase margin of from $+40°$ to $+60°$ is considered ideal in providing a damping factor which approaches unity.

The theory of gain-factor compensation is that any adjustment in the various cascaded factors contributing to the open-loop gain *does not affect* the Bode plot of phase angle. All that occurs is that the gain response is either raised or lowered on the dB scale. For example, if the gain of the amplifier in Ex. 10-9 is doubled (from 2500 to 5000), the overall transfer function is doubled. Thus, a gain of 2.0 corresponds to $+6$ dB. Consequently, every point on the dB gain curve would be raised by 6 dB and a second curve could be redrawn which is 6 dB higher. This is equivalent to shifting the vertical (ordinate) dB scale *downward* by 6 dB. The advantage of shifting the vertical *scale* is that the entire gain-response curve does not have to be redrawn!

Let us now examine Fig. 10-28 in the light of gain-factor compensation. The gain margin is negative at -42 dB, which allows an extensive latitude for gain-factor compensation. Ideally, a 50° phase margin provides unity damping factor when operated closed loop. Figure 10-28 shows a phase margin of only 30°, indicating that the servosystem is somewhat underdamped, even if it is stable. Example 10-10 shows how gain-factor compensation may be used to improve the stability of the system shown in Fig. 10-28.

EXAMPLE 10-10

For the open-loop transfer-function gain and phase response shown in Fig. 10-28, determine

a. The frequency at which a phase margin of $+50°$ is obtained
b. The dB value by which the gain response is raised or lowered
c. The change in gain margin and its effect on stability
d. Which element in the error loop should be varied and the change in gain of this element

Solution

(see Fig. 10-28):

a. Since the phase-response curve is unchanged by gain-factor compensation, a $+50°$ phase margin is produced at $-130°$ at a frequency of *38 rad/s*.
b. At a frequency of 38 rad/s, the gain is $+12$ dB. Thus, the gain-response curve could be *lowered* or the dB scale *raised* by 12 dB to provide a gain crossover at 38 rad/s. This corresponds to a reduction in gain of $-12\,dB$.
c. With the dB scale raised, the phase crossover of $-180°$ produces a gain margin of -42 dB $+ (-12$ dB$) = -54$ dB. A more negative gain margin implies greater stability. The gain margin has increased from -42 dB to $-54\,dB$. The phase margin is now $+50°$ indicating ideal damping.
d. Changing either the gear train or the motor affects the inertia reflected from the load. Ideally, the servoamplifier which is relatively linear can sustain a gain change without affecting the response curve of the system. The -12-dB change in gain corresponds to a gain reduction by a factor of $\frac{1}{4}$. The *servo amplifier* gain may be reduced to $\frac{2500}{4} = 625$.

10-17.
FEEDBACK
COMPENSATION

In Section 10-13, Rule (10-2) stated that any closed-loop system may be replaced by an equivalent open-loop system. If compensation networks are used in a feedback loop, the effect of such compensation is reflected equivalently to the open-loop transfer function, in accordance with Eq. (10-3) and the techniques shown in Section 10-13. Appendix Table A-10 gives the transfer functions of some common servo networks.

The three types of *cascade* compensation (lead, lag, and lag-lead) discussed in Section 10-16 have the disadvantage of sensitivity to changes in parameters of the fixed elements due to nonlinear behavior of the system. On the other hand, feedback compensation using high feedback loop gain is much less sensitive to fixed-element parameter variation. Furthermore, the networks used in feedback compensation may be much simpler in form than those used in cascade compensation.

The general configuration for feedback compensation is shown in Fig. 10-29a, where H_1 represents the overall transfer function of the *feedback* loop and G_1 is the overall transfer function of all cascaded elements in the *error* loop.

Using negative feedback, in accordance with Eq. (10-3) and the techniques given in Section 10-13, the equivalent open-loop transfer function,

325

H, as shown in Fig. 10-29b is

$$H = \frac{G_1}{1 + G_1 H_1} \qquad (10\text{-}17)$$

When the product $G_1 H_1$ in the denominator of Eq. (10-17) is much greater than 1, the equivalent open-loop transfer function, $H = 1/H_1$. Thus, in the frequency range where G_1 and H_1 are both reasonably large, the open-loop transfer function coincides with the reciprocal of the feedback function, H_1 in the error loop.

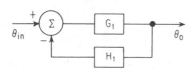

a. General configuration for feedback compensation

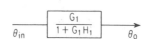

b. Equivalent open-loop configuration

Figure 10-29

Negative feedback stabilization.

When the product $G_1 H_1$ in Eq. (10-17) is small compared to 1, the equivalent open-loop transfer function, $H, = G_1$. Thus, in the frequency range where G_1 falls off and H_1 is attenuated, as well, the open-loop transfer function coincides with the transfer function of the cascaded elements in the error loop. In designing for feedback compensation, the frequency scale may be divided into several response regions, based on the relative magnitudes of the feedback transfer function and the forward-loop transfer function. In this way, the closed-loop response may be shaped as the feedback network compensation is adjusted. Ideally, the feedback network should provide those properties desirable for equivalent open-loop response: high gain at low frequencies, unity gain at a phase of approximately $+50°$, and low gain at high frequencies (negative gain margin).

While cascade compensation (Section 10-16) usually involves networks consisting of passive elements (resistance and capacitance), feedback compensation may take a variety of forms. For example, if there is an existing signal available at the output (motor armature voltage, tachometer voltage) which is proportional to output velocity, this feedback voltage may be fed directly back to the error detector to act as a compensation signal. Alternatively, the voltage may be fed back through a filter (passive network) or an amplifier (active network). The graphical representation of feedback compensation using Bode plots cannot be expressed as concisely as with cascade compensation. For this reason other techniques, such as the Root-Locus method or Nichol's diagrams, are used.* Transfer functions of some of the more common passive networks and their gain curves are given in Appendix Table A-10.

BIBLIOGRAPHY

Ahrendt, W. R., and C. J. Savant, Jr. *Servomechanism Practice* (New York: McGraw-Hill Book Company, 1965).

Baeck, H. S. *Practical Servomechanism Design* (New York: McGraw-Hill Book Company, 1968).

Blackburn, J. F., J. L. Shearer, and G. Reethof. *Fluid Power Control* (New York: John Wiley & Sons, Inc., 1960).

Bulliet, L. J. *Servomechanisms* (Reading, Mass.: Addison-Wesley Publishing Company, 1967).

Canfield, E. B. "Gearless Drive for Power Servo," *Electrical Manufacturing* (January 1959).

Chestnut, H., and R. W. Mayer. *Servomechanisms and Regulating System Design* (New York: John Wiley & Sons, Inc., 1959).

Clark, R. N. *Introduction to Automatic Control Systems* (New York: John Wiley & Sons, Inc., 1962).

Davis, S. A., and B. K. Ledgerwood, *Electromechanical Components for Servomechanisms* (New York: McGraw-Hill Book Company, 1961).

D'Azzo, J. J., and C. H. Houpis. *Control System Analysis and Synthesis* (New York: McGraw-Hill Book Company, 1960).

DeRoy, B. E. *Automatic Control Theory* (New York: John Wiley & Sons, Inc., 1966).

Eckman, D. P. *Automatic Process Control* (New York: John Wiley & Sons, Inc., 1958).

James, H. M., N. B. Nichols, and R. S. Phillips. *Theory of Servomechanisms* (New York: Dover Publications, Inc., 1965).

Johnson, E., *Servomechanisms* (Englewood Cliffs, N.J.: Prentice-Hall, Inc., 1963).

Kosow, I. L. *Electric Machinery and Transformers* (Englewood Cliffs, N.J.: Prentice-Hall, Inc., 1972).

Lewis, E. E. and H. Stern. *Design of Hydraulic Control Systems* (New York: McGraw-Hill Book Company, 1962).

Machol, R. E. *Systems Engineering Handbook* (New York: McGraw Hill, 1965).

---

* A discussion of these techniques is beyond the scope of this chapter. For more information see the references at the end of this chapter.

Melsa, J. L., and D. G. Schultz. *Linear Control Systems* (New York: McGraw-Hill Book Company, 1969).

Minorsky, N. *Journal of American Society of Naval Engineers*, vol. 34 (1922), p. 280.

Montgomery, T. B. "Regulex + Instability in Harness," *Allis-Chalmers Electrical Review* (Second and Third Quarters 1946).

Murphy, G. J. *Control Engineering* (Princeton, N.J.: D. Van Nostrand, 1959).

Nyquist, H. *Bell System Technical Journal*, Vol. 11 (1932), pp. 125–147.

O'Brien, D. G. "D-C Torque Motors for Servo Applications," *Electrical Manufacturing* (July 1959).

Ritow, I. "Specifying Servo Drives," *Electrical Manufacturing* (March 1956).

———. *Automatic Control System Design* (New York: Conover Mast, 1959).

———. *Advanced Servomechanism Design* (New York: Dolphin Books, 1963).

Truxal, J. G. *Automatic Feedback Control System Synthesis* (New York: McGraw-Hill Book Company, 1955).

———. *Control Engineers' Handbook* (New York: McGraw-Hill Book Company, 1958).

Watkins, B. *Introduction to Control Systems* (New York: The Macmillan Company, 1969).

QUESTIONS

10-1. Defining the closed-loop gain as $(G/1 + \beta G)$, prove mathematically that the closed-loop gain is
 a. The same as the amplifier gain, G, when the open-loop gain, βG, is much less than unity
 b. The same as $1/\beta$, when the open loop gain is much greater than unity
 c. The same as unity when β is unity
 d. Greater than the amplifier gain when positive feedback causes a negative loop gain between 0 and -1
 e. Approaching infinity as βG approaches -1

10-2. a. Draw a block diagram showing five basic elements of a servomechanism
 b. Indicate the purpose of each element
 c. Give five specific similarities possessed by all servomechanisms

10-3. a. List seven factors contributing to servomechanism error
 b. Describe each of these error factors
 c. Indicate one means of reducing each of these errors

10-4. a. Define instability
 b. Is instability due to error reduction?
 c. Is instability due to insufficient damping? Explain

10-5. Define underdamping, overdamping, and critical damping in terms of
  a. A damping factor ratio
  b. The roots of Eq. (10-6)
  c. Graphical representation of the variation of output with respect to time

10-6. It is desired to increase the rate of response of a critically damped servo-mechanism using a larger amplifier and servomotor to increase the output torque per unit error angle, K.
  a. Under what conditions may the rate of response be increased?
  b. What effect does this have on the steady-state error?
  c. Is it possible to achieve a steady-state zero error? Explain

10-7. Describe three classes of viscous dampers. Give one device serving as an example and discuss advantages and disadvantages of each type.

10-8. a. Describe viscous damping, its function, and explain why it is accompanied by a steady-state error
  b. Describe error-rate damping, its function, and explain why it is possible with this form of damping to reduce the steady-state error appreciably

10-9. a. Define a transducer
  b. Explain why the use of transducers extends the possibilities of closed-loop control over a variety of processes
  c. Give four inherent advantages resulting from the use of transducers

10-10. Discuss the necessity for modulators, explaining
  a. Purpose
  b. Application
  c. Types
  d. The advantages of electronic over mechanical-contact modulators

10-11. Modify the design of Fig. 10-11 using photovoltaic cells as transducers (dc output) to control the color density of the mixture. Draw a block diagram of the servomechanism, labeling and specifying all parts.

10-12. a. Repeat Question 10-11, using the motor pumping current as an indication of viscosity
  b. Show how a transduced voltage is obtained from the pump motor which serves as an analog of viscosity
  c. Draw a block diagram of the complete servomechanism, labeling and specifying all parts

P R O B L E M S

10-1. For the simple voltage divider shown below,
  a. Write the equation for the output voltage, v_2
  b. Write the open-loop transfer-function equation
  c. Write the equation for v_2 in closed-loop form
  d. Write the closed-loop transfer-function equation
  e. Show that the closed-loop transfer-function equation is equivalent to the open-loop equation.

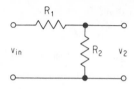

Problem 10-1

10-2. For the equation $y = G_1x_1 - G_2x_2 - 10$, draw a block diagram for the summing point inputs and the output, y.

10-3. For the simple voltage divider given in Problem 10-1 and equations developed, draw
a. The open-loop block diagram
b. The closed-loop block diagram

10-4. An oscillator crystal is located in an oven designed to maintain the crystal at a constant temperature. The voltage to a heating element in the oven is controlled by a thermostatic switch which serves as an error detector to sense the difference between the desired temperature and the actual oven temperature. Draw a block diagram representing the closed-loop system for maintaining constant crystal temperature.

10-5. In the human body, the nervous system is the error detector (or sensor) which maintains body temperature constant by sensing the actual skin temperature and controlling the output of the sweat glands at the skin surface. Draw a block diagram to represent the closed-loop system by which normal skin temperature is maintained.

10-6. In driving a car on a winding road, the eyes are the sensors and error detector between the road direction and the automobile's direction. Amplification of the error differences in the human brain controls the muscles of the hands and arms, which control the steering wheel and the power steering and wheels of the car. Draw a block diagram to represent the closed-loop system by which a human operator controls the automobile's direction to correspond to the road direction.

10-7. a. Draw the block diagram for a linear, unity, negative feedback control system having open-loop gain G, controlled output C, reference input R, and error E.
b. Write the open-loop transfer function for the block diagram.

10-8. a. Draw the block diagram for a feedback control system having a linear negative feedback element H in the feedback loop, a gain G in the error loop, controlled output C, reference input R, and error E.
b. Write the open-loop transfer function for this system.

10-9. a. Convert the block diagram drawn for Problem 10-8 to a unity feedback system
b. Prove that the open-loop transfer function for the block drawn in part a is equivalent to the open-loop transfer function obtained in your answer to Problem 10-8b

10-10. Reduce the feedback system shown in the block diagram below to a unity feedback system.

10-11. Prove the validity of Rule (10-3) using Fig. 10-14d.

10-12. Prove the validity of Rule (10-4) using Fig. 10-14e.

10-13. Simplify and determine the open-loop transfer function for the network shown.

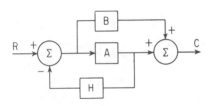

Problem 10-10

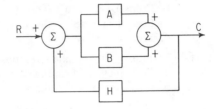

Problem 10-13

10-14. Simplify and determine the open-loop transfer function for the network shown.

10-15. Simplify and determine the open-loop transfer function for the network shown.

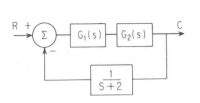

Problem 10-14

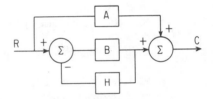

Problem 10-15

10-16. Convert the following dB values to relative voltage gain ratios:
   a. 25 dB
   b. 64 dB
   c. 100 dB
   d. 2 dB

10-17. Convert the following dB values to relative voltage gain ratios:
   a. -2 dB
   b. -10 dB
   c. -18 dB
   d. -30 dB

10-18. Convert the following relative voltage gain ratios to dB:
   a. 5000
   b. 2000
   c. 500
   d. 20

331

10-19. Convert the following relative voltage gain ratios to dB:
  a. 0.8
  b. 0.2
  c. 0.08
  d. 0.002

10-20. On three-cycle semilogarithmic paper plot the transfer function for the following open-loop conditions, showing both phase and gain response:
  a. From 0.1 to 0.4 rad/s, $K =$ gain ratio of 100
  b. From 0.4 rad/s to 1 rad/s, $G = 1/s$
  c. From 1 rad/s to 4 rad/s, $G = 1/s^2$
  d. From 4 rad/s to 40 rad/s, $G = 1/s^3$

10-21. From the gain and phase response plotted in Problem 10-20, determine
  a. Frequency at which the gain response is unity (0 dB)
  b. Frequency at which the phase response is 180°
  c. Phase margin
  d. Gain margin
  e. Whether the overall system is stable if connected as a closed-loop unity feedback system.

10-22. Closed-loop stability of the system of Problem 10-20 is achieved when the gain margin is +60°. For the open-loop transfer function plotted in Problem 10-20 using gain-factor compensation only, determine
  a. Frequency at which the phase margin is +60°
  b. Frequency at which the gain is unity for stability as a closed-loop unity feedback system
  c. Reduction or increase in amplifier gain constant, K, in gain ratio and dB
  d. Gain and phase margin of the system using gain-factor compensation, only
  e. Whether the overall system is stable if connected as a closed-loop unity feedback system

10-23. A typical servomotor has a stall torque of 0.33 oz-in, a no-load speed of 6800 rpm, and a rotor inertia of 0.46 g-cm². Assuming that the motor load has negligible inertia and/or viscous friction, calculate
  a. Stall torque in dyne-cm (see Appendix Table A-11)
  b. No load speed in rad/s (see Appendix Table A-11)
  c. Motor damping in dyne-cm per rad/s
  d. Motor time constant (ratio of motor inertia to motor damping), $\tau_m = J/D$
  e. Motor velocity constant, $K_v = 1/\tau_m$

10-24. A servomechanism has a controller producing a torque gradient, K, of 2×10^{-3} lb-ft/radian. The moment of inertia of the load, J_L, is 20×10^{-6} slug-ft² and F, the retarding output friction torque per unit output speed, is 2×10^{-4} lb-ft/rad/s. Calculate
  a. The natural frequency ($\omega_n = \sqrt{K/J}$) of the servomechanism in both rad/s and hertz
  b. The value of output damping for which the servo is critically damped, F_c

c. The damping factor, using the value of critical damping computed in part b

ANSWERS

10-1a. $v_{in}R_2/(R_1 + R_2)$ b. $R_2/(R_1 + R_2)$ c. $(R_2/R_1)v_{in} - (R_2/R_1)v_2$
   d. $H = (R_2/R_1)/(1 + R_2/R_1)$ e. $R_2/(R_1 + R_2)$.

10-7b. $C/R = \dfrac{G}{1 + G}$.

10-8b. $C/R = \dfrac{G}{1 + GH}$.

10-13. $C/R = (A + B)/[1 - H(A + B)]$.

10-14. $C/R = (A + B)/(1 + AH)$.

10-15. $C/R = [A(1 + HB) + B]/(1 + HB)$.

10-16a. 18 b. 1600 c. 10^5 d. 1.25.

10-17a. 0.8 b. 0.31 c. 0.125 d. 0.031.

10-18a. 61.78 dB b. 61.25 dB c. 41.78 dB d. 26 dB.

10-19a. -2 dB b. -14 dB c. -22 dB d. -54 dB.

10-21a. 5.3 rad/s b. 4.2 rad/s c. $-20°$ d. $+5$ dB e. unstable.

10-22a. 2 rad/s b. 2 rad/s c. $K = 10$ (or 20 dB) d. $+60°$ and -15 dB
   e. stable.

10-23a. 23,200 dyne-cm b. 712 rad/s c. $32.6\dfrac{\text{dyne-cm}}{\text{rad/s}}$ d. 14.1 ms
   e. 71 s^{-1}.

10-24a. 10 rad/s or 1.59 Hz b. $4 \times 10^{-4}\dfrac{\text{lb-ft}}{\text{rad/s}}$ c. 0.5.

ratings, selection, and maintenance of electric machinery

11-1.
FACTORS
AFFECTING
RATINGS OF
MACHINES

Electric machines are rated in terms of their *output* capacities. Generators and alternators are rated in terms of their output kilowatt capacity (kW) or kilovolt amperes (kVA) at a given rated prime-mover speed and a rated terminal voltage. Motors (dc and ac) are rated in terms of output capacity in **shaft horsepower** at rated speed, full-load current, and applied voltage.

When the electric machines are operated under these nameplate conditions, there is an implication that the temperature rise will not be excessive and that the machines will not overheat. While the manufacturer is aware that temporary overloads may be sustained, the rotating dynamos are not expected to carry sustained overloads for long periods. The consumer who, for reasons of economy, purchases a 10-hp motor to drive a 12- or 15-hp load continuously, runs the risk of *caveat emptor* in purchasing a product which (1) will deliver rated speed at rated load but not rated speed at an overload, (2) will overheat badly and (as a result) have a generally shorter life, and (3) will operate at a lower efficiency at the overload for the duration of its life. Thus, the lower initial cost is offset by poorer and more expensive running performance, coupled with the necessity for earlier replacement. For this reason, therefore, an indication is provided on all (electric rotating machinery) nameplates of the allowable temperature rise and duty cycle, as well as rated voltage, current, frequency, and speed.

11-2.
TEMPERATURE RISE
The standard allowable temperature rise of currently manufactured electric machinery is 40°C above ambient temperature.

Thus, if the room temperature is 70°F or 21°C, a 40°C rise in temperature implies that the motor temperature may be as high as 61°C or 142°F. It goes without saying that this matter of "allowable" temperature rise cannot be carried to a compulsive absurdity, however. A motor located in a confined area next to a high-temperature device, such as a boiler or heater, may have an "ambient" temperature* of 140°F or 60°C. A 40°C rise over such an "ambient" temperature brings the motor temperature to 100°C, the boiling point of water. Furthermore, even when the motor is not operating, the 140°F (60°C) "ambient" temperature of such a motor may be greater than its maximum limiting temperature based on the type of insulation employed. While any electric rotating machine will operate satisfactorily for some time at excessive temperatures, its life is shortened for precisely the *same* reason as if it were overloaded electrically.

Empirical studies show that for every 10°C increase in motor operating temperature over the recommended hottest-spot temperature limit (see Table 11-1), the winding life is cut in half. Conversely, for every 10°C reduction in motor operating temperature under the rated limit, the winding life is doubled.

Table 11-1 uses the standard maximum 40°C ambient temperature rise to assign a maximum allowable final temperature based on the limiting hottest-spot temperatures permitted for various classes of material.

---

* Ambient temperature, as defined by AIEE Standard No. 1, June 1947, is "the temperature of the medium used for cooling, either directly or indirectly, and is to be subtracted from the measured temperature of the machine to determine the temperature rise under specified test conditions." It is defined for particular cases as follows:

1. For self-ventilating apparatus, the ambient temperature is the average temperature of the air in the immediate neighborhood of the apparatus.
2. For air- or gas-cooled machines with forced ventilation or secondary water cooling, the ambient temperature is taken as that of the ingoing air or cooling gas.
3. For apparatus with oil or other liquid immersion of the heated parts where water cooling is employed, the ambient temperature is taken as that of the ingoing cooling water.

For the purpose of assigning a rating, 40°C is taken as the limiting ambient temperature of the cooling air or other gas.

335

The hottest-spot temperature allowed in a given dynamo, using a specific class of insulation, may be difficult to determine, since that spot may be buried in the stator or armature windings or may not be accessible for some other reason. Thermometers of the thermocouple or liquid-bulb type (mercury or alcohol) have to be located on the more accessible or outer portions of the dynamo, and such a temperature value depends on

TABLE 11-1.
LIMITING TEMPERATURES OF INSULATING
MATERIALS

DESCRIPTION OF MATERIAL	INSULA-TION CLASS	ALLOWABLE TEMP. FOR 40°C AMBIENT TEMP. STD.	HOTTEST-SPOT MAXIMUM LIMITING TEMPERATURE
Cotton, silk, paper, or other organic materials neither impregnated nor immersed in liquid insulating materials.	O	50°C	90°C
1. Any of the above materials immersed or impregnated in liquid dielectrics. 2. Enamels and varnishes applied to conductors. 3. Films and sheets of cellulose acetate or other cellulose products. 4. Molded and laminated materials having cellulose filler or phenolic resins or other resins of similar property.	A	65°C	105°C
Mica, asbestos, fiberglass, or other inorganic minerals with a small proportion of Class A materials as binders and fillers.	B	90°C	130°C
1. Mica, asbestos, fiberglass, and similar inorganic materials with binding substances composed of silicone compounds. 2. Silicone compounds in rubbery or resinous form, or materials with equivalent dielectric and temperature properties.	H	140°C	180°C
Pure mica, porcelain, glass, quartz, and similar inorganic material in pure form (glass wool, spun tapes, etc.)	C	No limit selected	

the thermodynamic temperature gradient created by the physical makeup of the machine. It is customary to add a correction of 15°C to the surface temperature to determine the hot-spot temperature. A higher and truer value of the internal maximum temperature is usually obtained by cold versus hot resistance measurements of the copper stator or rotor windings before and immediately after operation, using the temperature coefficient

CHAP. ELEVEN / *Rating, Selection, and Maintenance of Electric Machinery*

of copper.* A "hot-spot" correction of 10°C is generally added to the temperatures computed by resistance measurements.

Perhaps the best method of obtaining the hottest-spot temperature is by means of several embedded temperature detectors. These are either thermocouples or temperature-sensitive resistive material which are permanently built into the machine and whose leads are brought out for temperature monitoring purposes. Well-placed detectors usually yield higher values of temperature than either winding-resistance or contact-thermometer techniques.

On the basis of Table 11-1 the reader might be inclined to conclude that, if even the poorest insulation, Class O, can withstand a maximum temperature of 90°C, there is nothing to be concerned about. Unfortunately, there is, since most temperature measurements are made at the surface or by winding-resistance techniques. It is precisely for this reason that the American Standards Association (ASA) sets limiting "observable" temperature designations.† This is in recognition of the fact that, under most conditions, surface temperature rather than internal temperature is recorded.

In addition to a temperature rating, other rating factors, such as voltage, duty cycle, and speed, are assigned to dynamos.

11-3.
VOLTAGE RATING

The standard voltage ratings which have been adopted by NEMA are given in Table 11-2.

Note that, in Table 11-2, the voltage difference between dc generators and motors allows for a line voltage drop in the conductors supplying the motor. This is also true in the case of ac alternators and ac

TABLE 11-2.
STANDARD VOLTAGE RATINGS FOR ELECTRIC
ROTATING MACHINERY

MACHINE	STANDARD VOLTAGE RATINGS
dc generators	125, 250, 275, 600 V
dc motors	120, 240, 550 V
ac single-phase motors	115, 230, 440 V
ac polyphase motors	110, 208, 220, 440, 550, 2300, 4000, 4600, 6600 V
ac alternators	120, 240, 480, 600, 2400, 2500, 4160, 4330, 6990, 11,500, 13,800, 23,000 V

* The equation

$$\frac{R_2}{R_1} = \frac{(1/\alpha) + t_2}{(1/\alpha) + t_1}$$

may also be used for materials other than copper. For copper, $1/\alpha$ equals 234.5; and the equation is solved for t_2, where t_1 is the ambient temperature.

† The allowable 40°C rating given general-purpose machines (rather than those values given in the first temperature column of Table 11-1) is a safety factor based on "observable" temperature rises for different types of machines. See *Rotating Electrical Machines*, ASA Standard C-50.

337

polyphase or single-phase motors. These voltage ratings also correspond to Table A-3, A-4, and A-5 in the Appendix for dc and ac motors, although the tables do not include rated line current for the higher-voltage polyphase or synchronous motors shown above. As indicated in the appended tables, ratings are not available in the entire range of voltages. The higher-voltage dynamos usually are reserved for the higher-capacity ratings.

11-4.
EFFECT OF DUTY CYCLE AND AMBIENT TEMPERATURE ON RATING

In addition to temperature and voltage ratings, another rating factor is the *duty cycle*. The duty cycle of currently manufactured electric machinery is stated as either *continuous* duty, *intermittent* duty, *periodic* duty, or *varying* duty.*

For the same horsepower or kVA rating capacity, the continuous-duty machine will be *larger* in size, physically, than the intermittent-duty machine. The larger size results from conductors of larger diameter and heavier insulation. Furthermore, a larger frame size presents a larger surface area from which heat may be dissipated; and this, too, results in a lower operating temperature for the same duration of operation. In general, a 10-hp *continuous*-duty motor may be considered a 12- or 13-hp *intermittent*-duty motor (although the rated speed may be somewhat less), since the temperature rise is not excessive if intermittently operated; see Ex. 11-1, Sec. 11-7.

The duty cycle is closely related to temperature, therefore, and is generally taken to include environmental factors also. A 100-kVA alternator (intermittent rating) might be converted to a 200-kVA alternator if continuously operated at the North Pole at an ambient temperature of $-80°C$, since all the heat generated would still not be sufficient to overheat the alternator under such ambient conditions.

Just as the capacity rating and duty cycle are *reduced* by an *increase* in *ambient temperature*, so too are the capacity rating and duty cycle *increased* by an extreme *decrease* in ambient temperature.

In the same way, *totally enclosed* machines (without auxiliary forced ventilation, which do not permit ventilation and replacement of internal air) do not have as high a capacity rating as similar machines which are *not* totally enclosed and which are ventilated in such a manner that fresh air is drawn across the stator and rotor windings. (See Section 11-5.)

11-5.
TYPES OF ENCLOSURES

The National Electric Manufacturers Association (NEMA) recognizes and defines the type of motor enclosures listed below. Both cost and physical size for totally enclosed motors is higher than open motors of the same hp rating, duty cycle and ambient temperature rise.

Waterproof enclosure. A totally enclosed enclosure so constructed as to exclude water applied in the form of a stream from a hose, except that

* See Sec. 11-7 for definitions and calculations of rating based on duty cycle.

338

leakage may occur around the shaft providing it is prevented from entering the oil reservoir and provision is made for automatically draining such water. The means for the latter may be a check valve or a tapped hole at the lowest part of the frame for a drain pipe.

Dust-ignition-proof enclosure. A totally enclosed enclosure so designed and constructed as to exclude ignitable amounts of dust or such amounts that might affect the performance rating.

Explosion-proof enclosure. A totally enclosed enclosure so designed and constructed as to withstand an explosion of a specified gas or vapor which may occur within it, and to also prevent ignition of specified gas or vapor surrounding it by sparks, flashes, or explosions which may occur within the enclosure.

Totally enclosed enclosure. An enclosure which prevents the free exchange of air between the inside and outside of the enclosure but not sufficiently enclosed to be considered airtight.

Weather-protected enclosure. An open enclosure whose ventilating passages are so designed as to minimize entrance of rain, snow, and airborne particles to the electric parts.

Guarded enclosure. An open enclosure in which all openings giving direct access to live or rotating parts (except smooth motor shafts) are limited in size by the design of the structural parts or by screens, grills, expanded metal, etc., to prevent accidental contact with such parts. Such openings will not permit passage of a cylindrical rod $\frac{1}{2}$ inch in diameter.

Splash-proof enclosure. An open enclosure in which the ventilating openings are so constructed that drops of liquid or solid particles, falling on it, at any angle not greater than 100° from the vertical, cannot enter either directly or by striking and running along a horizontal or inwardly inclined surface.

Drip-proof enclosure. An open enclosure in which the ventilating openings are so constructed that drops of liquid or solid particles, falling on it, at any angle not greater than 15° from the vertical, cannot enter either directly or by striking and running along a horizontal or inwardly inclined surface.

Open enclosure. An enclosure having ventilating openings which permit passage of external cooling air over and around the windings of the machine. When an internal fan is provided, such machines are called self-ventilating.

11-6.
SPEED RATING;
CLASSIFICATIONS
OF SPEED AND
REVERSIBILITY

Generators, converters, and alternators are all designed for a given constant speed whose value or rating is expressed on the nameplate. When driven by a prime mover at this rated speed, the generator, alternator, or converter should deliver constant (rated) voltage at the rated load.

Motors, however, are subject to speed change. A reduced speed will produce poor ventilation and overheating. Motors, therefore, are rated at the speed at which they will deliver their rated output horse-

339

power at the rated voltage. When speed control is used on a motor, therefore, it cannot be expected that, for the same rated load current, a lower speed produces the rated horsepower output. In general, as the speed decreases, the motor rating should be decreased or *derated* proportionately.

Table 11-3 lists various groups of dc and ac motors, first, by their

TABLE 11-3.
MOTOR CLASSIFICATION BASED ON SPEED REGULATION AND SPEED VARIATION

GROUP	MOTOR TYPE	SPEED-REGULATION CHARACTERISTICS

1	Synchronous motor
	a. Polyphase
	b. Single-phase
	  1. Reluctance motor
	  2. Hysteresis motor
		Absolutely constant at synchronous speed, $S = 120f/P$

2	Asynchronous SCIM *a.* Polyphase *b.* Single-phase Shunt motor, dc	Relatively constant speed from no load to full load, with somewhat higher no-load speed
3	Polyphase SCIM, class D compound motor, dc	Moderate decrease in speed from no load to full load
4	Repulsion motor, repulsion-induction motor, series motor, dc and ac universal motor	Extremely large decrease in speed from no-load to full load; high speed at low or no load; very high starting torque, and low speed at high torque

ADJUSTABLE SPEED-VARIATION CHARACTERISTICS

1	Polyphase SCIM or synchronous motor using adjustable-frequency alternator	Speed-variation range up to 6 : 1
2	Dc motor using armature voltage and field rheostat control	Speed-variation range up to 200 : 1
3	Single-phase and polyphase motors using mechanical speed-adjustment systems or eddy-current clutches	Up to 25 hp, speed variation up to 16 : 1 Up to 100 hp, speed variation up to 100 : 1
4	Dc motor using solid-state control of input waveform	Speed variation up to 200 : 1
5	Polyphase WRIM using *a.* Secondary resistance control *b.* Concatenation (foreign voltage control) *c.* Leblanc system *d.* Kramer control system *e.* Scherbius system *f.* Solid-state foreign voltage control	Speed variation from 10 : 1 up to 200 : 1
6	Schrage brush-shifting (BTA) motor	Speed variation up to 4 : 1
7	Brush-shifting repulsion motor	Speed variation up to 6 : 1
8	Multispeed SCIMs, polyphase and single-phase	Speed ratios of 2 : 1 or 4 : 1 but not adjustable in these ranges; speed is definite with little change due to load

CHAP. ELEVEN / *Rating, Selection, and Maintenance of Electric Machinery*

speed-regulation characteristics and then by their speed-variation characteristics.

11-7. FACTORS AFFECTING GENERATOR AND MOTOR SELECTION

In addition to some of the factors mentioned above, other factors are of importance in the selection of generators or motors for specific use.

In the case of a generator, synchronous converter, or alternator, such factors include the type of prime mover; the method of mounting to be employed; direction of rotation; whether it is to be located in the open or in a totally enclosed building; the type of control which will be employed; the maintenance conditions in terms of accessibility; whether directly coupled, geared, or belted to the prime mover; and general humidity, atmospheric, or environmental conditions to which it will be subjected.

In the case of a motor, *duty service* (see Ex. 11-1) is perhaps the most important factor to be considered. The nature of load and overload frequency is a serious consideration; also, the type of mounting, whether horizontal or vertical, and whether floor, ceiling, or wall-mounted; the type of speed control to be employed; the method of coupling to the load; and how frequently it will be stopped, started, and reversed are factors affecting the type of motor to be selected and the horsepower-rating capacity. Wherever possible, data based on tests with a temporary motor or by calculation should be used. Both the average and maximum *load* conditions should be considered, both in tests and in calculations. In some instances, the maximum load requirements may occur only at starting; whereas, in other cases, periodic overloads of short duration may exceed the starting requirements. Other factors include power source available, frequency, voltage fluctuations, reversing characteristics, speed range, method of mounting, space available, lubrication provisions, accessibility of brushes (if any), maintenance, coupling provisions, speed-reduction techniques, type of enclosure (Section 11-5), cost per hp, starting and running torque, acceleration time, and breakdown torque.

Yet, as stated at the outset, duty cycle is perhaps most important. Four different types of duty cycles are classified by NEMA:

1. *Continuous duty*—dynamo use requiring operation at fairly constant load for reasonably long periods of time.
2. *Periodic duty*—load requirements recur regularly at periodic intervals over a reasonably long period of time.
3. *Intermittent duty*—irregular occurrence of load requirements, including fairly long periods of rest at which no load occurs.
4. *Varying duty*—both the loads and the periods of time at which the load requirements occur may be subject to a wide variation, without rest, over a reasonably long period of time without any regularity whatever.

Example 11-1 shows the method of calculating hp rating, in terms of **rms hp**, for intermittent-, varying-, and periodic-duty motors.

As a general rule, for all dynamos the capacity selected should be such that the dynamo will be operating between three-quarters to full load most of the time. A dynamo which is larger than necessary will have a low running efficiency and higher operating cost in addition to higher initial cost. In the case of a generator, with a possible increased anticipated load, this may not be a problem. In the case of a motor driving a specific load, such as an induction motor, not only is the efficiency (of a larger than necessary motor) poor, but the power factor is also poor. Similarly, a dynamo which is too small has a lower operating efficiency, and is subject to overheating, shorter life, and increased maintenance and repair costs.

In a number of applications, it may be required to select a motor for service conditions to drive a load which varies widely over continuously repeated cycles. A drill, for example, may be used with various drill bits and may be driven into various thicknesses of various metals. The heating of the motor is determined not by the peak but by the **rms** values of current under various load conditions. Furthermore, the cooling period during standstill or idle time is *less* effective than when the motor is running, and, therefore, it is customary to divide idle periods by an empirical factor of approximately 3. The required horsepower capacity rating, therefore, is the rms "average" of the various instantaneous horsepower ratings throughout a given test cycle, as shown by Ex. 11-1.

EXAMPLE 11-1

EXAMPLE 11-1. A 200-hp test motor was employed to determine the best capacity rating for a varying-load-requirement duty cycle over a 30-min period. The test motor operated at 200 hp for 5 min, 20 hp for 5 min, and a rest period of 10 min followed by 100 hp for 10 min. Calculate the horsepower required for such an intermittent varying load.

Solution

$$\text{rms hp} = \sqrt{\frac{[(200)^2 \times 5] + [(20)^2 \times 5] + ((100)^2 \times 10]}{5 + 5 + 10 + 10/3}} = \textbf{114 hp}$$

A 125-hp motor would be selected because that is the nearest larger commercial standard rating. This means that the motor would operate with a 160 per cent overload (at 200 hp) for 5 min, or one-sixth of its total duty cycle.

In general, most manufacturers of electric machinery employ applications engineers in their field service organizations to assist the consumer in the selection of the proper size and type of electric machine for a given load requirement. It is well to consult with one or more of these groups before purchasing a large piece of equipment which, if improperly selected,

CHAP. ELEVEN / *Rating, Selection, and Maintenance of Electric Machinery*

will result in high energy cost, inefficiency, poor service, overheating, breakdown, and increased maintenance costs.

11-8.
MAINTENANCE

Preventive maintenance and routine inspection techniques conserve and prolong the life of electric machinery. Induction-type machines require only periodic lubrication, while some, equipped with self-lubricating "lifetime" bearings, require no lubrication whatever. Dynamos equipped with brushes require periodic brush, commutator, or slip-ring maintenance in addition to lubrication. High-speed series-wound (dc, ac or universal) motors should not be selected for long and continuous-duty cycles because the severe brush sparking may require frequent commutator cleaning and brush replacement.

In lubricating electric machinery, excessive oiling is just as damaging as insufficient lubrication. Oil-gummed commutators and oil-soaked brushes may result in severe sparking of commutator machines. Oil leaking onto the stator may cause insulation breakdown of ac and dc stator windings.

Most types of electrical machinery require a minimum of maintenance confined only to minor lubrication. But many types of single-phase fractional-hp motors of the split-phase and repulsion types are equipped with centrifugal switches which may be a source of trouble that may damage the motor severely. If a centrifugal-switch mechanism is "stuck in its running position," the motor fails to start. If "stuck in its starting position," the starting winding overheats and the motor fails to reach rated speed. The contacts of the switches may be gummed or oxidized or worn out, as well. Such mechanisms should be replaced rather than repaired.

Because maintenance is usually confined merely to routine lubrication, inspection becomes an important factor in prolonging life of machinery and should not be ignored. Four of the five senses are of extreme importance here: sight, sound, smell, and touch. Visual inspection will reveal a number of troubles listed in Table 11-4. A noisy motor is an indication of worn bearing, overloading, or single-phasing. A burnt odor, characteristic of burning insulation, is an indication of overload or breakdown. An overheated bearing or winding is detected by touch (the surface should not be so hot that one cannot hold one's hand on it).

Further, in troubleshooting, certain symptoms if identified (see Table 11-4) automatically eliminate others. If heating occurs and the temperature rises significantly, it automatically eliminates other probabilities, such as blown fuse or failure to start.

Table 11-4 shows 19 common motor troubles and their causes. These are not exclusive but should be useful in diagnosing most difficulties.*

* Bodine Electric Company, *Fractional hp Motor and Control Handbook*, 3rd ed., 1968, p. 99.

TABLE 11-4.
REFERENCE GUIDE TO PROBABLE CAUSES OF MOTOR TROUBLES

MOTOR TYPE SYMPTOM OR TROUBLE	AC SINGLE PHASE				AC POLYPHASE (TWO OR THREE PHASE)	BRUSH-TYPE (UNIVERSAL, SERIES, SHUNT, OR COMPOUND)
	SPLIT-PHASE	CAPACITOR START	CAPACITOR START AND RUN	SHADED-POLE		
	PROBABLE CAUSES					
Will not start	1, 2, 3, 5	1, 2, 3, 4, 5	1, 2, 4, 7, 17	1, 2, 7, 16, 17	1, 2, 9	1, 2, 12, 13
Will not always start, even with no load, but will run in either direction when started manually	3, 5	3, 4, 5	4, 9		9	
Starts, but heats rapidly	6, 8	6, 8	4, 8	8	8	8
Starts, but runs too hot	8	8	4, 8	8	8	8
Will not start, but will run in either direction when started manually—overheats	3, 5, 8	3, 4, 5, 8	4, 8, 9		8, 9	
Sluggish—sparks severely at the brushes						10, 11, 12, 13, 14
Abnormally high speed—sparks severely at the brushes						15
Reduction in power—motor gets too hot	8, 16, 17	8, 16, 17	8, 16, 17	8, 16, 17	8, 16, 17	13, 16, 17
Motor blows fuse, or will not stop when switch is turned to off position	8, 18	8, 18	8, 18	8, 18	8, 18	18, 19
Jerky operation—severe vibration						10, 11, 12, 13, 19

1. Open in connection to line.
2. Open circuit in motor winding.
3. Contacts of centrifugal switch not closed.
4. Defective capacitor.
5. Starting winding open.
6. Centrifugal starting switch not opening.
7. Motor overloaded.
8. Winding short-circuited or grounded.
9. One or more windings open.
10. High mica between commutator bars.
11. Dirty commutator or commutator is out of round.
12. Worn brushes and/or annealed brush springs.
13. Open circuit or short circuit in the armature winding.
14. Oil-soaked brushes.
15. Open circuit in shunt winding.
16. Sticky or tight bearings.
17. Interference between stationary and rotating members.
18. Grounded near switch end of winding.
19. Shorted or grounded armature winding.

11-9.

FUTURE TRENDS
GOVERNING THE
SELECTION OF
MOTORS AND
DRIVES

Table 6-1 and its related text material showed that solid-state packages have begun (and will continue) to dominate as power drives for dc motors ranging from fractional up to 1000 hp. Similarly, Secs. 7-2, 7-10 and 8-5, respectively, indicate that solid-state packages are emerging for adjustable frequency/voltage control of small to medium polyphase and single-phase motors, respectively.

In selecting a motor drive system for a particular application, certain criteria tend to dictate the type of motor (whether dc or ac) and (consequently) the drive system required. Three major considerations are (1) initial overall system cost, (2) those factors governing the reliability and maintenance of the overall system (both drive and motor) and (3) relative accuracy with which the motor must be speed controlled.

Table 11-5 compares the various advantages of dc and ac solid-state systems* dictating the choice between a dc motor (and dc solid-state drive) versus an induction (or synchronous-induction or synchronous) motor (and an ac solid-state drive).

As of this writing, it is clear that the inherently greater accuracy of

---

TABLE 11-5.
ADVANTAGES OF DC AND AC SOLID-STATE
PACKAGED DRIVES AND MOTORS, RESPECTIVELY

ADVANTAGES OF DC DRIVES	ADVANTAGES OF AC DRIVES
1. Lower initial overall system cost for the same hp rating.	1. Lower maintenance (and initial) costs for motors of the same hp rating, resulting in higher reliability of operation.
2. Higher torque at low speeds with speed control accuracy ranging from 2 to 5% of desired speed (open loop).	2. Higher overall system reliability under severe ambient and environmental conditions.
3. Motors can be brought down to zero speed.	3. Higher accuracy of speed control (open loop) from high to low speeds (but not to zero) ranging from 0.25% down to 0.05% (for synchronous motors).
4. Motors can be reversed rapidly or effect a change in speed with little delay.	4. Even higher accuracy (open loop) at higher initial costs available using digital techniques (down to 0.001%).
5. Solid-state drive package has relatively high efficiency in converting ac to dc.	5. Several motors and their loads may be phase locked to the same identical speed for process control applications.
6. A wider selection of dc drive packages and motors, in all hp ranges, are (presently) available at lower cost.	6. New developments in solid-state devices and circuits indicate possibility of reduced costs and increasingly higher reliability.
7. Solid-state drive system is approximately half the overall initial system cost.	7. Motor cost only about 10% of overall initial system cost.

* For a more detailed discussion of the subject, cf., P. G. Mesniaeff, "Solid State Adjustable-Frequency AC Drives," *Control Engineering*, Nov. 1971, pp. 57–70.

345

the ac solid-state packaged drives indicate increased use of these drives as the cost of higher power SCRs and solid-state regulator packages continues to decrease. Simultaneously, however, the development of higher power brushless dc motor packages (Sec. 6-14) may result in lower maintenance costs of dc motor packages, tending to offset some of the maintenance and reliability advantages of ac motor drives. Clearly, the engineering designer is faced with a variety of tradeoffs in Table 11-5 which are themselves continuously changing with rapid changes in the state-of-the-art.

Both the dc and ac drive systems may be controlled with greater accuracy using closed loop servosystems. But since the ac drive tends to have greater inherent open-loop accuracy, its closed loop accuracy is correspondingly higher. Future trends may see extension of solid-state drives to wider hp ranges as well as techniques providing extremely accurate control of speed, even down to standstill.

BIBLIOGRAPHY

Buchanan, C. H. "Duty-Cycle Calculations for Wound-Rotor Motors," *Electrical Manufacturing* (November 1959).

Cook, J. W. "Squirrel-Cage Induction Motors Under Duty-Cycle Conditions," *Electrical Manufacturing* (February 1956).

Daniels, S. *The Performance of Electrical Machines* (New York: McGraw-Hill Book Company, 1968).

General Principles Upon Which Temperature Limits Are Based in the Rating of Electrical Equipment (New York: American Institute of Electrical Engineers, Publication No. 1).

Heumann, G. W. "Motor Protection," in *Magnetic Control of Industrial Motors* part 2 (New York: John Wiley & Sons, Inc., 1961), Chap. 6.

Industrial Control Equipment (Group 25) (ASA C42.25) (New York: American Standards Association).

Industrial Control Equipment (UL508) (Chicago: Underwriters Laboratories, Inc.). "Inherent Motor Overheat Protection Moves Inside the Field Coils," *Electrical Manufacturing*, November 1959.

Karr, F. R. "Squirrel-Cage Motor Characteristics Useful in Setting Protective Devices," AIEE Paper 59–13.

Kosow, I. L. *Electric Machinery and Transformers* (Englewood Cliffs, N.J.: Prentice-Hall, Inc., 1972).

Lebens, J. C. "Positive Over-Temperature Protection—With Heat Limiters," *Electrical Manufacturing* (January 1958).

Libby, C. C. *Motor Selection and Application* (New York: McGraw-Hill Book Company, 1960).

Machine Tool Electrical Standards (Cleveland: National Machine Tool Builders Association).

Morris G. C. "Duty-Cycle Motor Selection," *Electrical Manufacturing* (November 1958).

Motor and General Standards (MG1) (New York: National Electrical Manufacturers Association).

Slaymaker, R. R. *Bearing Lubrication Analysis* (New York: John Wiley & Sons, Inc., 1955).

Smeaton, R. W. *Motor Application and Maintenance Handbook* (New York: McGraw-Hill Book Company 1969).

Vaughan, V. G., and Glidden, R. M. "Built-In Overheat Protection for Three-Phase Motors," *Electrical Manufacturing* (August 1958).

—— and White, A. P. "New Hotter Motors Demand Thermal Protection," *Electrical Manufacturing* (February 1959).

Veniott, C. G. *Fractional and Subfractional Horsepower Electric Motors*, 3rd ed. (New York: McGraw-Hill Book Company, 1970).

Wilcock, D. F., and Booser, E. R. *Bearing Design and Application* (New York: McGraw-Hill Book Company, 1957).

Wilt, H. J. "Circuit Factors in Motor Protection," *Electrical Manufacturing* (June 1959).

PROBLEMS AND QUESTIONS

11-1.　a. What disadvantages occur from buying a dynamo which has a lower output rating than the average load for which it is intended?

b. What information is provided on the nameplates of most dynamos?

11-2.　a. What is the meaning of the rating "40°C temperature rise" above ambient temperature"?

b. How is the temperature rating affected by the class of insulation employed on the dynamo windings?

c. What is the disadvantage of high ambient temperatures with respect to insulation and dynamo life?

d. What is meant by hottest spot temperature?

e. Give three methods of measuring it practically.

11-3.　a. Why is the voltage rating of a dc motor lower than for a dc generator?

b. Repeat part a for the ac dynamo

c. Why is the standard voltage range of available ac alternators higher than for ac motors?

d. Why is the voltage range of available ac polyphase motors higher than single-phase motors?

11-4.　a. Why is the continuous-duty larger than an intermittent-duty dynamo?

347

b. How is the capacity of a given rated dynamo affected by (1) duty cycle, (2) ambient temperature, (3) enclosure, and (4) forced ventilation?

11-5. a. Which types of enclosures are suitable for outdoor dynamos exposed to the weather?

b. In what locations are open enclosures permitted?

c. Under what actual physical circumstances (give examples) would a consumer desire a waterproof motor enclosure?

11-6. a. How is the rating of a dynamo affected by a decrease in speed?

b. Why are motors more subject to a change in rating due to speed than generators?

c. Distinguish between a varying-speed motor and an adjustable-speed motor

d. Distinguish between an adjustable varying-speed motor and the types defined in part c.

11-7. a. Explain why Table 11-3 distinguishes between the speed-regulation characteristics and the adjustable speed-variation characteristics

b. What class of motors provides the best speed regulation with widest speed variation?

c. Can the speed of a hysteresis motor be varied? How?

d. What class of motors provides the lowest speed variation? Why?

e. Why is the polyphase alternator driven by a variable-speed prime mover ideally suited for speed variation of synchronous and asynchronous motors?

f. What effect does increase in frequency and speed have on the rating in part e?

11-8. In addition to the factors of rated voltage, rated current, frequency, speed, duty cycle, and temperature rise, list other considerations affecting selection of

a. generators and alternators

b. motors.

11-9. a. Define four types of duty service

b. What is meant by rms hp?

c. In the equation for calculating rms hp, why is the idle period divided by a factor of 3?

d. Assuming all other factors equal, on the basis of rms hp, arrange the following $\frac{1}{4}$-hp motors in order of descending physical size: intermittent duty, varying duty, continuous duty, and periodic duty.

11-10. a. Why is periodic inspection of dynamo operation an important factor in preventive maintenance and life of the dynamo?

b. Does a self-lubricating, explosion-proof SCIM require periodic inspection? Explain

c. Why is overlubrication just as harmful as underlubrication or no lubrication?

d. In making routine inspections, how are the human senses involved?

e. What are the limitations of human senses and what possible causes of troubles could only be revealed by instruments? (See Table 11-4)

11-11.   With reference to Table 11-4, for each motor type listed in the columns,
  a. total the number of *different* probable causes of failure
  b. total the number of different symptoms of trouble
  c. On the basis of part a and b, which single-phase motors are most trouble-free?
  d. Compare your answer in part c with polyphase motors and brush-type motors. Draw conclusions.

11-12.   A 440 V, 30 hp SCIM in a commercial laundry washer drives a load of 15 hp for 5 min, followed by a 60 hp load for 5 min, followed by a no load (drying) period of 15 min. If this cycle is repeated continuously for a 24 hr period, calculate
  a. the factor by which the motor rating is exceeded
  b. (a) again, if the idle period is increased to 30 min.

ANSWERS

11-12 a. 118.7% b. 103%

appendix

TABLE A-1. NATURAL TRIGONOMETRIC FUNCTIONS

Angle, °	sin	tan	cot	cos	Angle, °	Angle, °	sin	tan	cot	cos	Angle, °
0.0	.00000	.00000	∞	1.00000	90.0	6.0	.10453	.10510	9.5144	.99452	84.0
.1	.00175	.00175	572.96	1.00000	.9	.1	.10626	.10687	9.3572	.99434	.9
.2	.00349	.00349	286.48	0.99999	.8	.2	.10800	.10863	9.2052	.99415	.8
.3	.00524	.00524	190.98	.99999	.7	.3	.10973	.11040	9.0579	.99396	.7
.4	.00698	.00698	143.24	.99993	.6	.4	.11147	.11217	8.9152	.99377	.6
.5	.00873	.00873	114.59	.99996	.5	.5	.11320	.11394	8.7769	.99357	.5
.6	.01047	.01047	95.489	.99995	.4	.6	.11494	.11570	8.6427	.99337	.4
.7	.01222	.01222	81.847	.99993	.3	.7	.11667	.11747	8.5126	.99317	.3
.8	.01396	.01396	71.615	.99990	.2	.8	.11840	.11924	8.3863	.99297	.2
.9	.01571	.01571	63.657	.99988	.1	.9	.12014	.12101	8.2636	.99276	.1
1.0	.01745	.01746	57.290	.99985	89.0	7.0	.12187	.12278	8.1443	.99255	83.0
.1	.01920	.01920	52.081	.99982	.9	.1	.12360	.12456	8.0285	.99233	.9
.2	.02094	.02095	47.740	.99978	.8	.2	.12533	.12633	7.9158	.99211	.8
.3	.02269	.02269	44.066	.99974	.7	.3	.12706	.12810	7.8062	.99189	.7
.4	.02443	.02444	40.917	.99970	.6	.4	.12880	.12988	7.6996	.99167	.6
.5	.02618	.02619	38.188	.99966	.5	.5	.13053	.13165	7.5958	.99144	.5
.6	.02792	.02793	35.801	.99961	.4	.6	.13226	.13343	7.4947	.99122	.4
.7	.02967	.02968	33.694	.99956	.3	.7	.13399	.13521	7.3962	.99098	.3
.8	.03141	.03143	31.821	.99951	.2	.8	.13572	.13698	7.3002	.99075	.2
.9	.03316	.03317	30.145	.99945	.1	.9	.13744	.13876	7.2066	.99051	.1
2.0	.03490	.03492	28.636	.99939	88.0	8.0	.13917	.14054	7.1154	.99027	82.0
.1	.03664	.03667	27.271	.99933	.9	.1	.14090	.14232	7.0264	.99002	.9
.2	.03839	.03842	26.031	.99926	.8	.2	.14263	.14410	6.9395	.98978	.8
.3	.04013	.04016	24.898	.99919	.7	.3	.14436	.14588	6.8548	.98953	.7
.4	.04188	.04191	23.859	.99912	.6	.4	.14608	.14767	6.7720	.98927	.6
.5	.04362	.04366	22.904	.99905	.5	.5	.14781	.14945	6.6912	.98902	.5
.6	.04536	.04541	22.022	.99897	.4	.6	.14954	.15124	6.6122	.98876	.4
.7	.04711	.04716	21.205	.99889	.3	.7	.15126	.15302	6.5350	.98849	.3
.8	.04885	.04891	20.446	.99881	.2	.8	.15299	.15481	6.4596	.98823	.2
.9	.05059	.05066	19.740	.99872	.1	.9	.15471	.15660	6.3859	.98796	.1
3.0	.05234	.05241	19.081	.99863	87.0	9.0	.15643	.15838	6.3138	.98769	81.0
.1	.05408	.05416	18.464	.99854	.9	.1	.15816	.16017	6.2432	.98741	.9
.2	.05582	.05591	17.886	.99844	.8	.2	.15988	.16196	6.1742	.98714	.8
.3	.05756	.05766	17.343	.99834	.7	.3	.16160	.16376	6.1066	.98686	.7
.4	.05931	.05941	16.832	.99824	.6	.4	.16333	.16555	6.0405	.98657	.6
.5	.06105	.06116	16.350	.99813	.5	.5	.16505	.16734	5.9758	.98629	.5
.6	.06279	.06291	15.895	.99803	.4	.6	.16677	.16914	5.9124	.98600	.4
.7	.06453	.06467	15.464	.99792	.3	.7	.16849	.17093	5.8502	.98570	.3
.8	.06627	.06642	15.056	.99780	.2	.8	.17021	.17273	5.7894	.98541	.2
.9	.06802	.06817	14.669	.99768	.1	.9	.17193	.17453	5.7297	.98511	.1
4.0	.06976	.06993	14.301	.99756	86.0	10.0	.17365	.17633	5.6713	.98481	80.0
.1	.07150	.07168	13.951	.99744	.9	.1	.17537	.17813	5.6140	.98450	.9
.2	.07324	.07344	13.617	.99731	.8	.2	.17708	.17993	5.5578	.98420	.8
.3	.07498	.07519	13.300	.99719	.7	.3	.17880	.18173	5.5026	.98389	.7
.4	.07672	.07695	12.996	.99705	.6	.4	.18052	.18353	5.4486	.98357	.6
.5	.07846	.07870	12.706	.99692	.5	.5	.18224	.18534	5.3955	.98325	.5
.6	.08020	.08046	12.429	.99678	.4	.6	.18395	.18714	5.3435	.98294	.4
.7	.08194	.08221	12.163	.99664	.3	.7	.18567	.18895	5.2924	.98261	.3
.8	.08368	.08397	11.909	.99649	.2	.8	.18738	.19076	5.2422	.98229	.2
.9	.08542	.08573	11.664	.99635	.1	.9	.18910	.19257	5.1929	.98196	.1
5.0	.08716	.08749	11.430	.99619	85.0	11.0	.19081	.19438	5.1446	.98163	79.0
.1	.08889	.08925	11.205	.99604	.9	.1	.19252	.19619	5.0970	.98129	.9
.2	.09063	.09101	10.988	.99588	.8	.2	.19423	.19801	5.0504	.98096	.8
.3	.09237	.09277	10.780	.99572	.7	.3	.19595	.19982	5.0045	.98061	.7
.4	.09411	.09453	10.579	.99556	.6	.4	.19766	.20164	4.9594	.98027	.6
.5	.09585	.09629	10.385	.99540	.5	.5	.19937	.20345	4.9152	.97992	.5
.6	.09758	.09805	10.199	.99523	.4	.6	.20108	.20527	4.8716	.97958	.4
.7	.09932	.09981	10.019	.99506	.3	.7	.20279	.20709	4.8288	.97922	.3
.8	.10106	.10158	9.8448	.99488	.2	.8	.20450	.20891	4.7867	.97887	.2
.9	.10279	.10334	9.6768	.99470	.1	.9	.20620	.21073	4.7453	.97851	.1
6.0	.10453	.10510	9.5144	.99452	84.0	12.0	.20791	.21256	4.7046	.97815	78.0

Angle, °	cos	cot	tan	sin	Angle, °	Angle, °	cos	cot	tan	sin	Angle, °

Angle, °	sin	tan	cot	cos	Angle, °	Angle, °	sin	tan	cot	cos	Angle, °
12.0	.20791	.21256	4.7046	.97815	78.0	18.0	.30902	.32492	3.0777	.95106	72.0
.1	.20962	.21438	4.6646	.97778	.9	.1	.31068	.32685	3.0595	.95052	.9
.2	.21132	.21621	4.6252	.97742	.8	.2	.31233	.32378	3.0415	.94997	.8
.3	.21303	.21804	4.5864	.97705	.7	.3	.31399	.33072	3.0237	.94943	.7
.4	.21474	.21986	4.5483	.97667	.6	.4	.31565	.33266	3.0061	.94888	.6
.5	.21644	.22169	4.5107	.97630	.5	.5	.31730	.33460	2.9887	.94832	.5
.6	.21814	.22353	4.4737	.97592	.4	.6	.31896	.33654	2.9714	.94777	.4
.7	.21985	.22536	4.4373	.97553	.3	.7	.32061	.33848	2.9544	.94721	.3
.8	.22155	.22719	4.4015	.97515	.2	.8	.32227	.34043	2.9375	.94665	.2
.9	.22325	.22903	4.3662	.97476	.1	.9	.32392	.34238	2.9208	.94609	.1
13.0	.22495	.23087	4.3315	.97437	77.0	19.0	.32557	.34433	2.9042	.94552	71.0
.1	.22665	.23271	4.2972	.97398	.9	.1	.32722	.34628	2.8878	.94495	.9
.2	.22835	.23455	4.2635	.97358	.8	.2	.32887	.34824	2.8716	.94438	.8
.3	.23005	.23639	4.2303	.97318	.7	.3	.33051	.35020	2.8556	.94380	.7
.4	.23175	.23823	4.1976	.97278	.6	.4	.33216	.35216	2.8397	.94322	.6
.5	.23345	.24008	4.1653	.97237	.5	.5	.33381	.35412	2.8239	.94264	.5
.6	.23514	.24193	4.1335	.97196	.4	.6	.33545	.35608	2.8083	.94206	.4
.7	.23684	.24377	4.1022	.97155	.3	.7	.33710	.35805	2.7929	.94147	.3
.8	.23853	.24562	4.0713	.97113	.2	.8	.33874	.36002	2.7776	.94088	.2
.9	.24023	.24747	4.0408	.97072	.1	.9	.34038	.36199	2.7625	.94029	.1
14.0	.24192	.24933	4.0108	.97030	76.0	20.0	.34202	.36397	2.7475	.93969	70.0
.1	.24362	.25118	3.9812	.96987	.9	.1	.34366	.36595	2.7326	.93909	.9
.2	.24531	.25304	3.9520	.96945	.8	.2	.34530	.36793	2.7179	.93849	.8
.3	.24700	.25490	3.9232	.96902	.7	.3	.34694	.36991	2.7034	.93789	.7
.4	.24869	.25676	3.8947	.96858	.6	.4	.34857	.37190	2.6889	.93728	.6
.5	.25038	.25862	3.8667	.96815	.5	.5	.35021	.37388	2.6746	.93667	.5
.6	.25207	.26048	3.8391	.96771	.4	.6	.35184	.37588	2.6605	.93606	.4
.7	.25376	.26235	3.8118	.96727	.3	.7	.35347	.37787	2.6464	.93544	.3
.8	.25545	.26421	3.7848	.96682	.2	.8	.35511	.37986	2.6325	.93483	.2
.9	.25713	.26608	3.7583	.96638	.1	.9	.35674	.38186	2.6187	.93420	.1
15.0	.25882	.26795	3.7321	.96593	75.0	21.0	.35837	.38386	2.6051	.93358	69.0
.1	.26050	.26982	3.7062	.96547	.9	.1	.36000	.38587	2.5916	.93295	.9
.2	.26219	.27169	3.6806	.96502	.8	.2	.36162	.38787	2.5782	.93232	.8
.3	.26387	.27357	3.6554	.96456	.7	.3	.36325	.38988	2.5649	.93169	.7
.4	.26556	.27545	3.6305	.96410	.6	.4	.36488	.39190	2.5517	.93106	.6
.5	.26724	.27732	3.6059	.96363	.5	.5	.36650	.39391	2.5386	.93042	.5
.6	.26892	.27921	3.5816	.96316	.4	.6	.36812	.39593	2.5257	.92978	.4
.7	.27060	.28109	3.5576	.96269	.3	.7	.36975	.39795	2.5129	.92913	.3
.8	.27228	.28297	3.5339	.96222	.2	.8	.37137	.39997	2.5002	.92849	.2
.9	.27396	.28486	3.5105	.96174	.1	.9	.37299	.40200	2.4876	.92784	.1
16.0	.27564	.28675	3.4874	.96126	74.0	22.0	.37461	.40403	2.4751	.92718	68.0
.1	.27731	.28864	3.4646	.96078	.9	.1	.37622	.40606	2.4627	.92653	.9
.2	.27899	.29053	3.4420	.96029	.8	.2	.37784	.40809	2.4504	.92587	.8
.3	.28067	.29242	3.4197	.95981	.7	.3	.37946	.41013	2.4383	.92521	.7
.4	.28234	.29432	3.3977	.95931	.6	.4	.38107	.41217	2.4262	.92455	.6
.5	.28402	.29621	3.3759	.95882	.5	.5	.38268	.41421	2.4142	.92388	.5
.6	.28569	.29811	3.3544	.95832	.4	.6	.38430	.41626	2.4023	.92321	.4
.7	.28736	.30001	3.3332	.95782	.8	.7	.38591	.41831	2.3906	.92254	.3
.8	.28903	.30192	3.3122	.95732	.2	.8	.38752	.42036	2.3789	.92186	.2
.9	.29070	.30382	3.2914	.95681	.1	.9	.38912	.42242	2.3673	.92119	.1
17.0	.29237	.30573	3.2709	.95630	73.0	23.0	.39073	.42447	2.3559	.92050	67.0
.1	.29404	.30764	3.2506	.95579	.9	.1	.39234	.42654	2.3445	.91982	.9
.2	.29571	.30955	3.2305	.95528	.8	.2	.39394	.42860	2.3332	.91914	.8
.3	.29737	.31147	3.2106	.95476	.7	.3	.39555	.43067	2.3220	.91845	.7
.4	.29904	.31338	3.1910	.95424	.6	.4	.39715	.43274	2.3109	.91775	.6
.5	.30071	.31530	3.1716	.95372	.5	.5	.39875	.43481	2.2998	.91706	.5
.6	.30237	.31722	3.1524	.95319	.4	.6	.40035	.43689	2.2889	.91636	.4
.7	.30403	.31914	3.1334	.95266	.3	.7	.40195	.43897	2.2781	.91566	.3
.8	.30570	.32106	3.1146	.95213	.2	.8	.40355	.44105	2.2673	.91496	.2
.9	.30736	.32299	3.0961	.95159	.1	.9	.40514	.44314	2.2566	.91425	.1
18.0	.30902	.32492	3.0777	.95106	72.0	24.0	.40674	.44523	2.2460	.91355	66.0
Angle, °	cos	cot	tan	sin	Angle, °	Angle, °	cos	cot	tan	sin	Angle, °

Angle,°	sin	tan	cot	cos	Angle,°	Angle,°	sin	tan	cot	cos	Angle,°
24.0	.40674	.44523	2.2460	.91355	66.0	30.0	.50000	.57735	1.7321	.86603	60.0
.1	.40833	.44732	2.2355	.91283	.9	.1	.50151	.57968	1.7251	.86515	.9
.2	.40992	.44942	2.2251	.91212	.8	.2	.50302	.58201	1.7182	.86427	.8
.3	.41151	.45152	2.2148	.91140	.7	.3	.50453	.58435	1.7113	.86340	.7
.4	.41310	.45362	2.2045	.91068	.6	.4	.50603	.58670	1.7045	.86251	.6
.5	.41469	.45573	2.1943	.90996	.5	.5	.50754	.58905	1.6977	.86163	.5
.6	.41628	.45784	2.1842	.90924	.4	.6	.50904	.59140	1.6909	.86074	.4
.7	.41787	.45995	2.1742	.90851	.3	.7	.51054	.59376	1.6842	.85985	.3
.8	.41945	.46206	2.1642	.90778	.2	.8	.51204	.59612	1.6775	.85896	.2
.9	.42104	.46418	2.1543	.90704	.1	.9	.51354	.59849	1.6709	.85806	.1
25.0	.42262	.46631	2.1445	.90631	65.0	31.0	.51504	.60086	1.6643	.85717	59.0
.1	.42420	.46843	2.1348	.90557	.9	.1	.51653	.60324	1.6577	.85627	.9
.2	.42578	.47056	2.1251	.90483	.8	.2	.51803	.60562	1.6512	.85536	.8
.3	.42736	.47270	2.1155	.90408	.7	.3	.51952	.60801	1.6447	.85446	.7
.4	.42894	.47483	2.1060	.90334	.6	.4	.52101	.61040	1.6383	.85355	.6
.5	.43051	.47698	2.0965	.90259	.5	.5	.52250	.61280	1.6319	.85264	.5
.6	.43209	.47912	2.0872	.90183	.4	.6	.52399	.61520	1.6255	.85173	.4
.7	.43366	.48127	2.0778	.90108	.3	.7	.52547	.61761	1.6191	.85081	.3
.8	.43523	.48342	2.0686	.90032	.2	.8	.52696	.62003	1.6128	.84989	.2
.9	.43680	.48557	2.0594	.89956	.1	.9	.52844	.62245	1.6066	.84897	.1
26.0	.43837	.48773	2.0503	.89879	64.0	32.0	.52992	.62487	1.6003	.84805	58.0
.1	.43994	.48989	2.0413	.89803	.9	.1	.53140	.62730	1.5941	.84712	.9
.2	.44151	.49206	2.0323	.89726	.8	.2	.53288	.62973	1.5880	.84619	.8
.3	.44307	.49423	2.0233	.89649	.7	.3	.53435	.63217	1.5818	.84526	.7
.4	.44464	.49640	2.0145	.89571	.6	.4	.53583	.63462	1.5757	.84433	.6
.5	.44620	.49858	2.0057	.89493	.5	.5	.53730	.63707	1.5697	.84339	.5
.6	.44776	.50076	1.9970	.89415	.4	.6	.53877	.63953	1.5637	.84245	.4
.7	.44932	.50295	1.9883	.89337	.3	.7	.54024	.64199	1.5577	.84151	.3
.8	.45088	.50514	1.9797	.89259	.2	.8	.54171	.64446	1.5517	.84057	.2
.9	.45243	.50733	1.9711	.89180	.1	.9	.54317	.64693	1.5458	.83962	.1
27.0	.45399	.50953	1.9626	.89101	63.0	33.0	.54464	.64941	1.5399	.83867	57.0
.1	.45554	.51173	1.9542	.89021	.9	.1	.54610	.65189	1.5340	.83772	.9
.2	.45710	.51393	1.9458	.88942	.8	.2	.54756	.65438	1.5282	.83676	.8
.3	.45865	.51614	1.9375	.88862	.7	.3	.54902	.65688	1.5224	.83581	.7
.4	.46020	.51835	1.9292	.88782	.6	.4	.55048	.65938	1.5166	.83485	.6
.5	.46175	.52057	1.9210	.88701	.5	.5	.55194	.66189	1.5108	.83389	.5
.6	.46330	.52279	1.9128	.88620	.4	.6	.55339	.66440	1.5051	.83292	.4
.7	.46484	.52501	1.9047	.88539	.3	.7	.55484	.66692	1.4994	.83195	.3
.8	.46639	.52724	1.8967	.88458	.2	.8	.55630	.66944	1.4938	.83098	.2
.9	.46793	.52947	1.8887	.88377	.1	.9	.55775	.67197	1.4882	.83001	.1
28.0	.46947	.53171	1.8807	.88295	62.0	34.0	.55919	.67451	1.4826	.82904	56.0
.1	.47101	.53395	1.8728	.88213	.9	.1	.56064	.67705	1.4770	.82806	.9
.2	.47255	.53620	1.8650	.88130	.8	.2	.56208	.67960	1.4715	.82708	.8
.3	.47409	.53844	1.8572	.88048	.7	.3	.56353	.68215	1.4659	.82610	.7
.4	.47562	.54070	1.8495	.87965	.6	.4	.56497	.68471	1.4605	.82511	.6
.5	.47716	.54296	1.8418	.87882	.5	.5	.56641	.68728	1.4550	.82413	.5
.6	.47869	.54522	1.8341	.87798	.4	.6	.56784	.68985	1.4496	.82314	.4
.7	.48022	.54748	1.8265	.87715	.3	.7	.56928	.69243	1.4442	.82214	.3
.8	.48175	.54975	1.8190	.87631	.2	.8	.57071	.69502	1.4388	.82115	.2
.9	.48328	.55203	1.8115	.87546	.1	.9	.57215	.69761	1.4335	.82015	.1
29.0	.48481	.55431	1.8040	.87462	61.0	35.0	.57358	.70021	1.4281	.81915	55.0
.1	.48634	.55659	1.7966	.87377	.9	.1	.57501	.70281	1.4229	.81815	.9
.2	.48786	.55888	1.7893	.87292	.8	.2	.57643	.70542	1.4176	.81714	.8
.3	.48938	.56117	1.7820	.87207	.7	.3	.57786	.70804	1.4124	.81614	.7
.4	.49090	.56347	1.7747	.87121	.6	.4	.57928	.71066	1.4071	.81513	.6
.5	.49242	.56577	1.7675	.87036	.5	.5	.58070	.71329	1.4019	.81412	.5
.6	.49394	.56808	1.7603	.86949	.4	.6	.58212	.71593	1.3968	.81310	.4
.7	.49546	.57039	1.7532	.86863	.3	.7	.58354	.71857	1.3916	.81208	.3
.8	.49697	.57271	1.7461	.86777	.2	.8	.58496	.72122	1.3865	.81106	.2
.9	.49849	.57503	1.7391	.86690	.1	.9	.58637	.72388	1.3814	.81004	.1
30.0	.50000	.57735	1.7321	.86603	60.0	36.0	.58779	.72654	1.3764	.80902	54.0
Angle,°	cos	cot	tan	sin	Angle,°	Angle,°	cos	cot	tan	sin	Angle,°

Angle, °	sin	tan	cot	cos	Angle, °	Angle, °	sin	tan	cot	cos	Angle, °
36.0	.58779	.72654	1.3764	.80902	**54.0**	**40.5**	.64945	.85408	1.1708	.76041	**49.5**
.1	.58920	.72921	1.3713	.80799	.9	.6	.65077	.85710	1.1667	.75927	.4
.2	.59061	.73189	1.3663	.80696	.8	.7	.65210	.86014	1.1626	.75813	.3
.3	.59201	.73457	1.3613	.80593	.7	.8	.65342	.86318	1.1585	.75700	.2
.4	.59342	.73726	1.3564	.80489	.6	.9	.65474	.86623	1.1544	.75585	.1
.5	.59482	.73996	1.3514	.80386	.5	**41.0**	.65606	.86929	1.1504	.75471	**49.0**
.6	.59622	.74267	1.3465	.80232	.4	.1	.65738	.87236	1.1463	.75356	.9
.7	.59763	.74538	1.3416	.80178	.3	.2	.65869	.87543	1.1423	.75241	.8
.8	.59902	.74810	1.3367	.80073	.2	.3	.66000	.87852	1.1383	.75126	.7
.9	.60042	.75082	1.3319	.79968	.1	.4	.66131	.88162	1.1343	.75011	.6
37.0	.60182	.75355	1.3270	.79864	**53.0**	.5	.66262	.88473	1.1303	.74896	.5
.1	.60321	.75629	1.3222	.79758	.9	.6	.66393	.88784	1.1263	.74780	.4
.2	.60460	.75904	1.3175	.79653	.8	.7	.66523	.89097	1.1224	.74664	.3
.3	.60599	.76180	1.3127	.79547	.7	.8	.66653	.89410	1.1184	.74548	.2
.4	.60738	.76456	1.3079	.79441	.6	.9	.66783	.89725	1.1145	.74431	.1
.5	.60876	.76733	1.3032	.79335	.5	**42.0**	.66913	.90040	1.1106	.74314	**48.0**
.6	.61015	.77010	1.2985	.79229	.4	.1	.67043	.90357	1.1067	.74198	.9
.7	.61153	.77289	1.2938	.79122	.3	.2	.67172	.90674	1.1028	.74080	.8
.8	.61291	.77568	1.2892	.79016	.2	.3	.67301	.90993	1.0990	.73963	.7
.9	.61429	.77848	1.2846	.78908	.1	.4	.67430	.91313	1.0951	.73846	.6
38.0	.61566	.78129	1.2799	.78801	**52.0**	.5	.67559	.91633	1.0913	.73728	.5
.1	.61704	.78410	1.2753	.78694	.9	.6	.67688	.91955	1.0875	.73610	.4
.2	.61841	.78692	1.2708	.78586	.8	.7	.67816	.92277	1.0837	.73491	.3
.3	.61978	.78975	1.2662	.78478	.7	.8	.67944	.92601	1.0799	.73373	.2
.4	.62115	.79259	1.2617	.78369	.6	.9	.68072	.92926	1.0761	.73254	.1
.5	.62251	.79544	1.2572	.78261	.5	**43.0**	.68200	.93252	1.0724	.73135	**47.0**
.6	.62388	.79829	1.2527	.78152	.4	.1	.68327	.93578	1.0686	.73016	.9
.7	.62524	.80115	1.2482	.78043	.3	.2	.68455	.93906	1.0649	.72897	.8
.8	.62660	.80402	1.2437	.77934	.2	.3	.68582	.94235	1.0612	.72777	.7
.9	.62796	.80690	1.2393	.77824	.1	.4	.68709	.94565	1.0575	.72657	.6
39.0	.62932	.80978	1.2349	.77715	**51.0**	.5	.68835	.94896	1.0538	.72537	.5
.1	.63068	.81268	1.2305	.77605	.9	.6	.68962	.95229	1.0501	.72417	.4
.2	.63203	.81558	1.2261	.77494	.8	.7	.69088	.95562	1.0464	.72297	.3
.3	.63338	.81849	1.2218	.77384	.7	.8	.69214	.95897	1.0428	.72176	.2
.4	.63473	.82141	1.2174	.77273	.6	.9	.69340	.96232	1.0392	.72055	.1
.5	.63608	.82434	1.2131	.77162	.5	**44.0**	.69466	.96569	1.0355	.71934	**46.0**
.6	.63742	.82727	1.2088	.77051	.4	.1	.69591	.96907	1.0319	.71813	.9
.7	.63877	.83022	1.2045	.76940	.3	.2	.69717	.97246	1.0283	.71691	.8
.8	.64011	.83317	1.2002	.76828	.2	.3	.69842	.97586	1.0247	.71569	.7
.9	.64145	.83613	1.1960	.76717	.1	.4	.69966	.97927	1.0212	.71447	.6
40.0	.64279	.83910	1.1918	.76604	**50.0**	.5	.70091	.98270	1.0176	.71325	.5
.1	.64412	.84208	1.1875	.76492	.9	.6	.70215	.98613	1.0141	.71203	.4
.2	.64546	.84507	1.1833	.76380	.8	.7	.70339	.98958	1.0105	.71080	.3
.3	.64679	.84806	1.1792	.76267	.7	.8	.70463	.99304	1.0070	.70957	.2
.4	.64812	.85107	1.1750	.76154	.6	.9	.70587	.99652	1.0035	.70834	.1
40.5	.64945	.85408	1.1708	.76041	**49.5**	**45.0**	.70711	1.00000	1.0000	.70711	**45.0**
Angle, °	cos	cot	tan	sin	Angle, °	Angle, °	cos	cot	tan	sin	Angle, °

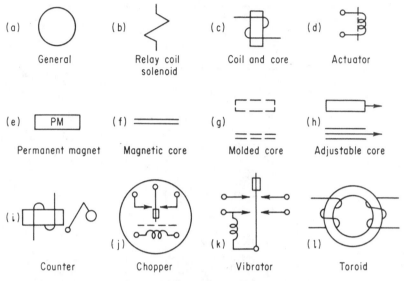

Figure A-1. Symbols for windings, cores, and magnetic devices.

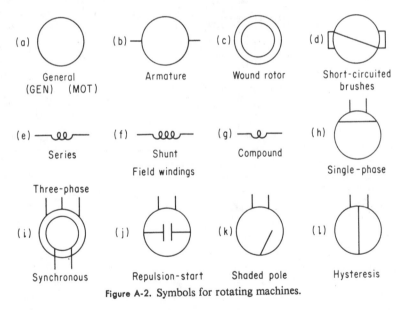

Figure A-2. Symbols for rotating machines.

* Excerpted from G. Shiers, *Electronic Drafting,* Prentice-Hall, Inc., 1962.

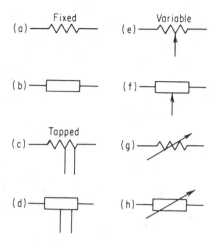

Figure A-3. Resistor symbols.

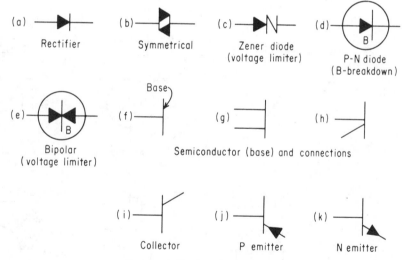

Figure A-4. Semiconductor symbols.

357

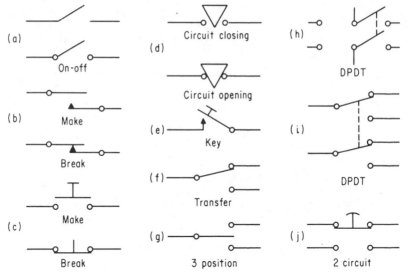

Figure A-5. Basic switch symbols.

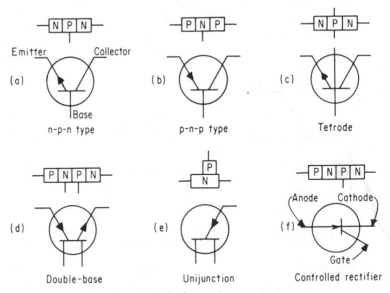

Figure A-6. Basic transistor symbols.

358

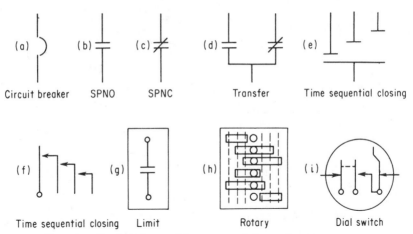

Figure A-7. Symbols for special-purpose switches and contactors.

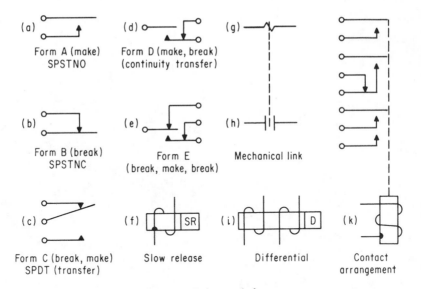

Figure A-8. Relay symbols.

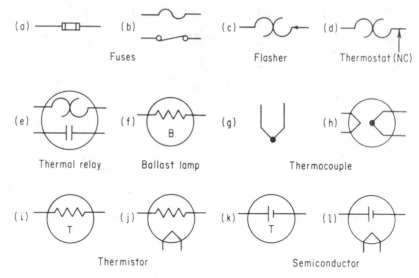

(a) ——[ ]—— (b) Fuses (c) Flasher (d) Thermostat (NC)

(e) Thermal relay (f) Ballast lamp (g) Thermocouple (h)

(i) Thermistor (j) (k) Semiconductor (l)

Figure A-9. Symbols used for thermally-operated devices.

TABLE A-3. FULL-LOAD CURRENTS IN AMPERES
DIRECT-CURRENT MOTORS (NEC 430-147)

The following values of full-load currents are for motors running at base speed.

HP	120 V	240 V
$\frac{1}{4}$	2.9	1.5
$\frac{1}{3}$	3.6	1.8
$\frac{1}{2}$	5.2	2.6
$\frac{3}{4}$	7.4	3.7
1	9.4	4.7
$1\frac{1}{2}$	13.2	6.6
2	17	8.5
3	25	12.2
5	40	20
$7\frac{1}{2}$	58	29
10	76	38
15		55
20		72
25		89
30		106
40		140
50		173
60		206
75		255
100		341
125		425
150		506
200		675

Appendix

TABLE A-4. FULL-LOAD CURRENTS IN AMPERES
SINGLE-PHASE ALTERNATING-CURRENT MOTORS (NEC 430-148)

The following values of full-load currents are for motors running at usual speeds and motors with normal torque characteristics. Motors built for especially low speeds or high torques may have higher full-load currents, in which case the nameplate current ratings should be used.

To obtain full-load currents of 208 and 200 V motors, increase corresponding 230 V motor full-load currents by 10 and 15 per cent, respectively.

The voltages listed are rated motor voltages. Corresponding nominal system voltages are 110 to 120, 220 to 240, 440 to 480.

HP	115 V	230 V	440 V
$\frac{1}{6}$	4.4	2.2	
$\frac{1}{4}$	5.8	2.9	
$\frac{1}{3}$	7.2	3.6	
$\frac{1}{2}$	9.8	4.9	
$\frac{3}{4}$	13.8	6.9	
1	16	8	
$1\frac{1}{2}$	20	10	
2	24	12	
3	34	17	
5	56	28	
$7\frac{1}{2}$	80	40	21
10	100	50	26

HP	\multicolumn{5}{c}{Induction Type Squirrel-Cage and Wound Rotor Amperes}	\multicolumn{4}{c}{Synchronous Type Unity Power Factor Amperes†}
	110 V	220 V	440 V	550 V	2300 V	220 V	440 V	550 V	2300 V
$\frac{1}{2}$	4	2	1	.8					
$\frac{3}{4}$	5.6	2.8	1.4	1.1					
1	7	3.5	1.8	1.4					
$1\frac{1}{2}$	10	5	2.5	2.0					
2	13	6.5	3.3	2.6					
3		9	4.5	4					
5		15	7.5	6					
$7\frac{1}{2}$		22	11	9					
10		27	14	11					
15		40	20	16					
20		52	26	21					
25		64	32	26	7	54	27	22	5.4
30		78	39	31	8.5	65	33	26	6.5
40		104	52	41	10.5	86	43	35	8
50		125	63	50	13	108	54	44	10
60		150	75	60	16	128	64	51	12
75		185	93	74	19	161	81	65	15
100		246	123	98	25	211	106	85	20
125		310	155	124	31	264	132	106	25
150		360	180	144	37		158	127	30
200		480	240	192	48		210	168	40

For full-load currents of 208 and 200 volt motors, increase the corresponding 220 volt motor full-load current by 6 and 10 per cent, respectively.

* These values of full-load current are for motors running at speeds usual for belted motors and motors with normal torque characteristics. Motors built for especially low speeds or high torques may require more running current, in which case the nameplate current rating should be used.

† For 90 and 80 per cent P. F., the above figures should be multiplied by 1.1 and 1.25 respectively.

The voltages listed are rated motor voltages. Corresponding nominal system voltages are 110 to 120, 220 to 240, 440 to 480 and 550 to 600 volts.

Code letter	Kilovolt-amperes per horsepower with locked rotor
A	0 — 3.14
B	3.15 — 3.54
C	3.55 — 3.99
D	4.0 — 4.49
E	4.5 — 4.99
F	5.0 — 5.59
G	5.6 — 6.29
H	6.3 — 7.09
J	7.1 — 7.99
K	8.0 — 8.99
L	9.0 — 9.99
M	10.0 — 11.19
N	11.2 — 12.49
P	12.5 — 13.99
R	14.0 — 15.99
S	16.0 — 17.99
T	18.0 — 19.99
U	20.0 — 22.39
V	22.4 — and up

The above table is an adopted standard of the National Electrical Manufacturers Association.

The code letter indicating motor input with locked rotor must be in an individual block on the nameplate, properly designated. This code letter is to be used for determining branch-circuit overcurrent protection by reference to Table A-7.

	Per cent of full-load current		
		Circuit-breaker Setting	
			Time limit
Type of motor	Fuse rating	Instantaneous type	type
All AC single-phase and polyphase squirrel cage and synchronous motors with full-voltage, resistor, or reactor starting:			
Code Letter A	150		150
Code Letters B to E	250		200
Code Letters F to V	300		250
All AC squirrel-cage and synchronous motors with auto-transformer starting:			
Code Letter A	150		150
Code Letters B to E	200		200
Code Letters F to V	250		200

For certain exceptions to the values specified see Sections 430-52 and 430-54 (NEC). The values given in the last column also cover the ratings of nonadjustable, time-limit types of circuit-breakers which may also be modified as in Section 430-52 (NEC).

Synchronous motors of the low-torque, low-speed type (usually 450 rpm or lower), such as are used to drive reciprocating compressors, pumps, etc., which start up unloaded, do not require a fuse rating or circuit-breaker setting in excess of 200 per cent of full-load current.

For motors not marked with a Code Letter, see Table A-8.

365

	Per cent of full-load current		
		Circuit-breaker Setting	
Type of motor	Fuse rating	Instantaneous type	Time limit type
Single-phase, all types	300		250
Squirrel-cage and synchronous (full-voltage, resistor and reactor starting)	300		250
Squirrel-cage and synchronous (auto-transformer starting)			
Not more than 30 amperes	250		200
More than 30 amperes	200		200
High-reactance squirrel-cage			
Not more than 30 amperes	250		250
More than 30 amperes	200		200
Wound-rotor	150		150
Direct-current			
Not more than 50 H.P.	150	250	150
More than 50 H.P.	150	175	150
Sealed (Hermetic Type) Refrigeration Compressor*			
400 KVA lacked-rotor or less	175†		†175

For certain exceptions to the values specified see Sections 430-52, and 430-59 (NEC). The values given in the last column also cover the ratings of non-adjustable, time-limit types of circuit-breakers which may also be modified as in Section 430-52 (NEC).

Synchronous motors of the low-torque low-speed type (usually 450 rpm or lower) such as are used to drive reciprocating compressors, pumps, etc., which start up unloaded, do not require a fuse rating or circuit-breaker setting in excess of 200 per cent of full-load current.

For motors marked with a Code Letter, see Table A-7.

* The locked-rotor KVA is the product of the motor voltage and the motor locked-rotor current (LRA), given on the motor nameplate, divided by 1000 for single-phase motors, or divided by 580 for 3-phase motors.

† This value may be increased to 225 per cent if necessary to permit starting.

Motor hp	Rated current amps	Starting current, amps Classes B, C, D	Starting current, amps Class F	Starting torque, per cent rated torque at rated voltage Classes A and B 4 pole	6 pole	8 pole	Class C 4 pole	6 pole	8 pole
0.5	2.0	12				150			
1.0	3.5	24		275	175	150			
1.5	5.0	35		265	175	150			
2	6.5	45		250	175	150			
3	9.0	60		250	175	150		250	225
5	15	90		185	160	130	250	250	225
7.5	22	120		175	150	125	250	225	200
10	27	150		175	150	125	250	225	200
15	40	220		165	140	125	225	200	200
20	52	290		150	135	125	200	200	200
25	64	365		150	135	125	200	200	200
30	78	435	270	150	135	125	200	200	200
40	104	580	360	150	135	125	200	200	200
50	125	725	450	150	135	125	200	200	200
60	150	870	540	150	135	125	200	200	200
75	185	1085	675	150	135	125	200	200	200
100	246	1450	900	125	125	125	200	200	200
125	310	1815	1125	125	125	125	200	200	200
150	360	2170	1350	125	125	125	200	200	200
200	480	2900	1800	125	125	125	200	200	200

* For changes in voltage, use the following equations:

a. Starting current $= \dfrac{V_l}{220} \times I_s$,

where V_l = new voltage applied to stator and I_s = starting current in above table.

b. Starting torque $= \left(\dfrac{V_l}{220}\right)^2 \times T_s$,

where T_s = starting torque in above table.

† Starting currents of class A motors are usually *higher* than corresponding class B, C, D motors.

Starting torques of class D motors are usually *higher* than corresponding class A, B, C motors.

Starting torques of class F motors are usually *lower* than corresponding class A, B, C motors.

TABLE A-10. TRANSFER FUNCTIONS OF VARIOUS SERVO NETWORKS

Network	Open-circuit transfer function	Gain curve
1. (R series, C shunt)	$H = \dfrac{1}{1+RCs}$	$\|H\|$dB, 0, Log ω; $\dfrac{1}{RC}$, -6 dB/oct
2. (C series, R shunt)	$H = \dfrac{RCs}{1+RCs}$	$\|H\|$dB, 0, Log ω; $+6$, $\dfrac{1}{RC}$
3.	$H = \dfrac{R_1}{R+R_1}\ \dfrac{1+RCs}{1+\dfrac{R_1}{R+R_1}RCs}$	$\|H\|$dB, $+6$ db, Log ω; $20\log\left(\dfrac{R}{R+R_1}\right)$, $\dfrac{1}{RC}$, $\dfrac{R_1R}{R+R_1}C$
4.	$H = \dfrac{1+R_1Cs}{1+(R+R_1)Cs}$	$\|H\|$dB, -6db, Log ω; $20\log\left(\dfrac{R_1}{R+R_1}\right)$, $\dfrac{1}{(R+R_1)C}$, $\dfrac{1}{R_1C}$
5.	$H = \dfrac{(1+RCs)(1+R_1C_1s)}{RR_1CC_1s^2+(RC+R_1C_1+RC_1)s+1}$	$\|H\|$dB, 0, Log ω
6.	$H=\dfrac{s^2(C_1CR_1R_2R)+s(C_1R_1R_2+CR_1R+R_2RC)+(R_1+R_2)}{s^2(C_1CR_1R_2R)+s(C_1R_1R_2+CR_1R+CR_2R+C_1RR_2)+(R+R_1+R_2)}$	$\|H\|$dB, Log ω; -6dB, $+6$dB oct; $20\log\left(\dfrac{R_1+R_2}{R+R_1+R_2}\right)$
7.	$H=\dfrac{R_1(R+R_1)CC_1s^2+(RC+R_2C+R_1C_1)s+1}{(RR_1+RR_2+R_1R_2)CC_1s^2+(RC+RC_1+R_1C_1+R_2C)s+1}$	$\|H\|$dB, Log ω; -6 dB, $+6$dB oct; $20\log\left[\dfrac{R_1(R+R_2)}{R_1(R+R_2)+RR_2}\right]$
8.	$H=\dfrac{1}{1+R_1C_1s}-\dfrac{R_2C_2s}{1+R_2C_2s}$	$\|H\|$dB, 0, Log ω; For $R_1C_1=R_2C_2$
9.	$H=\dfrac{s^2(R_1C_1R_2C_2)+s\left[R_1(C_1+C_2)\right]+1}{s^2(R_1C_1R_2C_2)+s\left[R_1(C_1+C_2)+C_2R_2\right]+1}$	$\omega_0=\dfrac{1}{\sqrt{R_1C_1R_2C_2}}$; $\|H\|$dB, 0, Log ω; $20\log\left[\dfrac{R_1(C_1+C_2)}{R_1(C_1+C_2)+R_2C_2}\right]$
10.	$H=\dfrac{1+\left(\frac{s}{\omega_0}\right)^2+\sqrt{2\frac{R_1}{R_2}}\left(\frac{s}{\omega_0}+\frac{s^3}{\omega_0^3}\right)}{1+\left(3+4\frac{R_1}{R_2}\right)\left(\frac{s}{\omega_0}\right)^2+\left(3\sqrt{\frac{2R_2}{R}}+\sqrt{\frac{2R_2}{R}}\right)\frac{s}{\omega_0}+\sqrt{\frac{2R_1}{R_0}}\left(\frac{s}{\omega_0}\right)^3}$ $C_1=\dfrac{\sqrt{\frac{2R_1}{R_2}}}{2\omega_0R_1}\qquad C_2=4\dfrac{R_1}{R_2}C_1$	$\omega_0=\dfrac{\sqrt{\frac{2R_1}{R_2}}}{2R\,C}$; $\|H\|$dB, 0, Log ω

	Multiply:	By:	To obtain:
Angular measure	degrees	17.45	mils
	degrees	60	minutes
	degrees	1.745×10^{-2}	radians
	mils	5.730×10^{-2}	degrees
	mils	3.438	minutes
	mils	1.000×10^{-3}	radians
	minutes	1.667×10^{-2}	degrees
	minutes	0.2909	mils
	minutes	2.909×10^{-4}	radians
	radians	57.30	degrees
	radians	1.000×10^{3}	mils
	radians	3.438×10^{3}	minutes
Angular velocity	deg/s	1.745×10^{-2}	rad/s
	deg/s	0.1667	rpm
	deg/s	2.778×10^{-3}	rps
	rad/s	57.30	deg/s
	rad/s	9.549	rpm
	rad/s	0.1592	rps
	rpm	6.0	deg/s
	rpm	0.1047	rad/s
	rpm	1.667×10^{-2}	rps
	rps	360	deg/s
	rps	6.283	rad/s
	rps	60	rpm
	ARMY MIL	1/6400	revolutions
Damping	$\dfrac{\text{ft-lb}}{\text{rad/s}}$	20.11	$\dfrac{\text{oz-in}}{\text{rpm}}$
	$\dfrac{\text{oz-in}}{\text{rpm}}$	4.974×10^{-2}	$\dfrac{\text{ft-lb}}{\text{rad/s}}$
	$\dfrac{\text{oz-in}}{\text{rpm}}$	6.75×10^{-2}	newton-m/rad/s
Density	g/cm³	10^{3}	kg/m³
	lb/ft³	16.018	kg/m³
Distance	cm	10^{-2}	meters (m)
	in	2.5400×10^{-2}	meters
	ft	0.30480	meters
	yd	0.91440	meters
	km	10^{3}	meters
	miles	1609.4	meters
Energy	ergs	10^{-7}	joules
	kwhr	3.6×10^{6}	joules
	calories	4.182	joules
	ft-lb	1.356	joules
	Btu	1655	joules

	Multiply:	By:	To obtain:
Force and weight	dynes	10^{-5}	newtons
	poundals	0.13826	newtons
	lb (force)	4.4482	newtons
Inertia	g-cm²	10^{-7}	kg-m²
	g-cm²	5.468×10^{-3}	oz-in²
	g-cm²	7.372×10^{-8}	slug-ft²
	oz-in²	1.829×10^{2}	g-cm²
	oz-in²	1.348×10^{-5}	slug-ft²
	slug-ft²	1.357×10^{7}	g-cm²
	(lb-ft-s²)	7.419×10^{4}	oz-in²
	slug-ft²	1.357	kg-m²
	lb-in²	2.925×10^{-4}	kg-m²
	oz-in²	1.829×10^{-5}	kg-m²
Mass	g	10^{-3}	kilograms
	slug	14.594	kilograms
Power	ergs/s	10^{-7}	watts
	cal/s	4.182	watts
	Btu/hr	0.2930	watts
	joules/s	1.00	watts
	horsepower	746	watts
	ft-lb/s	1.356	watts
Pressure	dynes/cm²	10^{-1}	newton/m²
	psi	6.895×10^{3}	newton/m²
	atmospheres	1.013×10^{5}	newton/m²
	cm Hg	1333	newton/m²
Torque	ft-lb	1.383×10^{4}	g-cm
	ft-lb	192	oz-in
	g-cm	7.235×10^{-5}	ft-lb
	g-cm	1.389×10^{-2}	oz-in
	oz-in	5.208×10^{-3}	ft-lb
	oz-in	72.01	g-cm
	oz-in	7.0612×10^{-3}	newton-m (joules)
Torque error	$\dfrac{\text{oz-in}}{\text{min}}$	0.0558	$\dfrac{\text{lb-ft}}{\text{rad}}$
	$\dfrac{\text{lb-ft}}{\text{rad}}$	17.9	$\dfrac{\text{oz-in}}{\text{min}}$
Velocity	ft/s	0.30480	m/s
	miles/hr	0.44704	m/s
	knots	1.152	miles/hr

Note: $HP = \dfrac{\text{ft-lb-rpm}}{5250}$

370

index

372

374